PHYSICS IN OXFORD
1839–1939

PHYSICS IN OXFORD 1839–1939

Laboratories, Learning, and College Life

EDITED BY

ROBERT FOX
Modern History Faculty,
University of Oxford

GRAEME GOODAY
Division of History and Philosophy of Science,
University of Leeds

OXFORD
UNIVERSITY PRESS

Great Clarendon Street, Oxford OX2 6DP

Oxford University Press is a department of the University of Oxford.
It furthers the University's objective of excellence in research, scholarship,
and education by publishing worldwide in

Oxford New York

Auckland Cape Town Dar es Salaam Hong Kong Karachi
Kuala Lumpur Madrid Melbourne Mexico City Nairobi
New Delhi Shanghai Taipei Toronto

With offices in

Argentina Austria Brazil Chile Czech Republic France Greece
Guatemala Hungary Italy Japan Poland Portugal Singapore
South Korea Switzerland Thailand Turkey Ukraine Vietnam

Published in the United States
by Oxford University Press Inc., New York

Oxford is a registered trade mark of Oxford University Press
in the UK and in certain other countries

© Oxford University Press 2005

The moral rights of the authors have been asserted
Database right Oxford University Press (maker)

First published 2005

All rights reserved. No part of this publication may be reproduced,
stored in a retrieval system, or transmitted, in any form or by any means,
without the prior permission in writing of Oxford University Press,
or as expressly permitted by law, or under terms agreed with the appropriate
reprographics rights organization. Enquiries concerning reproduction
outside the scope of the above should be sent to the Rights Department,
Oxford University Press, at the address above

You must not circulate this book in any other binding or cover
and you must impose the same condition on any acquirer

British Library Cataloguing in Publication Data

Data available

Library of Congress Cataloging in Publication Data

Data available

Typeset by Newgen Imaging Systems (P) Ltd., Chennai, India
Printed in Great Britain
on acid-free paper by
Biddles Ltd., King's Lynn

ISBN 0–19–856792–8 (Hbk)   978–0–19–856792–9

10 9 8 7 6 5 4 3 2 1

PREFACE

Physics in Oxford grew from the much larger project that culminated with the publication, in 2000, of the last of the eight volumes of *The History of the University of Oxford*. Several chapters of that *History*, as well as Jack Morrell's pioneering *Science at Oxford 1914–1939* (1997), hinted at both the desirability and the feasibility of a detailed study of physics in the University in the modern period. The challenge was an attractive but formidable one. Given the scale and scope of the project as we conceived it, a collaborative endeavour was obviously necessary, and this was the task on which the six authors of the present book embarked over a decade ago.

It was soon evident that the development of physics in Oxford between the mid-nineteenth century and the Second World War was a complex affair, pursued in contexts extending far beyond the confines of the University's core facility for physics since 1870, the Clarendon Laboratory. That point is all too often missed in the existing secondary literature, which tends to make much of the flagging state of the Clarendon in the later years of Robert Clifton's long occupancy of the chair of experimental philosophy and during the inevitable disruption of the First World War. Certainly, the decade or so before the appointment of Frederick Lindemann as Clifton's successor in 1919 marked a low point in the laboratory's history. But to state, as Lindemann did retrospectively in 1939, that at the time of his arrival 'the reputation of the Department had sunk almost to zero' is to focus, unduly as we argue, on one element within the bigger picture of Oxford physics. What of J. S. E. Townsend's attempts to establish research in ion physics in the Electrical Laboratory, for example? Or what, more strikingly, of the traditions of research and teaching that had taken root in less obvious settings since the mid-nineteenth century, notably in college laboratories for physics and physical chemistry? John Heilbron has shown how H. J. G. Moseley's time in the Balliol–Trinity Laboratories and in Oxford's undergraduate science clubs fashioned the researcher who, as a young graduate, went on to produce his mould-breaking X-ray analyses of atomic structure in the Electrical Laboratory during 1913–14 (Heilbron 1974). But few other historians have explored the significance of college laboratories or, more generally, sought to look beyond the Clarendon in search of a bigger picture of Oxford physics. As a result, too many studies have been partial, University-centred accounts of precisely the kind that *Physics in Oxford* seeks to review.

The diversity of contexts in which physics was pursued in Oxford is an important recurring theme of the chapters that follow. Sensitivity to the University's decentralized structure has led all of us, in treating different themes and periods, to open new perspectives on the peculiarly Oxonian manifestation of the discipline. In the process, it has helped to highlight a striking characteristic that distinguished Oxford from its most obvious counterpart, Cambridge. Whereas physics emerged in Oxford, often with difficulty though not infrequently with conspicuous success, in a predominantly arts-based community dominated by the diffused culture of collegiate life, in Cambridge the history of the University's engagement with physics can perhaps more properly be written as a history of the Cavendish Laboratory and its succession of powerful Cavendish professors from James Clerk Maxwell on.

In a collaboration that has been in train for so long, the authors of *Physics in Oxford* have accumulated many debts to organizations, librarians, archivists, and others who have helped their work. Among these, special mention must be made of Mr Simon Bailey, who has offered unfailingly efficient and friendly support in his capacity as Keeper of the Oxford University Archives, and Mr Chris Stammers of the Photographic Unit of the Department of Physics in Oxford, who has expertly prepared many of the photographs in the volume. As editors, we have also benefited from financial support from the Modern History Faculty in Oxford and from the Royal Society and have incurred numerous debts of a more personal kind. In the Modern History Faculty, Stephanie Jenkins has given invaluable help at all stages in the editorial process. She has prepared successive drafts of the book with meticulous care, good humour, and a keen eye for lapses from clarity. She has also given important assistance with the preparation of the bibliography and index. In Leeds, Leacha Veneer too has worked on the index, while Annie Jamieson has helped with the editing and proof-reading of several chapters and supplied patient and genial assistance in checking references.

We also wish to express our gratitude to Dr John Sanders, whose reading of all the chapters has steered us clear of numerous errors and infelicities. His creation of the Clarendon Laboratory Archive, following his retirement from the laboratory in 1991, has made available to us, as it has to other researchers, a rich collection of photographs, manuscripts, printed sources, recorded oral reminiscences, and instruments. The inventory of the collection, which was initially catalogued by Dr Katherine Watson (Watson 1994: 1–33), is now accessible on the website of the Oxford Department of Physics at http://www.physics.ox.ac.uk/history.asp.

Finally, we are indebted to Sönke Adlung, Anita Petrie, and their colleagues at Oxford University Press, who have given every possible assistance and helped us to maintain a brisk pace on the road to publication.

No history is definitive, and we do not offer *Physics in Oxford* as an exception to that obvious rule. There are key areas of research in our period that deserve more detailed analysis than would have been appropriate in a volume covering a hundred years and physics in all its aspects. Also, as we indicate in the Epilogue, there are especially rich opportunities for work on the decades since 1939, in which both teaching and research in the discipline have undergone dramatic expansion and endowed Oxford with one of the largest and most dynamic schools of physics in the world.

Robert Fox and Graeme Gooday
April 2005

CONTENTS

Notes on Contributors	xi
List of Illustrations	xii
Acknowledgements for Illustrations	xv
Abbreviations	xvii
Administrative Terminology	xx
Terms and Residence	xxi

1. Physics in Oxford: Problems and Perspectives — 1
   Robert Fox, Graeme Gooday, and Tony Simcock

2. The Context and Practices of Oxford Physics, 1839–77 — 24
   Robert Fox

3. Robert Bellamy Clifton and the 'Depressing Inheritance' of the Clarendon Laboratory, 1877–1919 — 80
   Graeme Gooday

4. Laboratories and Physics in Oxford Colleges, 1848–1947 — 119
   Tony Simcock

5. Mechanical Physicists, the Millard Laboratory, and the Transition from Physics to Engineering — 169
   Tony Simcock

6. Translating Ion Physics from Cambridge to Oxford: John Townsend and the Electrical Laboratory, 1900–24 — 209
   Benoit Lelong

7. The Lindemann Era — 233
   Jack Morrell

8. Redefining the Context: Oxford and the Wider World of British Physics, 1900–1940 — 267
   Jeff Hughes

9. Epilogue — 301
   Robert Fox and Graeme Gooday

Appendix I The Classification of the Oxford B.A.	311
Appendix II The Syllabus for the Oxford B.A., 1831–1872	312
Appendix III Letter from Robert Clifton to Sir William Thomson	315
Bibliography	317
Index	347

NOTES ON CONTRIBUTORS

Robert Fox is Professor of the History of Science at the University of Oxford. His main research interest is the history of the physical sciences and industrial technology in Europe since the eighteenth century.

Graeme Gooday is Senior Lecturer in the History and Philosophy of Science at the University of Leeds. His research focuses on electrical cultures, laboratories, and technologies since the mid-nineteenth century. He was a British Academy/Royal Society Research Fellow in the Modern History Faculty at Oxford from 1992 to 1994.

Jeff Hughes is Senior Lecturer in History of Science and Technology at the University of Manchester. His research interests focus mainly on the history of physics in Britain since c.1900.

Benoit Lelong works as an historian in association with the REHSEIS group (CNRS and University of Paris-7). He is also a researcher in France Telecom's Laboratoire de Sociologie des Usages et de Traitement statistique de l'Information. Both his historical and his sociological work is concerned with scientific and technical innovation and its social uses.

Jack Morrell was Reader in History of Science in the University of Bradford until 1991. He is currently Honorary Visiting Lecturer in History of Science at the University of Leeds. His publications include *Science at Oxford 1914–1939. Transforming an Arts University* (Oxford: Clarendon Press, 1997).

Tony Simcock is Archivist of the Museum of the History of Science, Oxford. He is the author of several articles and booklets on the history of Oxford science.

LIST OF ILLUSTRATIONS

2.1	Stephen Peter Rigaud	25
2.2	Robert Walker	25
2.3	The Clarendon Building	27
2.4	First floor of the Clarendon Building, plan by Robert Smirke	27
2.5	Model of Watt beam engine	28
2.6	Announcement of Robert Walker's lectures, c. 1840	29
2.7	The Clarendon dry pile	30
2.8	Pages from Robert Walker's record of lecture-fees	36
2.9	The University Museum	37
2.10	Plan of ground floor of the University Museum	39
2.11	Plan of first floor of the University Museum	40
2.12	Robert Bellamy Clifton	45
2.13	Ground plan and elevation for extension to the University Museum, 1867	52
2.14	Illuminated parchment commemorating the building of the Clarendon Laboratory	57
2.15	The Clarendon Laboratory, from *The Builder*, 1869	59
2.16	Sketch-plan of ground floor and first floor of the Clarendon Laboratory	61
2.17	Lazarus Fletcher	64
2.18	Arthur Rücker	66
2.19	Robert Clifton's private room, Clarendon Laboratory	67
2.20	Ground-floor hall, Clarendon Laboratory	68
2.21	First-floor gallery, Clarendon Laboratory	69
3.1	Workshops, Clarendon Laboratory	89
3.2	Instrument-shop, Clarendon Laboratory	89
3.3	John Viriamu Jones	91
3.4	Engine-room, Clarendon Laboratory	93
3.5	Switchboards, Clarendon Laboratory	94
3.6	The Clarendon Laboratory's 'Manchester' dynamo	94
3.7	Alternating-current Wilde dynamo, Clarendon Laboratory	95
3.8	Quadrant electrometer and miner's lamp, designed by Robert Clifton	105
3.9	Greenwich Board of Visitors, 1890s	106

List of Illustrations

3.10	Experiment for measurement of G, by C. V. Boys, showing octagon house surrounding apparatus	111
3.11	Boys apparatus, showing telescope for viewing deflections of lead balls	112
3.12	Boys apparatus, showing cathetometer	113
3.13	Chronograph drum used by Boys for recording time during his experiment	114
4.1	Thomas Henry Toovey Hopkins	122
4.2	Edward Chapman	124
4.3	John Job Manley	125
4.4	Henry Thomas Tizard	127
4.5	Benjamin Collins Brodie, junior	129
4.6	Map of parts of Balliol, Trinity, and St John's Colleges	131
4.7	Harold Baily Dixon	132
4.8	William Stroud	134
4.9	John Conroy	135
4.10	David Henry Nagel	136
4.11	Robert Tabor Lattey	138
4.12	Harold Brewer Hartley	139
4.13	Robert Edward Baynes	143
4.14	David Leonard Chapman	151
4.15	Carl Howard Collie	160
4.16	Sprengel mercury pump	163
4.17	Letter from Lord Berkeley's copy-book	165
5.1	Henry Reginald Arnulph Mallock	176
5.2	Mallock's device for measuring the growth-rate of trees	177
5.3	Frederick John Jervis-Smith	186
5.4	(a and b) Printed notices of courses offered by Jervis-Smith	190, 191
5.5	Printed notice issued by Jervis-Smith in response to the demand for instruction in electrical engineering	193
5.6	Earliest prototype of Jervis-Smith's tram chronograph	197
5.7	Second prototype of Jervis-Smith's tram chronograph	197
5.8	Final production model of Jervis-Smith's tram chronograph	198
5.9	Jervis-Smith's integrating ergometer	202
6.1	Paul Jerome Kirkby	216
6.2	Henry Gwyn Jeffreys Moseley in the Balliol–Trinity Laboratories	221
6.3	The Electrical Laboratory	223
6.4	Plaque commemorating the role of the Drapers' Company in the building of the Electrical Laboratory	224
6.5	John Sealy Edward Townsend	230

7.1	Group photograph taken at the second Solvay conference on physics, 1913	235
7.2	Gordon Miller Bourne Dobson	240
7.3	Thomas Ralph Merton	241
7.4	Derek Ainslie Jackson	242
7.5	Alfred Charles Glyn Egerton	242
7.6	Helium liquefier and cryostat, 1933	246
7.7	Franz Eugen Simon, with low-temperature apparatus	247
7.8	Franz Simon in his laboratory, Breslau	248
7.9	'Tug of war' between Bosch and Simon	248
7.10	Heinrich Gerhard Kuhn	249
7.11	Kurt Alfred Georg Mendelssohn, with John Gilbert Daunt and Rex Pontius Bush	253
7.12	Judith Rachel Moore	254
7.13	Bernard Vincent Rollin	255
7.14	Nicholas Kurti	256
7.15	Claude Hurst	257
7.16	Group photograph: Jackson, Keeley, Hurst, and Collie	258
7.17	James Howard Eagle Griffiths	260
7.18	Frederick Alexander Lindemann	261
7.19	Plan of the Science Area, 1937	264
7.20	The new Clarendon Laboratory	265
8.1	Moseley's graph mapping characteristic X-ray emission spectra	278
8.2	Moseley in military uniform, 1915	279
8.3	Robert Jemison van de Graaff	290
9.1	South front of the Clarendon Laboratory shortly before its demolition	302
9.2	Francis Simon, c.1950	304
9.3	Thomas Clews Keeley	305
9.4	Lindemann with one of Simon's daughters	306
9.5	Ida Busbridge and Madge Adam	307
9.6	Fellows of Brasenose College, late 1940s	310

ACKNOWLEDGEMENTS FOR ILLUSTRATIONS

We are grateful to the following for permission to reproduce the illustrations in this volume

Clarendon Laboratory Archive (by permission of the Archivist, Dr J. H. Sanders):
  Figures 2.2 (location of original unknown), 2.6, 2.7, 2.8, 2.9, 2.12, 2.15, 2.19, 2.20, 2.21, 3.1, 3.2, 3.4, 3.5, 3.6, 3.7, 3.8, 3.10, 3.11, 3.12, 3.13, 6.2, 6.5, 7.2, 7.4, 7.5, 7.6, 7.7, 7.8, 7.9, 7.10, 7.13, 7.14, 7.16, 7.17, 7.18, 8.1, 8.2, 8.3, 9.2, 9.3, 9.4, 9.6

Museum of the History of Science, University of Oxford (by permission of the Director, Dr J. A. Bennett):
  Figures 2.1, 2.17, 3.3, 3.9, 4.1, 4.2, 4.3, 4.4, 4.5, 4.7, 4.8, 4.10, 4.11, 4.12, 4.14, 4.15, 4.16, 4.17, 5.1, 5.2, 5.3, 5.5, 5.6, 5.7, 5.8, 5.9, 6.2, 7.3, 9.1

Bodleian Library, University of Oxford (by permission of the Bodleian Library, University of Oxford):
  Figures 7.11, 7.12

Oxford University Archives (by permission of the Keeper of the University Archives, Mr Simon Bailey):
  Figures 2.4, 2.13

Department of Physics, University of Oxford (by permission of the Chairman of Physics, Professor K. Burnett):
  Figure 2.14

Balliol College, Oxford (by permission of Balliol College):
  Figure 4.9

Christ Church, Oxford (by permission of the Senior Common Room, Christ Church):
  Figures 4.13, 7.15

St Hugh's College, Oxford (by permission of the Principal and Fellows of St Hugh's College, Oxford):
  Figure 9.5

Trinity College, Oxford (by permission of the President and Fellows of Trinity College):
  Figure 5.4 (*a* and *b*)

Cambridge University Library (by permission of the Syndics of Cambridge University Library):
> Figure 2.16

Chorley Parish Church of St Laurence (by permission of the Revd Dr John Cree, Rector of Chorley, and Mr Ed Fisher, Churchwarden):
> Figure 6.1

Francis Frith Collection (by permission of the Francis Frith Collection <www.francisfrith.co.uk>)
> Figure 2.3 (copy lent by Stephanie Jenkins)

Gilman & Soame, Limited, Oxford
> Figure 9.6 (from print in the Clarendon Laboratory Archive)

International Institute for Physics and Chemistry, founded by E. Solvay (Photographie Benjamin Couprie, by permission of American Institute of Physics, Emilio Segrè Visual Archives)
> Figure 7.1

Professor G. L'E. Turner and Museum of the History of Science:
> Figure 2.5

University of Leeds Library (by permission of Leeds University Archive):
> Figure 2.18

ABBREVIATIONS

Published works

Devonshire Commission. First, Supplementary, and Second Reports 1872	*Royal Commission on Scientific Instruction and the Advancement of Science. Vol. I. First, Supplementary, and Second Reports, with Minutes of Evidence and Appendices [Devonshire Commission]* [Cd 536] (London: HMSO, 1872)
Devonshire Commission. Third Report 1873	*The Third Report of the Royal Commission on Scientific Instruction and the Advancement of Science [Devonshire Commission]* [Cd. 868] (London: HMSO, 1873)
DNB	*The Dictionary of National Biography*, ed. L. Stephen and S. Lee. 22 volumes (Oxford: Oxford University Press, 1885–1901)
DNB 1951–1960	*The Dictionary of National Biography 1951–1960*, eds. E.T. Williams and H.M. Palmer (London: Oxford University Press, 1971)
DNB Missing Persons	*The Dictionary of National Biography. Missing Persons*, ed. C.S. Nicholls (Oxford: Oxford University Press, 1993)
DSB	*Dictionary of Scientific Biography*, ed. C. C. Gillispie. 16 volumes (New York: Charles Scribner's Sons, 1970–80)
Historical Register	*The Historical Register of the University of Oxford, being a Supplement to the Oxford University Calendar with an Alphabetical Record of University Honours and Distinctions to the End of Trinity Term 1900* (Oxford: Clarendon Press, 1900)
Historical Register First Supplement	*University of Oxford. First Supplement to the Historical Register of 1900 containing a Complete Record of University Honours and Distinctions for the Years 1900–1920* (Oxford: Clarendon Press, 1921)
Historical Register Supplement 1901–1930	*University of Oxford. Supplement to the Historical Register of 1900 including an Alphabetical Record of University Honours and Distinctions for the Years 1901–1930* (Oxford: Clarendon Press, 1931)
Historical Register Supplement 1931–1950	*University of Oxford. Supplement to the Historical Register of 1900 including an Alphabetical Record of University Honours and Distinctions for the Years 1931–1950* (Oxford: Clarendon Press, 1951)

New Statutes 1877	*New Statutes Made for the University by the University of Oxford Commissioners under the Universities of Oxford and Cambridge Act, 1877* (Oxford: Clarendon Press, 1882)
ODNB	*The Oxford Dictionary of National Biography*, eds. H. C. G. Matthew and B. H. Harrison. 60 volumes (Oxford: Oxford University Press, 2004)
OU Calendar	*Oxford University Calendar*, first published in 1810 and then annually from 1813, initially by Parker and subsequently by the Clarendon Press
OUG	*Oxford University Gazette*, published from 1870
Oxford University Commission 1852	*Oxford University Commission. Report of Her Majesty's Commissioners appointed to inquire into the State, Discipline, Studies, and Revenues of the University and Colleges of Oxford: together with the Evidence, and an Appendix* [P.P. 1852, XXII] (London: HMSO, 1852)
University of Oxford Commission 1881	*University of Oxford Commission. Part I. Minutes of Evidence taken by the Commissioners, together with an Appendix and Index* [Cd.2868] (London: HMSO, 1881)

Other abbreviations

BAAS	British Association for the Advancement of Science
Bodl.	Bodleian Library, Oxford (followed by shelf-mark of printed item)
CLA	Clarendon Laboratory Archive, in the Archive Room, Clarendon Laboratory, Parks Road, Oxford OX1 3PU. A regularly updated inventory of the Archive, with links to sources of historical interest, is available at http://www.physics.ox.ac.uk/history.asp. See also the printed inventory in Watson (1994: 1–33)
CUL	Cambridge University Library
double first	For the purpose of this volume, graduates described as having a double first had taken first-class honours in the Schools of both Mathematics and Natural Science
ET	Easter Term*
HCP	Hebdomadal Council papers, Oxford University Archives
HT	Hilary Term*
Lit. Hum.	Literae Humaniores
LP	Lindemann Papers, Nuffield College, Oxford OX1 1NF
MHS	Museum of the History of Science, Broad Street, Oxford OX1 3AZ
Moderations (Mods)	The First Public Examination, introduced by the statutes of 1850. Although candidates could be awarded Honours (in either Classics or Mathematics), most took the easier path to a Pass

Abbreviations

MP	Mendelssohn Papers, Bodleian Library, Oxford OX1 3BG
MT	Michaelmas Term*
NSS	School of Natural Science
OUA	Oxford University Archives
Preliminary examination (Prelim)	Preliminary Honour examination, introduced by decisions in Congregation and then Convocation in 1871. The examination, in chemistry and 'mechanics and physics', had to be taken by all candidates for Honours in the NSS at some time after they had taken Mods. The words 'Preliminary students (candidates) and 'Preliminary teaching' refer to such candidates and the instruction they received
SP	Simon papers, in Royal Society of London, 6–9 Carlton House Terrace, London, SW1Y 5AG
SPSL	Society for the Protection of Science and Learning papers, Bodleian Library, Oxford OX1 3BG
TCA	Trinity College archives
TT	Trinity (or Act) Term*

* On the nomenclature of the Oxford terms, see 'Terms and Residence', p. xxi

ADMINISTRATIVE TERMINOLOGY

The terms Council (or Hebdomadal Council), Congregation, and Convocation are used frequently in this book. All three bodies have origins going back to the seventeenth century and beyond, and over this long period their composition and responsibilities have changed in complex ways. The functions they had in the period 1839-1939 were defined chiefly by the university statutes ratified during the Chancellorship of Archbishop Laud in 1636 and by the Oxford University Act of 1854 (17 and 18 Vict., c.81). They may be summarized as follows:

Hebdomadal Council Instituted in 1631 as the Hebdomadal Board and reconstituted and renamed in accordance with the 1854 Act, Hebdomadal Council met weekly to conduct the regular business of the University. Its official members were the Chancellor, Vice-Chancellor, previous Vice-Chancellor, and two Proctors (fellows of colleges elected to be responsible for the discipline of the University). The remaining members (eighteen for most of the period) were elected by Congregation in equal numbers from among heads of colleges and halls, professors, and other members of Convocation.

Congregation Not to be confused with the 'Ancient House of Congregation', whose role was reduced by the 1854 Act to that of granting degrees, Congregation consisted of resident members of Convocation and was essentially the body of teaching members of the University. Statutes framed by Hebdomadal Council had to be promulgated in Congregation, which had a central role not only in legislation but also in administrative decisions affecting the running of the University and the various boards and delegacies through which authority on specific matters was devolved.

Convocation With a membership composed of all Masters of Arts and Doctors of Civil Law, Medicine, and Divinity, whether resident or not, Convocation was required to accept or reject statutes and decrees coming from Hebdomadal Council or Congregation. Unlike the other bodies, it had no power to initiate any statute or other measure or even to amend one that came before it.

For further details see the article 'Constitution of the University' in *Historical Register*: 9–15.

TERMS AND RESIDENCE

For most of the period treated in this book, the academic year was formally divided into four terms. These were normally determined as follows: Hilary (or Lent) Term, beginning on 14 January and ending on the day before Palm Sunday; Easter Term, beginning on the Wednesday in Easter Week and ending on the Friday before Whit Sunday; Trinity (or Act) Term, beginning on the Saturday before Whit Sunday and ending on the Saturday after the first Saturday in July; and Michaelmas Term, beginning on 10 October and ending on 17 December. By the mid-1860s it had become common for the Easter and Trinity Terms to be regarded as a single term, with Trinity Term being used to denote the combined period of the two terms. This distribution of the terms represented a total period of about 34 weeks in which residence would be recognized. However, undergraduates were required to reside for only eighteen weeks in the year, with a good deal of latitude on the distribution of their qualifying periods. From the time of the reforms of the early nineteenth century until 1883, examinations took place twice a year, one session being held in Easter or Trinity term, the other in Michaelmas Term. Thereafter, examinations for the Final Honour Schools were held in Trinity Term only. The present practice of concentrating the examinations for Honour Moderations in Classics in Hilary Term began in 1886. Honours in Mathematical Moderations, however, continued to be awarded in both Michaelmas and Trinity Terms until 1896, whereafter the examination took place in Trinity Term only.

What had long been normal practice with regard to the distribution of terms was formalized in 1917, when the Easter and Trinity Terms were united in a single Trinity Term. For details of this measure see the *OU Calendar* for 1918: 118.

1

Physics in Oxford: Problems and Perspectives

Robert Fox, Graeme Gooday, and Tony Simcock

Physics in Oxford aims to depict the complex trajectory by which physics evolved in Oxford from 1839, when the University appointed the Revd Robert Walker to its readership in experimental philosophy, until the beginning of the Second World War a century later. The choice of 1839 as the starting-point for the volume should not be interpreted as implying that Oxford physics began in that year. By then, the teaching of experimental philosophy in the University had already had a long history (Turner 1986; Simcock 1984, 1993b). It had been inaugurated in the early years of the eighteenth century by John Theophilus Desaguliers and John Keill, and thereafter, from about 1715, John Whiteside and then James Bradley had offered lectures and demonstrations on physical phenomena and applied mathematics on a more formal basis. The creation, in 1749, of the readership in Experimental Philosophy marked the steadily growing prominence of the subject. Held successively by Bradley, Thomas Hornsby, and Stephen Peter Rigaud, the readership became a focus not only for interested students and colleagues but also for benefactions. From 1786, it received modest but regular funding in the form of an annual allowance under Lord Leigh's benefaction for the purchase of apparatus, and in 1810 the Prince Regent, the future George IV, established a further annual grant of £100 from the revenues of the Crown. Bradley, Hornsby, and Rigaud were able and, in different ways, successful. But they all held the readership in association with another position; Rigaud, for example, held it in addition to the Savilian chairs of geometry (1810–27) and astronomy (1827–39). Hence the conversion of the post into an independent one bearing a complete salary (a total endowment of about £230), which occurred on Walker's appointment in 1839, marked an important new departure, essentially the beginning of the complete institutionalization of physics in the University.

The choice of 1839 and the volume's other chronological boundary of 1939 carries with it a challenge inherent in the existing secondary literature. For, while the work of Morrell (1997: esp. 381–432, 1992) has done much to force a reappraisal of Oxford science in the last twenty years or so of the period, our understanding of physics in the University before 1939 remains patchy. With regard to the period up to 1920 especially, it has been widely assumed that in physics little of interest occurred in Oxford by comparison with what was afoot in Cambridge, Glasgow, and the more progressive universities on the continents of Europe and North America. Typically, a main cause of this 'failing' has been found in Robert Clifton's apparent inactivity during his half century as professor of experimental philosophy from 1865 to 1915 and as the effective founder and first occupant of the Clarendon Laboratory from 1870. Yet the essential elements of this interpretation – whether concerning the supposed decades of somnolence or Clifton's responsibility for them – have never been subjected to the kind of systematic scrutiny that is attempted here.

Such studies as we have – Morrell's conspicuously excepted – tend to be based either on shakily elaborated comparisons with Cambridge or on assumptions insufficiently sensitive to the distinctive characteristics of the academic cultures within which science was practised in Oxford. In recent years, however, those cultures have been examined in unprecedented detail in the eight volumes of the *History of the University of Oxford*, not only by Morrell but also by Mark Curthoys and, with special reference to science, mathematics, and medicine, by Rupke (1997), Robb-Smith (1997), Harley (1997), Fox (1997), Howarth (2000), Hannabuss 2000c), Webster (1994 and 2000), and Roche (1994). As a result, the authors of this volume have been able to base their reconsideration of the particular case of physics on firmer foundations, including a far richer body of source materials than was available to earlier scholars. The period covered in the book is a long one, and the perspectives of the individual contributors are diverse. But the work converges to a reappraisal of physics and related areas of physical science in Oxford. In doing so, it offers the opportunity for a new look at the pervasive folklore of decline and inactivity under Clifton, followed by eventual, tardy redemption between the 1920s and the Second World War, when physics in the University achieved salvation through Clifton's successor in the chair of experimental philosophy, Frederick Lindemann.

While the chapters that follow do not break with every aspect of the received view, they do draw attention to circumstances, peculiar to Oxford, that compel us to modify the terms on which we judge the case. These circumstances include three main contextual issues. The first concerns the ways in which certain activities that were considered elsewhere to be a part of physics were pursued in Oxford within other domains, notably physical chemistry and mechanics. Secondly, a strong theme of the volume, most obviously in the contributions by Simcock, is the largely unrecognized level of activity in college laboratories, as opposed to the central facilities of the University. Finally, all the contributors have been sensitive to the locally defined goal of Oxford science, which for most of the century in question was to fashion a small elite of graduates for a variety of professional and clerical careers rather than large numbers of research-oriented physicists. It is part and parcel of this sensitivity that the volume engages with some all too familiar historiographical concerns about how evaluations of past science can be made without anachronistically imposing present-day criteria for judging a university's success or failure.

LEADING THEMES

The many faces of physics that emerge from the particularities of the case of Oxford force us to rethink easy assumptions about the discipline as a global enterprise united by consensus as to its nature and objectives. Too often, such assumptions have been regarded as so self-evident that they have not been subject to critical discussion. Broadly, the assumptions to which we refer can be summarized as follows:

- that physics is a necessary and prestigious part of university life
- that physics occupies a disciplinary territory common to all universities
- that physics is, or should be, characterized by a priority given to research rather than teaching

Take the first of these assumptions, one made by most historians and by physicists writing on the history of their discipline, who see physics as an essential element in any respectable university's offering in science. The assumption draws its strength from perceptions of academic prestige that have gone virtually unchallenged in the age of Big Science since the mid-twentieth century. The resulting hallmark of excellence has inevitably bestowed favour on physics departments that, by either contemporary or later standards, have supported large communities of teachers, pupils, and researchers and maintained flows of important publications and highly trained postgraduate students. On this evaluative framework, the Cavendish Laboratory in Cambridge has assumed the status of a supremely successful institution since its rise to eminence in the last quarter of the nineteenth century, while the Clarendon Laboratory has emerged as a comparative failure, at least until the 1930s.

But how justified are the criteria that lead to such evaluations? How far does a university need a sustained presence at whatever may be deemed to be the forefront of innovative physics in order for us to see it as successful? Clearly the status of universities was judged by very different standards before teaching in natural philosophy was instituted in the seventeenth and eighteenth centuries, and by different ones again for most of the nineteenth century, before the emergence of well-endowed research laboratories in the late nineteenth and early twentieth centuries. So, when and how did the prestige-enhancing traits of academic physics that are familiar to us now become so overwhelmingly important? They were incontestably in place by the time the Cold War was raging in the 1950s; then, as never before, the future prospects of civilization on both sides in the nuclear arms race seemed to depend on a high level of expertise in physics. Arguably, the evaluative framework that held sway in the years of the Cold War had already asserted its dominance in the 1920s and 1930s, elsewhere if not in Oxford. Before the First World War, however, it is not easy to identify such physics-oriented evaluations of university life, apart from those proffered self-servingly by physicists themselves. And it is harder still to locate them as we go back through the nineteenth century, when chemistry and astronomy were still the sciences most likely to be regarded as the *sine qua non* of a high-status university. The danger of anachronism in our understanding of success and failure is therefore a real one. It has left its mark across the secondary literature and has been very present in the minds of all the contributors to this volume.

The second assumption to which we refer above concerns the fluctuating definition of what in the nineteenth and early twentieth centuries constituted the realms of physics or (to use terms that still circulated in the period) natural philosophy, experimental philosophy, or mechanical philosophy. The boundaries between physics and the adjacent disciplines of chemistry, mechanics, mathematics, and engineering have been susceptible to challenge and change with a frequency that leaves little ground for believing that the scope of physics in our period was ever seen as being natural or in any sense given. Until the early nineteenth century, for example, electricity and heat were largely the province of chemistry; the motion of bodies was studied in mechanics, which itself formed part of the category of 'mixed mathematics';[1] and concepts of work and energy were elaborated mainly by steam engineers. Thereafter, one by one, these areas were partially or completely appropriated by the nascent community of physicists that began to emerge in the mid-nineteenth century. Similar questions of disciplinary alignment arose again at the end of the century with respect to the new topic of radioactivity and accounts of atomic structure that were developed using the new electron theory of matter. In Cambridge, these subjects immediately entered the domain of the Cavendish Laboratory as part of physics. In Oxford, on the other hand, radioactivity was assimilated to the existing territory of physical chemistry. Similarly, as Lelong's chapter shows, the study of electrons and ion physics was incorporated

---

[1] The term 'mixed mathematics' was commonly used in our period to denote the category of mathematics applied to physical phenomena, especially optics and mechanics. For a discussion of the category as it evolved in nineteenth-century Britain see Warwick (2003: 34–8 and 114–75).

into what was, in effect, the independent specialism of 'electrical science', pursued by John Townsend, the holder of the newly created Wykeham chair of physics from 1900, working from 1910 in the laboratory that the Drapers' Company built for him next to the Clarendon. The contrast underlines the elusiveness of any agreed definition of physics that would achieve consensus between one institution and another. Definitions were typically local and idiosyncratic. Even the Cambridge model started as such, and it was only the diaspora of students and teachers trained at the Cavendish from the late nineteenth century that eventually made it otherwise.

Of course, any emphasis on the local and the particular in the evolving definition of physics runs the risk of institutional solipsism. Analysis that touches insufficiently on contextual elements on the national or international scale will necessarily be flawed, even with regard to a university as apparently autonomous as Oxford. By the same token, an exaggerated preoccupation with idiosyncracies can make comparative study difficult. If we are to compare Oxford with Cambridge, for example, should we adopt the Cambridge model of physics, embracing radioactivity and electron theory as natural parts of the discipline? Or, using a dualistic framework, should we try to accommodate both models and strive for a degree of symmetry in our analysis of the two divergent conceptions? The chapters by Gooday and Hughes that follow come down firmly in favour of the latter approach. In this way, they are able to explore both how Cambridge physics of the 1890s onwards looked from Oxford – expansionist and eager to colonize the new areas of physics that presented themselves about the turn of the nineteenth century – and how Oxford physics looked from Cambridge – somewhat unadventurous in its readiness to cede emerging areas to adjacent disciplines. While this framework may strike some readers as troublesomely relativistic, it is helpful in the important respect that it encourages comparison on a topic-by-topic basis; from that type of comparison, Oxford emerges as not being anything like as quiescent as has commonly been supposed.

Turning to the third of the assumptions we mention above, the contributors to *Physics in Oxford* have reflected at length on the tendency for histories of physics to see research, especially as pursued collectively in research schools, as determining an institution's standing in the discipline. This tendency, with its associated perception of teaching as an auxiliary and relatively uninteresting activity, reflects the extent to which historical explanations of change in physics have focused on the making of new knowledge and on the intellectual and material resources relevant to discovery and the creative process. The common corollary of such a view is that all else comes to be regarded as following naturally in the wake of research; it is all too easily assumed, for example, that the syllabuses of schools and universities would in due course, and in a process all too seldom subjected to analysis, assimilate simplified versions of findings at the cutting edge of the subject. Of course, historians of science might well be justified in making research the main focus of their attention, but they overlook at their peril the extent to which the work of research physicists is moulded in some degree by their prior training at school or in higher education. What, after all, would have been the point of such training unless it were to have some directive effect on its recipients? Indeed, as Thomas Kuhn points out in his *The Structure of Scientific Revolutions*, there could scarcely have been a significant research community in physics in the absence of a formal university-level training in the existing practices and knowledge in the field (Kuhn 1962: 1–12). Exposure at that level also pre-supposes structures for the training and employment of teachers in secondary education. This is a pertinent observation with respect to the functions of the instruction in physics that Oxford offered. For, while none of the contributors to *Physics in Oxford* has undertaken a quantitative analysis of the production of physics teachers in the University, such a study would almost certainly show that from the 1860s and for at least half a century, probably on into the 1930s, secondary education was the commonest career destination for graduates in physics. That conclusion would point to clear demographic reasons why historians should attend to the quality of the training offered to potential recruits for schoolteaching. Not least, it would remind us that if a main aim of a university's provision for physics was to prepare graduates for careers other than that of research scientist, it would be inappropriate to dismiss such a university as underachieving on the basis of its performance in research alone.

Physics in Oxford, then, treats a period in which historians cannot appeal to an institution's aspirations to a high reputation in research as the self-evident motivation for its promotion of the subject. Consider, for example, the development of the first generation of university physics laboratories in the last quarter of the nineteenth century. As Gooday has shown, the financial and curricular rationale for such laboratories was strictly educational, both in Oxford and elsewhere, notably in London and Manchester (Gooday 1990). When the Cavendish Laboratory was inaugurated in 1874, James Clerk Maxwell was no more obliged to undertake innovative research than Robert Clifton had been when the Clarendon Laboratory first opened its doors in Oxford four years earlier. Certainly some of their students and demonstrators were active in producing original papers. But neither Maxwell nor Clifton set up new large-scale projects that drew substantial numbers of outsiders (national or international) to undertake research in their respective new laboratories. The predominant intake to their experimental domains was, in both cases, a modest flux of internally trained graduates in mathematics and/or physics. Both men were heavily burdened with the university and extramural duties that accrue to holders of institutional chairs. And, as far as longer-term endeavours were concerned, Maxwell's devotion to replicating Henry Cavendish's researches and managing the BAAS electrical standards programme (Schaffer 1995;

Harman 2002)[2] was matched in Clifton's obsessive preparation and supervision of undergraduate classes and his development of lecture demonstration apparatus. The highest institutional priority of both Maxwell and Clifton, in fact, was not so much the publication of scientific papers but the sustaining of past endeavours and the passing on of their expertise to the rising generation, whether wranglers in the Cavendish or the ablest Natural Science undergraduates in the Clarendon. The implications of this observation recur frequently in the chapters that follow, in which the authors strive for sensitivity to the priorities of a particular university's diverse community of physicists whose views of the proper balance between pedagogy and research might be unfamiliar to readers in the twenty-first century.

THE COLLEGIATE DIMENSION

A further question arises about the sites of activity that we should consider in examining the implantation of a new discipline, such as physics, in an ancient university whose structures had evolved over hundreds of years in response to challenges and incentives unrelated to the sustained presence of the sciences. The most pertinent question with regard to Oxford concerns the relative importance of the University's centralized resources and those in the quasi-autonomous colleges that made up the University. Too frequently, the default assumption has been that the former were more important. But the critical histories of public and popular science that have appeared over the last few decades have revealed the danger of an undue emphasis on the supposedly authoritative 'central' bodies that were once the stock in trade of social historians of science, often unduly influenced by the exaggerated claims of those who worked within them. The point has special relevance to any study of science in Oxford, where for some decades after the opening of the University Museum as a consolidated centre for science in 1860 and of the Clarendon Laboratory a decade later a large proportion of teaching continued to be pursued in the decentralized and largely self-sufficient world of the colleges, where it was already firmly entrenched.

A recurring theme of this volume is therefore that lectures, laboratories, and scientific activities in the constituent colleges should be treated as being as characteristic of the life of Oxford as the facilities managed by the University. By the 1860s, physical science was taught in the laboratories and lecture-rooms of Magdalen, Balliol, and (in a laboratory whose history went back to the eighteenth century)

---

[2] In an extensive secondary literature on the Cavendish see especially Crowther (1974); Schaffer (1992); Kim (2002).

Christ Church. Far from reducing their engagement in science, all three colleges kept pace with developments and maintained their self-sufficiency, offering teaching that was complementary, and not subservient, to that of the University Museum. Indeed, from the 1870s through to the First World War, the collegiate presence in the panorama of Oxford science became even more prominent. The college laboratories at Jesus and Queen's were founded in this period, and Trinity College's Millard Laboratory, as Simcock shows in Chapter 5, became the setting for a disciplinary new departure of great significance. It was here, in 1885, that the Revd Frederick Jervis-Smith instituted a tradition of practical instruction in mechanics that filled a long-felt gap in the Clarendon's teaching and led in due course to the founding of the University's department of engineering science in 1908. With the colleges providing such a menu of complementary and alternative provision, the Clarendon Laboratory has to be seen as just one of several locations for the pursuit of physics. The contrast with the central role that the Cavendish Laboratory played in Cambridge from the 1890s could scarcely be more striking.

In the history of college-based science, physics began strongly. In mid-nineteenth-century Oxford, it was if anything more prominent in the colleges' offerings than the two other core sciences of the new School of Natural Science (NSS): chemistry and biology.[3] This was partly because of the substantial dose of physics in the existing syllabus of the Mathematical School, which in turn ensured the relatively generous availability of mathematical fellows and lecturers, at least one of whom was to be found in most colleges. By comparison, both chemistry and biology in the 1850s had ground to make up as fields of undergraduate study. As it happened, they did so with spectacular success, leaving physics, by the late 1880s, the least popular of the three chief specialisms at Honours level. This may well have been so because chemistry and biology did not require the solid grounding in mathematics that Clifton demanded for advanced study in the Clarendon. Clifton's apparent aloofness also played its part; he was not easily approachable, except to those who were already his students, and there were few others to whom undergraduates could turn for instruction in physics at any but an elementary level, such as was required for the Preliminary Honour examination. Indeed, some who went on to become physicists, such as Sir John Conroy (who graduated in 1868) and George Burch (in 1889), opted to specialize in chemistry as undergraduates since tuition in the subject was more readily available to them. Conroy, for example, found his mentor in A. G. Vernon Harcourt, Dr Lee's reader in chemistry at Christ Church, and it was Harcourt who introduced Conroy to the emerging hybrid of 'physical chemistry', a distinctive feature of the Oxford collegiate scene that occupied a good deal of common disciplinary territory with physics.

[3] On Charles Daubeny's conception of the three 'primary' sciences see Fox, Chapter 2, 31–3.

Balliol and Trinity too supported teachers and students of physics as well as of chemistry, and in the combined Balliol–Trinity laboratories, as at Christ Church, local cohabitation engendered a fruitful fusion of the two disciplines, one that, to Oxford's advantage, coincided with the rise of physical chemistry in the outside world. New arrangements for practical instruction in both physics and chemistry at the level of the Preliminary examination in 1904–5 had the effect of strengthening still further the position of physical chemistry by largely transferring responsibility for laboratory teaching in physics from the colleges to the central facilities under Townsend's control.[4] The transfer left college-based chemistry to inherit much of the tradition and expertise that had hitherto been fostered at the interface between physics and chemistry. In these circumstances, it was easy for a range of topics that would elsewhere have been more naturally incorporated into physics, such as radioactivity, thermodynamics, and spectroscopy, to find a home within chemistry. During the 1920s and 1930s, the intellectual breadth of Oxford chemistry consolidated what had begun to occur before the war. In a period when Thomas Merton and the chemists Cyril Hinshelwood, D. L. Chapman, and C. H. Collie were pursuing physics-related research in college laboratories, it might even be argued that some of the University's best physicists were to be found in such laboratories, all formally designated for chemistry.[5] The point was not lost on Lindemann, who on his arrival in 1919 did what he could to reclaim some of these individuals and their areas of research for the Clarendon. Collie, for example, was put in charge of the Clarendon's involvement in nuclear physics, a responsibility he retained until the 1950s. Merton too was appropriated. After working at Balliol before the war and returning after war work to become reader in spectroscopy in the Department of Chemistry under Soddy, he was moved to the Clarendon in 1920.

The diffuseness of the provision for physics was taken still further by the practice of extracollegiate coaching for examinations. The practice is more commonly associated with Cambridge, where until the 1880s teaching by private coaches specializing in the preparation of ambitious candidates for the Mathematical Tripos was an essential adjunct to the tuition offered by the colleges. As Warwick has shown (2003: 89), the terms 'coach' and 'coaching' appear to have originated as student slang in Oxford during the 1830s. And certainly many Oxford undergraduates in the mid-nineteenth century depended on coaches as heavily as their Cambridge counterparts did, and the services of coaches were copiously advertised. One such coach was Bartholomew Price, who offered private tuition in addition to his work

---

[4] On the Preliminary Honour examination see Fox, ibid., 69–71. The effects of the changes are evident in the annual reports of Clifton and Townsend for 1904, in *OUG*, 35 (1904–5): 572–3 (16 May 1905). Although Clifton was glad to be relieved of elementary practical teaching, he was clearly concerned about the loss of the income for the laboratory that the instruction of preliminary students had brought with it through the 1890s.

[5] Nevil Sidgwick was another college-based chemist whose research concerned aspects of physics, although his college, Lincoln, had no laboratory.

as lecturer in mathematics at Pembroke College (1845–72) and Sedleian professor of natural philosophy (1853–98). In the words of Clifton's obituary of Price, 'a considerable number of private pupils obtained from him the more detailed tuition required in the higher branches of mathematics, so that for many years he took a large share of the teaching in these subjects in Oxford' (Clifton 1905: 31). And to mathematics, of course, Clifton could well have added the branches of physics that were examined in the Mathematical School as parts of mixed mathematics. All this said, the contribution of coaches seems to have diminished somewhat earlier in Oxford than it did in Cambridge. In the process, which went hand in hand with the greater prominence and seriousness of mathematical lectures organized within the colleges, William Esson, a fellow of Merton from 1860, college lecturer and tutor in mathematics from 1865, and finally, from 1897, Savilian professor of geometry, played a central role. In an obituary of him in 1917, his former pupil Edwin Bailey Elliott (by now Waynflete professor of mathematics) saw Esson's contribution to the changes as marking something of a turning-point in the history of mathematical teaching in Oxford. As he observed, rather archly, 'things did not move quite so fast in Cambridge' (Elliott 1916–17: lv).[6]

THE PLACE OF MATHEMATICS

While a leading purpose of this volume is to question the tendency to take the Cavendish Laboratory in Cambridge as the main touchstone of excellence in physics, there is one point of comparison between Oxford and Cambridge that has to be broached. This is the role of university mathematics in nurturing the skills of undergraduates seeking to specialize in physics. With regard to Cambridge, it has commonly been argued that that university's eminence in physics owed much to the strength of the Mathematical Tripos. Most recently, Warwick (2003) has shown that such a thesis works well to explain the vibrant nineteenth-century Cambridge culture of mixed mathematics rooted in Newtonian mechanics and, latterly, theoretical electromagnetism. Yet, as both he and Schaffer have argued, this particularist interpretation of Cambridge University mathematics cannot be extended so easily to account for the growth of experimental physics, as epitomized in the Cavendish Laboratory, opened in 1874. A significant caucus of Cambridge dons was hostile to

---

[6] By the time of his appointment to the Savilian chair in 1897, Esson had already deputized for his ailing predecessor, J. J. Sylvester, for three years. It is not entirely clear what Elliott regarded as Esson's main contribution. But it seems likely to have been Esson's role in promoting the system of 'combined' college lectures, by which college lecturers divided the syllabus between them, so harnessing college-based (and college-funded) expertise in preparing undergraduates for the University's examinations.

this development, and even Maxwell, writing in 1871, feared that laboratory work would be hard to integrate into the university system and might even harm an undergraduate's prospects of success in the Mathematical Tripos (Schaffer 1992: 32–3 and 46; Warwick 1994).[7] Indeed, as Cavendish professor from 1870 until his death in 1879, Maxwell (Second Wrangler in 1854) pre-empted this unfortunate outcome by admitting to the Cavendish only recent graduates of the Mathematical Tripos and a select few others.[8] In Oxford, by contrast, the entry of practical physics into the undergraduate curriculum was a relatively uncontroversial matter. The common practice for Oxford's high-flyers in physics of reading first for the Mathematical School and then for the quite separate School of Natural Science meant that mathematicians there did not feel threatened by the possible incursion of practical physics into their syllabus. Consequently, with laboratory exercises belonging unequivocally with the NSS, most Oxford mathematicians (the Revd Charles Dodgson being a conspicuous exception) were supportive of the early Clarendon Laboratory. In Cambridge, it was not until well into the 1880s, under Maxwell's successor, Lord Rayleigh, that laboratory work in physics became a central part of the experience of undergraduate scientists (initially of those reading for the recently introduced Part II of the Natural Sciences Tripos).

This instance of institutional contingency in the relations between mathematics and physics challenges the stereotypical contrast between Cambridge as *the* place for the study of mathematics and physics, and Oxford as the university for studying the classical languages, literature, and ancient and (to a lesser extent) modern philosophy, united in the prestigious degree of Literae Humaniores or 'Greats'. While in demographic terms this institutional contrast retains a core of validity, it must be qualified by a recognition that mathematics had a degree of centrality in Oxford learning that has only recently been examined with due scholarly care, notably by the contributors to *Oxford Figures* (especially Fauvel 2000; Hannabuss 2000a, 2000b, and 2000c) and to the *History of the University of Oxford*.

---

[7] On 15 March 1871 Maxwell wrote to Lord Rayleigh asking him to come to Cambridge occasionally 'for it will need a good deal of effort to make Exp.[erimental] Physics bite into our University system which is so continuous and complete without it'. As Maxwell saw it, the danger was that candidates for the Mathematical Tripos might become so preoccupied with laboratory experimentation that they might graduate as second-class senior optimes rather than first-class wranglers – a consequence that would appal both their parents and the University. See the full text of Maxwell's letter in Harman (1995: 614–16).

[8] Maxwell's decision not to encourage undergraduates to work in the Cavendish might superficially seem akin to Clifton's similar policy of limiting access to advanced study in the Clarendon to those who possessed a high level of competence in mathematics, ideally obtained through study for the Mathematical School. But, as Gooday shows in his chapter (Chapter 3, 108), Clifton explicitly disavowed this view in 1888 in seeking to encourage more senior students to pursue experimental physics at the Clarendon. Towards the end of his very long career, however, Clifton seems to have reverted to his initial position of discouraging any but the mathematically trained elite from handling the costly and delicate apparatus in the Clarendon.

The evidence about the introduction of mathematics into the degree structure in Oxford reveals a number of instances in which the University followed Cambridge chronologically but with real commitment and not necessarily mimetically. It is true that the Cambridge Mathematical Tripos emerged early, in the 1720s, but it only achieved formally recognized status in the 1750s, just a decade before Oxford embarked on an invigoration of the teaching of mathematics, most notably at Christ Church (Fauvel 2000). In the Oxford reforms, spherical and plane geometry were added to the existing diet of Euclid and Maclaurin's algebra in the later terms of study, and from the 1780s Newton's *Principia* was incorporated into the syllabus. Oxford's major restructuring of the examination system at the beginning of the nineteenth century further strengthened the position of mathematics. Following an important reform dating from 1807, candidates for Honours in Mathematics were examined in a separate School of Mathematics, formally designated the Schola Disciplinarum Mathematicarum et Physicarum. This did not allow such candidates to bypass the classical demands of the core syllabus; it was rather a voluntary additional hurdle for ambitious undergraduates who wished to supplement their required Pass or Honours in Lit. Hum. by displaying a special competence in mathematics. The challenge was not an unattractive one, and in the first half of the nineteenth century, between 10 and 15 per cent of Oxford undergraduates who took Honours in Lit. Hum. also did so in Mathematics. This was always regarded as a notable intellectual exploit, and exceptional performances, such as those of Robert Peel and William Gladstone who took firsts in the same term in both schools (in 1808 and 1831, respectively) were guarantees of celebrity (Brock 1997: 17).

In the format of its examinations too, Oxford did not lag significantly behind Cambridge. From about 1830, the gruelling procedures of the early years of the century were gradually amended in ways that invested examining in mathematics with a new visibility and seriousness (Curthoys 1997: 342–52). Until the mid-1820s, candidates in both Lit. Hum. and Mathematics, whether for a Pass or for Honours, appeared before a single board of examiners, which always included at least one mathematician. The examination, which was conducted for the most part orally and in the unnerving public arena of the Schools (situated round the Bodleian quadrangle, now assimilated into the Bodleian Library), was evidently a fearsome test of undergraduates' stamina, and it soon became one for examiners too, as the number of candidates (only six of whom could be examined in any one day) rose from 188 in 1808 to a peak of 404 in 1827. Stress, however, did not necessarily equate with intellectual rigour, especially in the mathematical part of the examination, which was left a somewhat perfunctory exercise. Convocation's decision of 1825 to appoint separate boards of examiners for Lit. Hum. and Mathematics, therefore, was a response both to the problem of superficiality and to the need to ease the burden of the examination process. The measure, implemented in the following year, of separating the two examinations by a period of three weeks was undertaken for the same ends.

The move towards greater autonomy for the Mathematical School reflected the widespread, though by no means universal, sense in Oxford that the University would gain by moving closer to Cambridge in the status it accorded to mathematics. The Revd Baden Powell, as Savilian professor of geometry from 1827 until his death in 1860, worked hard to import into Oxford something of Cambridge's practices in both the examining and the teaching of the subject. Early in his tenure, in 1828, he gained a notable victory in winning approval for written mathematical papers, but he was not always so successful. As Pietro Corsi has shown, Convocation displayed a conservatism in this area that was hardened by the sectarian turmoil that followed Peel's departure from his Oxford parliamentary seat in 1829 (Corsi 1988: 110–14; Curthoys 1997: 352). In the following year, for example, it rejected the moves of Powell and other members of the mathematical lobby who argued that the examinations in Lit. Hum. and mathematics should be separated by terms rather than the recently instituted norm of three weeks and that those taking Honours in the Mathematical School should be listed in order of merit, as they were in the Cambridge Tripos. Some opponents of Powell's reforms, Hurrell Froude among them, went even further, maintaining that the Mathematical School should be abolished altogether and that the books of Euclid should be restored to their original place in a single classics-dominated syllabus, where for them all ancient learning belonged.

Despite the obstacles, Powell did succeed in getting some elements of Cambridge-style calculus into the Oxford syllabus; in this, a new textbook, his *Short Treatise on the Principles of the Differential and Integral Calculus* (1829), was an important weapon. Powell's efforts to promote calculus in the undergraduate curriculum were followed up by Bartholomew Price during his forty-five years as Sedleian professor of natural philosophy from 1853. Lampooned by one of his pupils in the lines 'Twinkle, twinkle little bat, how I wonder what you're at',[9] 'Bat' Price did much to promote and modernize mathematics as an examination subject, especially during his time as an examiner in the Mathematical School in 1847–8 and 1853–5, an experience that stimulated him to write two important textbooks of calculus (Price 1848 and 1852–60).

These books to a large extent defined the place that calculus had for many years in the Oxford mathematical curriculum, alongside the traditional place that had

---

[9] Lewis Carroll [Charles Lutwidge Dodgson], *Alice's Adventures in Wonderland* (1865), chapter 7. Questions remain as to whether this was a reference to Price's loftier excursions into calculus or to his financial wheeler-dealing on behalf of the Clarendon Press; see Cohen (1995: 44, 59, 136, and 255). Dodgson was not an especially popular mathematics teacher; see Green (1953–4:66–7 and 176).

long been accorded to geometry. While geometrical techniques, defended with special vigour by Powell's successor as Savilian professor, Henry J. S. Smith, were long regarded as constituting the core of the syllabus, calculus was used in conjunction with geometry in tackling the physical subjects of the Mathematical School.[10] In this context, physics and mathematics found themselves in intimate alliance, with common ground not only in astronomy and geometry (the subjects of the two Savilian chairs, Oxford's senior appointments in mathematics) but also in some of the classic themes of 'experimental philosophy': mechanics, hydrostatics, dynamics, optics, pneumatics, and electricity. The point is illustrated by the syllabus for the Final Honour School of Mathematics in 1872, which consisted of eight topics in 'pure mathematics' and four – solid and fluid mechanics, geometrical and physical optics, Newton's *Principia*, and astronomy – in 'mixed mathematics'.[11]

From this vantage-point, we can see how successfully, from the early nineteenth century, Oxford fostered a vigorous if not fully independent mathematical tradition that was not greatly unlike that at Cambridge. How, then, are we to judge the relative influence of mathematics in Cambridge and that of classics at Oxford? The analysis of the Honours degrees taken between 1855 and 1900 in Table 1.1 shows that the disciplines peaked at different times in the two universities. In Cambridge, the numbers attempting the Mathematical Tripos reached a maximum

Table 1.1. Classified 'Finalists' at five-year intervals: Mathematical School, Oxford; Mathematical Tripos, Cambridge; Literae Humaniores, Oxford

Year	Oxford Mathematical School		Cambridge Mathematical Tripos		Oxford Literae Humaniores	
	Male	Female	Male	Female	Male	Female
1855	18	–	143	–	78	–
1860	29	–	124	–	68	–
1865	10	–	94	–	97	–
1870	20	–	113	–	83	–
1875	23	–	86	–	100	–
1880	16	–	100	–	103	–
1885	22	–	118	–	140	–
1890	23	–	108/11[a]	17/0[a]	124	–
1895	23	0[b]	101/8[a]	16/2[a]	152	1
1900	26	0	64/7[a]	19/1[a]	159	2

[a] Part I/Part II   [b] 2 female students in 1894

Sources: *Historical Register* and J. R. Tanner (ed.), *The Historical Register of the University of Cambridge. Being a Supplement to the Calendar with a Record of University Offices, Honours and Distinctions to the Year 1910* (Cambridge, 1917)

[10] Even in Mods, candidates for Honours had to possess a good command of calculus. For a comment bearing on the particular case of H. G. J. Moseley, who was better schooled in mathematics than has sometimes been supposed, see Howarth (2000: 341, note 1).   [11] *OUG*, 3 (1872): 10 (19 January 1872).

Table 1.2. Classified 'Finalists' at five-year intervals: School of Natural Science, Oxford (1855–1900); Natural Sciences Tripos, Cambridge (1855–1900)

Year	Oxford School of Natural Science		Cambridge Natural Sciences Tripos	
	Male	Female	Male	Female
1855	12	–	5	–
1860	10	–	6	–
1865	10	–	11	–
1870	15	–	17	–
1875	20	–	27	–
1880	27	–	32	–
1885	22	–	20[a]	5[a]
1890	26	–	20[a]	3[a]
1895	43	3	18[a]	3[a]
1900	42	4	23[a]	2[a]

[a] Part II only from 1881

Sources: *Historical Register* and J. R. Tanner (ed.), *The Historical Register of the University of Cambridge. Being a Supplement to the Calendar with a Record of University Offices, Honours and Distinctions to the Year 1910* (Cambridge, 1917)

in the 1850s, whereas those taking Honours in Lit. Hum. reached their peak at the very end of the nineteenth century. Yet both institutions launched Bachelor's degree schemes in natural science (including physics) at roughly the same time, about 1850. Also, as Table 1.2 indicates, both experienced a roughly comparable growth in the numbers taking the degrees up to 1882 (when the Cambridge degree was split into a Part I and a Part II). This suggests that any attempt to understand the comparative development of physics, or for that matter of the other sciences, in the two institutions cannot build on any simple correlation with the relative popularity of Cambridge mathematics and Oxford classics. Certainly the strength of the Mathematical Tripos does not appear to have been as influential in fostering science in Cambridge, as has commonly been supposed. This conclusion is reinforced by the fact that at the time when the School of Natural Science was inaugurated in the early 1850s, Oxford's leading practitioners of activities recognizable as physics all held posts, independently, in one or another area of mathematics: this was true not only of Price and Powell but also of Robert Walker, who, as well as being the University's reader in experimental philosophy, was mathematical lecturer at Wadham College, and of William Donkin, who was both Savilian professor of astronomy and mathematical lecturer at University College.

Despite its growing prominence, however, mathematics in Oxford did have difficulty in emerging from the shadow of a tenacious classical tradition in teaching and scholarship. Even the new examination statutes of 1850 had only a limited effect in adjusting the relative status of the disciplines. Under the statutes, undergraduates were obliged to pass in two of the four schools that were now available: Lit. Hum. (which was compulsory) and either mathematics or one of the new schools of Natural Science or Law and Modern History (Curthoys 1997: 352–5; Fox, Chapter 2, 32; Appendix II: 312–14). Thereafter, an almost inevitable relaxing of the classical requirement occurred, but it took place slowly. Only from 1864 was it possible for candidates to proceed to a degree via any one of the schools alone (on condition that in the school concerned a minimum of third-class honours, or from 1870 fourth-class honours, was achieved). And even then classics remained a powerful presence in the experience of most Oxford undergraduates, at least until they took the overwhelmingly classical Moderations examination, usually at the end of their first or second year.

On this last count, there was a contrast with Cambridge that may well have been important for the ways in which the sciences evolved in the two universities; while Oxford undergraduates required a thorough knowledge of Homer and Virgil in order to succeed in Mods, their peers in Cambridge needed only enough rudimentary Greek and Latin to get through the 'Previous' Examination after their first term. It is true that in 1886 the requirement of a Pass in Mods was removed for Oxford undergraduates going on to the School of Natural Science; the Preliminary examination now became the only hurdle that such students had to face between the still unavoidable Responsions and the Final examination. But the residual classical requirement in Responsions remained tougher than anything that Cambridge scientists had to face. And even the modest liberalization of 1887 was won in the face of opposition by conservative dons across the disciplinary spectrum who argued that a narrowing of the syllabus allowing this amount of specialization in non-classical subjects would devalue the Oxford degree. As the Revd Charles Dodgson and other conservatives always maintained, it was the breadth of the University's curriculum, with its option of working for two schools, that had fashioned a distinctively Oxonian elite eminently suited for politics and public life.[12]

By the late nineteenth century, such defences of humane learning were beginning to smack of nostalgia for a world that was disappearing, in Oxford as

---

[12] Like nearly half the prime ministers of the nineteenth century, Peel, Gladstone, and Cecil had been undergraduates at Christ Church, a college that, more than any other, actively supported mathematical learning from the 1760s to the mid-nineteenth century. Thereafter, the centre of both mathematical and political power shifted to Balliol. But Oxford continued its tradition of fashioning political and civil service elites. Seven chancellors of the exchequer educated at Oxford before 1850 attained honours in mathematics, Gladstone included; see Curthoys 1997: 352. MacLeod and Moseley note that of the 93 Cabinet ministers in office between 1868 and 1955, 49 were from Oxford whilst only 26 were from Cambridge (MacLeod and Moseley 1980: 192, note 17). This is, of course, entirely compatible with the drift of MacLeod and Moseley's thesis that Cambridge Natural Sciences Tripos was closely associated with elites in academic and science-related professions rather than (by implication) in politics.

elsewhere, before the onslaught of specialization. In physics, Clifton had consistently championed the cause of modernity, while always insisting on a modest element of classical study in the early stages of an undergraduate's career. For him, as a former Cambridge wrangler (in 1859) who had learned his experimental physics from the Lucasian professor of mathematics, G. G. Stokes, mathematics was far more important than ancient languages as a foundation for physics, and in Oxford he continued to teach relevant areas of mathematics for some time after his appointment to the chair of experimental philosophy in 1865. In this perception, Clifton had influential allies in Henry Smith and Bartholomew Price until their deaths in 1883 and 1898, respectively. It is no coincidence that, until the introduction of the Preliminary Honour examination in the 1870s, the candidates for Honours in the School of Natural Science whom Clifton accepted for advanced study in the Clarendon had commonly been high flyers in the Mathematical School. This trait is especially apparent among those of Clifton's pupils who went on to become demonstrators in the laboratory: the cases of Arnold Reinold, Arthur Rücker, and Lazarus Fletcher, all of whom took firsts in the Mathematical School before going on, as physicists, to firsts in Natural Science, illustrate the point.[13]

Such career-patterns fed the common belief that mathematical prowess was a prerequisite for entry to specialized work in the Clarendon, and that belief in turn helped to create an aura that surrounded Oxford physics in the first quarter of a century or so of Clifton's tenure. John Viriamu Jones was one student who was attracted by it. As the son of a Welsh dissenting preacher and a graduate of University College, London, who was entirely unversed in Greek, Jones was, by Oxonian standards, an unusual candidate (Jones 1915: esp. 19–50 and 255–6; Poulton 1911: 5–109). Yet he applied to Oxford, probably on the advice of George Carey Foster, his professor of physics at University College and an acquaintance of Clifton. The fact that he went up to Balliol College in 1876 as the holder of a prestigious Brackenbury scholarship did not allow him to avoid the classical requirement, and he had to call on Alfred Goodwin, a Fellow of the college who

---

[13] As an undergraduate at Brasenose between 1863 and 1866, Reinold won the University's junior and senior mathematical scholarships and took firsts in Mathematics and Natural Science in 1866 and 1867, respectively. Appointed to a fellowship at Merton in 1866 and the newly created readership in physics at Christ Church in 1869, he served from 1870 to 1873 as Clifton's first demonstrator in the Clarendon (Stocker 1921). Rücker entered Brasenose with a mathematical scholarship in 1867, subsequently winning the University's junior mathematical scholarship and taking firsts in Mathematics in 1870 and Natural Science in 1871. He was elected to a fellowship and college lectureship in mathematics in 1871 and held those positions while serving from 1871 to 1874 as demonstrator in the Clarendon (Thorpe 1921). A similar pattern can be seen in the career of Lazarus Fletcher, whose firsts in Mathematics (1875) and Natural Science (1876) led on to service as demonstrator in the Clarendon from 1875 to 1877; see A. L., 'Sir Lazarus Fletcher, 1854–1921', *Proceedings of the Royal Society of London*, series A, 99 (1921), 'Obituary notices of Fellows deceased', ix–xi. The publications of Reinold, Rücker, and Fletcher bore unmistakable marks of their high-level exposure to both mathematics and physics. Fletcher's characteristically Oxonian expertise in geometry, for example, is evident in his book *The Optical Indicatrix and the Transmission of Light in Crystals* (Fletcher 1892).

was also professor of Greek at UC, for tuition in his preparation for Mods. But in 1879 Jones took a first in the Mathematical School, entered the Clarendon, and in the following year took another first, in Natural Science, specializing (as was now possible) in physics. Clifton's approval of Jones's ability and his demonstrated success in both mathematics and physics made him an obvious candidate for the demonstratorship in the Clarendon that fell vacant in 1881. Jones was duly appointed, although in the event he stayed in the post for only two terms, before becoming Principal of Firth College, Sheffield, and professor of physics there, at the remarkably early age of 25. From Sheffield, in later years, he spoke as a friendly critic of Oxford physics, in particular for its lack of sustained commitment to research and, following a spell as examiner in the School of Natural Science between 1887 and 1889, for the somewhat lax standards of undergraduate examining. It was as a result of his, and his fellow-examiners', comments that the University implemented moves to make examinations more rigorous by introducing a practical exercise at the Preliminary Honour level (Jones 1915: 255–6).

TOWARDS REDEMPTION?

In this chapter, we have insisted on the broad perspective that is required if we are to explore the full extent of the diffused community of physics in Oxford, a community whose members did not necessarily owe a primary allegiance either to the Clarendon Laboratory or, from its inauguration in 1910, to the Electrical Laboratory under Townsend. In such a decentralized structure, it was hard for any one person to develop a unified culture of teaching and research that would rival that which became the hallmark of the Cavendish in the 1880s and 1890s or of Manchester after Rutherford's arrival there in 1907. Clifton, who initially aspired to the authority that was essential if such a culture was to be established in Oxford, must on this count be judged a failure, though in a context that would have defeated any but the most resolute and deft academic politician. To many potential recruits to the Clarendon, the college laboratories were more welcoming and easier of access, and, after an initial surge of interest, serious students working under Clifton became, from the 1880s, rare birds. By 1887, as a result, Convocation's rejection of Clifton's request for £4800 to pay for a dedicated electrical laboratory was widely considered to be reasonable given the scarcity of undergraduate physicists working for Honours. But, reasonable though the decision was in the Oxford context, it served only to highlight the Cavendish's surge to international pre-eminence in one of its flagship areas of research, on electrical discharges, and to foster the view beyond Oxford that physics there was in need of a profound shake-up.

It is not hard to understand the attraction of the chair of experimental philosophy to a man of Frederick Lindemann's ambition in the immediate aftermath of the First World War. Oxford still stood high in the international hierarchy of universities, but its potential for reform on many fronts, soon to be revealed in the report of the Asquith Commission in 1922, was beyond question. Lindemann arrived, therefore, with a vision for his discipline that took little account of the established ways of Oxford and with an expectation that the wind of change would blow in his favour. From the start, he rested his plan on the promotion of research in general and of low-temperature physics in particular. It was a bold choice, even bolder than Lindemann, as an outsider to Oxford, may have realized at the time. But it squared well with his own intellectual trajectory, which had led him to the highest reaches of European physics. Born in Germany, though British by nationality and cultural inclination, he came at the age of 33 armed with the doctorate he had obtained after four years in Walther Nernst's laboratory in Berlin, wartime experience at the Royal Aircraft Factory at Farnborough, and a wide range of contacts gained in Berlin and Farnborough and as joint-secretary of the First Solvay Conference in Physics, held in Brussels in 1911.

Predictably for those who knew the University well, the peculiarities of Oxford soon brought home to Lindemann the difficulty of transforming the Clarendon into a well-organized and firmly led laboratory of international standing comparable with the Cavendish or one of the leading German institutes. Townsend, in particular, proved a disappointment to him. After more than a decade of destructive rivalry in which Townsend had mounted lectures at the time traditionally reserved for Clifton, he was soon at odds with Lindemann too.[14] With the Electrical Laboratory and the Clarendon going their largely separate ways, with Townsend becoming, through the late 1920s and 1930s, increasingly detached from the cutting edge of physics (and proud of it), and with another potential ally, the Dr Lee's professor of chemistry Frederick Soddy, withdrawing into cantankerous isolation, Lindemann had to pursue his aspirations with minimal support.

The promotion of science, especially with research as its core activity, was always going to be an uphill struggle in what remained primarily, as Morrell (1997 *passim*, esp. 433–51) has shown, an arts university. But the mosaic of often conflicting allegiances that characterized the local physics community exacerbated Lindemann's problems, which were already grave enough in the cash-starved post-war years. One strategy that came naturally to someone of his personal wealth was the recruitment of men of independent means: the colourful horse-riding spectroscopist Derek Jackson was such a recruit, as were Thomas Merton,

---

[14] On the relations between Lindemann and Townsend, see the chapters by Lelong, Morrell, and Hughes.

also a spectroscopist, and Gordon Dobson, a pioneer of atmospheric physics who (like Merton) eventually moved the base for his research from the Clarendon to his home.[15] Lindemann's success in securing the appointment of such colleagues testified to his ingenuity and resolve. But in a decade in which (to use Jeff Hughes's term) physics elsewhere was being 'reconstituted', Oxford felt little of the reinvigoration that came in many other universities in the wake of wartime and post-war appointments (such as those of Rutherford at the Cavendish in 1919) and the embracing of new departures in the discipline. Developments in atomic physics, wireless communications, and quantum mechanics (an approach to which Lindemann's mind remained resolutely closed) all swept over Oxford in the 1920s, but they left little trace. And Lindemann himself seems to have lost heart even with regard to his chosen priority in research, low-temperature physics. He published nothing on the subject after 1924, and while he continued to campaign for science in the University and to take a constructive interest in the work of younger colleagues, he seemed by the late 1920s to be set on leading the existence of a gentleman-don and frequenter of elevated political circles and high society. Had he settled definitively into this leisurely way of life, Oxford might have faced the prospect of another two or three decades of moderate productivity in research in the Clarendon, while Cambridge, Bristol (with its state-of-the-art Henry Herbert Wills Physics Laboratory, opened in 1927), and the still very strong laboratory in Manchester, surged ahead.

Chance, however, was soon to play its part. Just as in 1870 the Clarendon had owed its birth to the fortuitous availability of a generous allocation from the almost forgotten funds held by the Clarendon trustees, so now events unrelated to physics helped rekindle the zeal of Lindemann's earliest years in the chair. The signs of his renewed interest in low-temperature research from 1930 would probably not have been enough to revitalize his laboratory. But that interest, which brought the distinguished low-temperature physicist Kurt Mendelssohn from the Technische Hochschule in Breslau to Oxford on a Rockefeller grant in the winter of 1932–3 (during which time Mendelssohn became the first person to produce liquid helium in Britain), soon drew unanticipated strength from the political events that were now escalating in continental Europe. By the spring of 1933, Lindemann was being driven about Germany (by his chauffeur) to offer a home in Oxford to physicists who were affected or likely to be affected by the Nazi laws on race.[16] With the aid of short-term funding, mainly from ICI, Lindemann was able to implant in the Clarendon not only key figures in the

---

[15] On Jackson, Merton, and Dobson, see Morrell, Chapter 7, 238–44.

[16] Lindemann's initiatives probably owed more to his opportunism than they did to any deep humanitarian conviction. As Morrell has pointed out, the Academic Assistance Council was only just being formed when Lindemann made his successful sorties to Breslau and Göttingen, and (in contrast with Rutherford, for example) he took little part in the council's activities (Morrell 1997: 369–72). Other Oxford scientists, however, were deeply moved by the plight of Jewish colleagues in Germany.

Breslau low-temperature team – Mendelssohn, Franz Simon, Heinz London, and Nicholas Kurti – but also Heinrich Kuhn, a spectroscopist from the University of Göttingen.[17]

By his skilful exploitation of the opportunities proffered by events in Germany, and on what was, by international standards, a shoestring, Lindemann was able at last to push ahead with establishing the vibrant research culture to which he had aspired on his arrival in the chair and hence with changing the structures and ideals of Oxford physics. A steady flow of doctoral students who had graduated in Oxford – Bernard Rollin, Brebis Bleaney, Claude Hurst, James Griffiths, and Richard Hull, among them – started on the road to Oxford careers at this time. And while work on low temperatures and spectroscopy remained the staples, new specialisms cautiously made their mark. With an eye on the spectacular developments in Cambridge in the early 1930s, for example, Lindemann dipped his toe in the water of atomic 'machine' physics by installing a high-voltage machine for nuclear experiments and setting some of the young products of the research training in the Clarendon, notably Griffiths and Hurst, along with C. H. Collie (who thus completed his passage from an undergraduate career as a chemist), to work with it.

As Hughes points out in Chapter 8, there is a danger in analysing the vigour of the Clarendon in the 1930s as the personal achievement of Lindemann alone; perhaps, as he suggests, it is ICI rather than Lindemann that should be credited with setting up the Oxford low-temperature group. The fact remains, however, that in this phase of his leadership of the Clarendon Lindemann displayed an astuteness and energy that bore rich fruit. By the eve of the Second World War, his decision to use research as the corner-stone of his campaign to rehabilitate his discipline in Oxford had been amply vindicated. It is true that the number of undergraduates, which had not risen significantly since the early 1920s, remained at about twenty Honours candidates a year throughout the interwar period. But stagnation on the undergraduate front was far outweighed in Lindemann's mind by the growing number of research students and the research teams he put in place. By 1934, as a result, the ageing original Clarendon was already feeling cramped, and Lindemann's thoughts turned to the securing of funds for new premises.

The completion of the new Clarendon Laboratory in the autumn of 1939 is a measure both of Lindemann's powers of persuasion and of the changing attitudes to research-led disciplines in a university that had traditionally given priority to broad liberal education purveyed through the collegiate system. The milestone was a notable one not only for physics but also for the wider project for an expanded

---

[17] Lindemann also brought a number of physicists who did not stay long in Oxford. These included Leo Szilard, Erwin Schrödinger, Fritz London, and (on a particularly brief visit) Albert Einstein. On the impact of the refugees on Oxford physics generally and on the development of low-temperature physics in particular see Morrell, Chapter 7, 244–56, and Gavroglu (1995: 96–138).

Science Area along the edges of the Parks that had been a goal of reform-minded Oxonians in the sciences since the mid-1930s. Yet the old idiosyncracies of Oxford still weighed heavily. Few physicists had college fellowships, for example: apart from Lindemann at both Wadham and Christ Church and Townsend at New College, only four of them were fellows of colleges.[18] And within the Clarendon there were only two university-funded demonstratorships, a situation that left most of the twenty-three researchers to depend on short-term external grants (Morrell 1997: 431–2). The imbalance between the Clarendon and the Electrical Laboratory on this score was especially galling to Lindemann. For, as Lelong shows, Townsend had managed to preserve the gains of his more active years, to the point of having four demonstrators under him in the late 1930s, all paid by the University, in a laboratory that, although still contributing to undergraduate teaching, now lagged some way behind the Clarendon in its research reputation.

So had Oxford physics achieved 'redemption' by the eve of the Second World War? Were we to judge by the criterion of the Cavendish, the answer would be a virtually unqualified No. The purist cast of atomic and nuclear physics that contemporaries saw as the hallmark of Cambridge physics by 1939 had not established itself to anything like the same extent in Oxford. But, as we have argued, and as Hughes does more fully in his chapter, using the touchstone of the Cavendish (and, more particularly, the touchstone of historians' traditional perceptions of the Cavendish) has a narrowing effect on our evaluations of the achievements of other universities. Perhaps Lindemann was right to favour low-temperature physics (with its potential links to the wider world of industry and government, in which he was very much at home) rather than research in the Cambridge style on the subatomic level. Perhaps, too, Oxford-educated physicists, with their solid grounding in 'old-fashioned physics' and geometrical methods, really were more useful in industry than Cambridge graduates nurtured on the headier diet of differential equations and cutting-edge developments of nuclear physics (Gooday 2005). To raise such questions is not, of course, to undervalue the Cavendish's towering achievements in its areas of strength. It is merely to question the canonical status, as the unique criterion of excellence, of what Hughes defines as the characteristic 'reductionism' of Cambridge physics.

The sheer range of research traditions and the number of physics laboratories that had established themselves in British universities by 1939 reinforce the warning we have advanced against any selective vision founded on judgement against a single model, however ennobled by the accretions of history. With that warning,

---

[18] Lindemann's position deserves comment. Elected a fellow of Wadham on his appointment, he moved to Christ Church in 1922, when the chair was renamed the Dr Lee's professorship of experimental philosophy following the death of the Dr Lee's reader, R. E. Baynes. Despite the move, he remained a member of the governing body at Wadham; university statutes precluded his becoming a member of that of Christ Church as well. He resided in Christ Church until his death in 1957.

of course, there come the obvious challenges to historians of physics in our period. The fact that the contextual analysis, as opposed to the celebration or apologetic dismissal, of traditions and laboratories, is now recognized as a leading focus for scholarly research only highlights the need for work on the swathes of problems and materials that lie beyond the areas that have traditionally won historians' attention. *Physics in Oxford* is one response to this challenge. It attempts no sentimental rehabilitation of Oxford's reputation in physics. Rather, it seeks to lay bare the particular constraints, incentives, and shifting profile of opportunities that fashioned the world in which the discipline's practitioners in the University pursued their diverse and, in large measure, locally determined aspirations.

This is not to deny that there are criteria by which Clifton, Townsend, and Lindemann might be deemed to have failed, in different respects and in relation to certain specific roles. But such judgements can only be properly made in the context of a serious engagement with the objectives that such figures set for themselves and with the pressures, traditions, and opportunities that aided or impeded their aspirations. Oxford physics, in other words, has to be judged in its own terms, rather than as a case of manifest inadequacy, and with a sensitivity to personal rivalries, chance, the vagaries of university politics and finance, and the special difficulty of promoting science in a university in which classics occupied the summits of prestige, all of which played their part, against a wider national and international backcloth of disciplinary change and political and economic circumstances. It is just such complexity that emerges from the chapters that follow. We do not see this as a weakness. Certainly, the alternative of easy characterization is not one to which any of the authors would wish to subscribe.

2

The Context and Practices of Oxford Physics, 1839–77

Robert Fox

The appointment of Robert Walker to the University's readership in experimental philosophy in 1839 marked the beginning of four decades in which physics came to occupy an unprecedentedly prominent place among the disciplines represented within Oxford.[1] Succeeding Stephen Peter Rigaud, who had held the title of reader in experimental philosophy in conjunction with his tenure of the Savilian chairs of, first, geometry (1810–27) and then astronomy (1827–39), Walker came to the post with a background that was at once mathematical and clerical. In his late thirties at the time of his appointment, he had taken a second in Literae Humaniores and a first in Mathematics in the same term, Michaelmas 1822. He had become a deacon and junior chaplain of his college, Wadham, in 1826 (a position he held until his marriage in 1831) and gone on to ordination in 1827 (Boase 1892–1921: vol. 3, col. 1151–2; Gardiner 1895: 282–3; Simcock 1993a). His academic career had begun in 1828, when he was appointed mathematics tutor at Wadham (although, as a married man, he was ineligible to become a fellow), and it was as a mathematician and the author of a well-informed and up-to-date textbook of mechanics that he was elected a Fellow of the Royal Society in 1831.[2] Despite his appointment to the readership, Walker maintained his engagement with mathematics (though without publishing) and he continued to teach the subject at Wadham and to examine regularly

---

This chapter has been written over a period of several years. In that time, I have benefited greatly from advice generously given by Mark Curthoys. I am also indebted to John Sanders, who has helped me to make the most of sources in his care in the Clarendon Laboratory Archive, and Simon Bailey, who has been unfailingly supportive as Keeper of the University Archives. I wish to express my gratitude to Stephanie Jenkins, who has not only typed successive drafts of the chapter but also read them with a critical eye and drawn my attention to sources (cited in notes 25 and 57) of which I was unaware. Finally, the text has benefited from the comments of other authors in this volume who, in the spirit of our collaborative venture, have read it at various stages of its preparation.

[1] On the earlier history of the readership and the decision to appoint Walker with a salary that did not necessitate his holding another post see Fox, Gooday, and Simcock, Chapter 1, 1.

[2] The textbook, which displayed Walker's familiarity with modern French mathematical techniques, appeared as *The Elements of the Theory of Mechanics* (Oxford, 1830).

Fig. 2.1. The Revd Stephen Peter Rigaud, reader in experimental philosophy, 1810–39. Rigaud also held the Savilian chairs of geometry (1810–27) and astronomy (1827–39) and was Radcliffe observer from 1827. From an original silhouette in the Museum of the History of Science (inventory no. 15701).

Fig. 2.2. The Revd Robert Walker, reader in experimental philosophy, then professor, 1839–65. As an Anglican clergyman, Walker was also chaplain of his college, Wadham, from 1826 until his marriage in 1831, and vicar of Culham from 1848 until he resigned the living in 1862.

in the School of Mathematics until 1853. From 1848, he also served as vicar of Culham, a few miles to the south of Oxford, and lived in the vicarage there.

The decision to convert the readership into a fully independent post carrying the salary of £230 p.a. that Walker received was motivated more by a rather dimly perceived potential for the position than by evidence of an immediate unmet demand. Indeed, it was taken at a time when the lectures of most professors and readers in the sciences were conspicuously less attractive than they had been in the years from 1814 to 1830, when William Buckland in particular had drawn consistently healthy audiences, averaging 34 in mineralogy and 61 in geology. The general decline of interest in scientific lectures in the 1830s and 1840s, which saw Buckland having difficulty in even securing audiences in double figures (*Oxford University Commission* 1852: *Evidence* 285), may well have owed something to the displacement of the interests of dons and their pupils towards the debates on Anglican doctrine raised by the Oxford Movement. But it probably had more to do with a growing preoccupation with the gaining of examination-based distinctions among undergraduates, the ablest of whom were drawn to the disciplines, supremely the classical languages, literature, and philosophy, in which scholarships and fellowships were most plentiful. In this respect, it is significant that lectures in certain non-scientific disciplines, such as law, also suffered through being unrepresented in the examining process.

Whatever the reasons for decline, the fact remains that the science professors who addressed the Commission of Inquiry into the University in 1850–1 made much of the particular difficulty they faced in attracting significant numbers to their lectures. The Aldrichian professor of chemistry (Charles Daubeny) and the Savilian professors of geometry (Baden Powell) and astronomy (William Donkin) spoke as those who seem to have been most directly affected. Daubeny, for example, noted a fall in the attendances at his lectures from an average of 31 a year between 1822 and 1830 to 16 between 1831 and 1838 and finally to 12 between 1838 and 1850, while Powell and Donkin reported audiences that were much smaller still, not infrequently falling to zero (*Oxford University Commission* 1852: *Evidence* 257–62 and 267–8). Walker, for his part, could draw some comfort from the relative popularity of his predecessor's lectures, attendances for which over the previous eighteen years had averaged almost forty and been boosted since the early 1830s by the decision of the Dean of Christ Church, Thomas Gaisford, to make attendance at a course of lectures on experimental philosophy compulsory for all the college's undergraduates (*Oxford University Commission* 1852: *Evidence* 259).[3] But the climate remained unpromising.

Walker occupied the rooms on the first floor of the Clarendon Building that had been allocated to Rigaud in 1832, when geology and mineralogy, whose collections and lecture-room were installed on the same floor as those for experimental philosophy,

---

[3] Rigaud's sustained drawing power in experimental philosophy stands in sharp contrast with his dwindling audiences as Savilian professor of astronomy after his appointment in 1827.

Fig. 2.3. The Clarendon Building from the south-east, c.1922. Rigaud (from 1832) and Walker (from 1839) lectured on the first floor of this early-eighteenth-century building until the opening of the University Museum in 1860. Francis Frith Collection, postcard 72736.

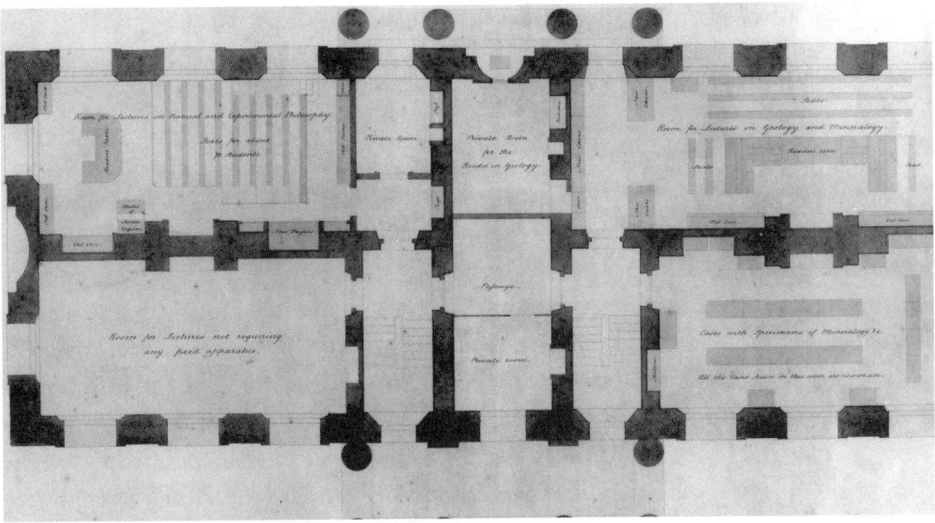

Fig. 2.4. Plan of the first floor of the Clarendon Building, drawn by Robert Smirke in May 1831 when acting as architect for the refurbishment that followed the departure of the Press to its new premises in Walton Street. The area designated for the lecture-room for experimental philosophy, in the top left-hand (south-eastern) quarter of the building, was first used by Rigaud, when he moved his teaching from the Old Ashmolean Building in 1832. The drawing, signed by Smirke, is in the Oxford University Archives, UD/11/1/3.

Fig. 2.5. Model of a James Watt beam engine, one of the more spectacular items acquired by the Oxford physics lecturers in the early nineteenth century from the London manufacturers Watkins & Hill, who were the leading contemporary suppliers of physics teaching apparatus. It is a fully working model, burning charcoal. The 1831 plan of the physics lecture-room in the Clarendon Building (see Figure 2.4) shows the position for 'Model of Steam Engine'. The engine is now in the Museum of the History of Science (inventory no. 26418).

had also moved in to occupy the space left by the departure of the University Press. Here, the apparatus for experimental philosophy was housed, and Walker gave his lectures – one series in Hilary Term and one in Michaelmas Term – in a room with space for 70 students at the south-eastern corner of the building. As a lecturer, he was evidently more engaging than most of his scientific contemporaries. For A. G. Vernon Harcourt, who took a first in Natural Science in Trinity Term 1858, he was an 'excellent' lecturer (Vernon Harcourt 1910: 358). And William Tuckwell, in his reminiscences of his undergraduate days at New College in the midcentury, spoke just as glowingly, recalling 'a cheery Mr Walker, who constructed and exploded gases, laid bare the viscera of pumps and steam engines, forced mercury through wood blocks in a vacuum, manipulated galvanic batteries, magic-lanterns, air-guns' (Tuckwell 1900: 41). The range of Walker's lectures, covering mechanics, hydrostatics, meteorology, optics, electricity, and magnetism (see Figure 2.6), was impressive, and correspondence, instruction-leaflets, and a notebook that he kept during his time

EXPERIMENTAL PHILOSOPHY.

INTRODUCTION—The laws by which inquiries in Natural Philofophy fhould be governed—matter and its properties—the laws of motion—the attraction of cohefion—gravitation.

MECHANICS.

The lever—center of gravity—the balance—pully—wheel and axle—wedge—inclined plane—fcrew—the means of altering the direction of motion—compound machines—friction—ftrength of materials.

Uniformly accelerated and retarded motion—projectiles—central forces—the Solar fyftem.

Bodies moving down inclined planes—pendulums—weights and meafures.

Collifion.

Springs—watches.

HYDROSTATICS, PNEUMATICS, HYDRAULICS.

Properties of fluids in general—thofe of incompreffible fluids—fpecific gravities—properties of elaftic fluids—height and preffure of the atmofphere—found—fteam—the motion of water through pipes—fiphons—water fpouting from an orifice—time in which veffels will be emptied—refiftance of fluids.

METEOROLOGY.

Barometer—Thermometer—Hygrometer.

OPTICS.

Dioptrics—refraction—foci of lenfes—ftructure of the eye—microfcopes—telefcopes.

Catoptrics—reflection—foci of reflecting furfaces—aberration produced by fpherical furfaces—reflecting telefcopes.

Perfpective—camera obfcura—camera lucida—anamorphofes.

Colours—rainbow—difperfion of light—achromatic lenfes.

ELECTRICITY.

MAGNETISM.

Terms of attending the Lectures—For the Firft and Second Courfes, each Two Guineas, and for any fubfequent Courfe, One Guinea.

‸ All Courfes attended in the time of the late Reader in Experimental Philofophy to be reckoned.

S. Collingwood, Printer, Oxford.

Fig. 2.6. Announcement of Robert Walker's lectures on experimental philosophy, c. 1840.

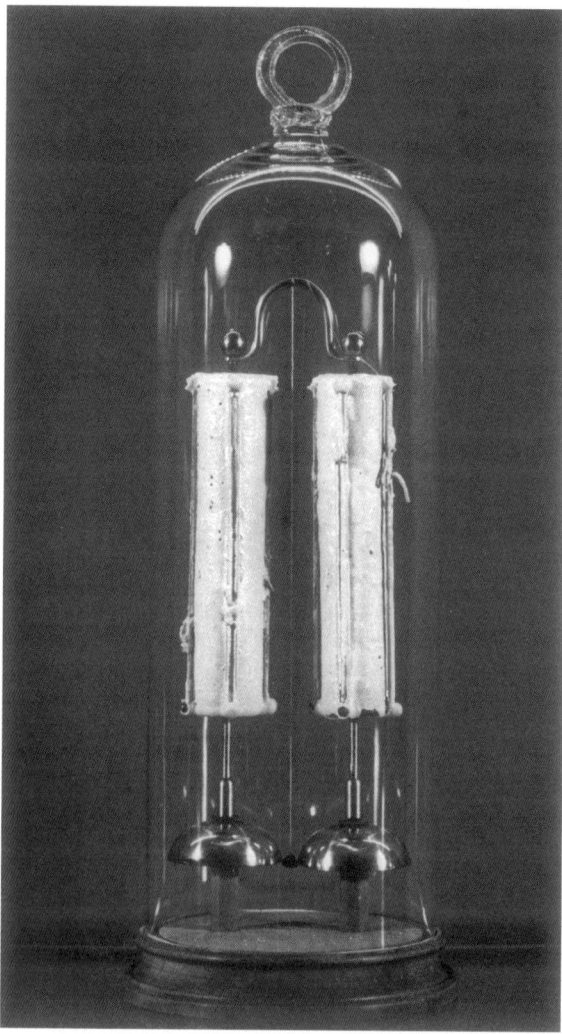

Fig. 2.7. The Clarendon dry pile. This pair of voltaic 'dry-piles' bears a label in Robert Walker's handwriting: 'Set up in 1840'. Since then, the potential difference generated by it has rung two bells, probably making it the world's most durable battery. Its sound is now muffled in the display cabinet that it occupies near the main entrance of the present Clarendon Laboratory's Lindemann Building.

in the post shows that, until the onset of his protracted last illness in the early 1860s, he purchased equipment, within the constraints of a limited budget, and made every effort to care for it and master its use.[4] The equipment was almost exclusively for demonstration purposes; there is no evidence that it served either for Walker's own research or for any but the most occasional practical exercises by undergraduates and others attending his lectures.

Despite his capacity to draw a respectable audience and a positive, outgoing personality that found characteristic expression in an active engagement in ballooning, Walker was at one with Daubeny in his zeal for the advancement of science as a

---

[4] The correspondence and notebook, with some miscellaneous manuscript material dating from his and Clifton's time, are preserved in the Clarendon Laboratory Archive.

subject for serious academic study in the University. The measure, taken in 1839, requiring all candidates for the B.A. to attend two courses of lectures by professors or readers had given some fleeting encouragement to the disciplines, non-scientific as well as scientific, that were not examined in the existing Schools of Literae Humaniores and Mathematics. But the imperfect monitoring of attendance, by written certificates issued after each course or each lecture, and the absence of attempts to test what had been assimilated, had undermined any hope of lasting benefit.[5] The only possible remedy, as both Walker and Daubeny agreed by the late 1840s, lay in the formal integration of science in the syllabus for the bachelor's degree.

Walker's views on syllabus reform were, if anything, more radical than Daubeny's. Whereas Daubeny sought to make the study of the three 'primary sciences' of mechanical or natural philosophy, chemistry, and physiology (essentially biology) a normal part of what would remain, for the most part, the existing undifferentiated curriculum, Walker argued for a new and more demanding degree structure. In this structure, in order to take a B.A., it would be necessary to pass through two schools: one in moral philosophy and history (in which candidates would be examined on Old and New Testament history, one or two Pauline epistles, the historical and religious significance of the Thirty-Nine Articles, and a philosophical or historical book in both Greek and Latin), and the other in either mathematics or 'physics', a term that, as Walker used it in this context, embraced all of Daubeny's primary sciences (Walker 1848). Candidates for a Pass would be judged on this basis alone; those aspiring to Honours in one of the schools would undergo a further examination, for which a deeper knowledge (of more difficult authors or an individual chosen science) would be required.

In the event, Walker's concrete proposal gathered little support, and it was quickly lost from view in the gathering campaign on behalf of science. In this campaign, Walker had no difficulty in standing shoulder to shoulder not only with Daubeny but also with the young Henry Acland, recently appointed Dr Lee's reader in anatomy at Christ Church, whose views were close to and probably in large measure derived from Daubeny's. The three men drew strength from the knowledge that the tide was already beginning to turn in ways that would inevitably favour the sciences. One sign lay in Oxford's sense of mounting pressure to initiate reforms of the kind that were to be urged on the University following the Oxford University Commission of 1850–1. Another, more closely related to science, lay in the encouragement that had followed the meeting of the British Association for the Advancement of Science in Oxford in June 1847. It was

---

[5] The potential for the abuse of the system is illustrated by the abnormally large number of undergraduates attending two lectures by Daubeny (a notoriously poor lecturer) on rural economy. It had been spotted that the two lectures could be deemed to constitute a course and hence to offer a short cut to the fulfilment of the requirement. See Vernon Harcourt (1910: 352–3).

in the wake of that meeting that Walker joined Daubeny, Acland, and the keeper of the Ashmolean Museum, Philip Bury Duncan, in signing a very visible memorandum supporting the cause of Oxford science in the University. Dated 12 July 1847, the memorandum urged the University to erect a building that would house the scientific collections currently dispersed between the Clarendon Building, the Ashmolean Museum, and Christ Church and offer space for lectures, meetings, and a library.[6] Buckland's refusal to sign the memorandum, on the grounds that the cause was 'utterly hopeless', was not unexpected, since his hopes of achieving a better status for science had now succumbed to the profound disenchantment that was to mark his declining years.[7] But while Buckland's attitude came as a setback, it was a minor one, and the champions of science turned their immediate energies to the more promising area of the reform of the undergraduate curriculum.

The reform duly came about with the adoption of the examination statutes of 1850, which brought into being the School of Natural Science (hereafter NSS) along with the School of Law and Modern History. The move, which left Oxford with a complement of four schools – the two new ones in addition to those of Literae Humaniores and Mathematics – stood as a victory for modernization over the considerable resistance of certain Heads of House and in Convocation. It was, in its general thrust, a bold new departure, since candidates for the B.A. now had to achieve a Pass or Honours in two of the four schools (see Appendix II and Curthoys 1997: 325–5). But its effect was limited by the requirement that one of the schools had to be the still dominant Literae Humaniores. As far as the NSS was concerned, there were other constraints too. One that Walker immediately addressed was the inadequacy of existing textbooks; his textbook of mechanics, hydrostatics, and pneumatics covered at least part of the syllabus in Mechanical Philosophy up to Honours level (Walker 1851–2). But a far graver constraint was the weakness of science teaching in the public schools and grammar schools. This made science appear, in the eyes of the vast majority of undergraduates, as of their classically trained tutors, a forbidding field of study.

However daunting the NSS seemed to undergraduates, it is hard to see its syllabus as difficult in any respect other than breadth. The regulations for the School established a syllabus covering Daubeny's three primary sciences.[8] For

---

[6] A manuscript copy of the memorandum is in the archives of the University Museum (now part of the Oxford University Archives): 'History of the building of the Museum' (hereafter OUA UMA), box 5, folder 1. A printed version is reproduced in Acland's evidence to the Devonshire Commission: *Devonshire Commission. First, Supplementary, and Second Reports* 1872: *Minutes of Evidence* 172.

[7] A manuscript copy of the letter in which Buckland expressed his hopelessness is in OUA UMA, box 5, folder 1.

[8] The regulations are reproduced in many places, including the annual *OU Calendar* and the *Oxford University Commission* 1852: *Report* 66–7.

Honours, candidates were to be examined in all three of these sciences and, at a more advanced level, either in one of the primary sciences or in another scientific discipline chosen by the candidate. In this way, abler students could venture into such subjects as geology, physical geography, botany, zoology, and mineralogy, all of them 'special' or 'subordinate' sciences (in Daubeny's terminology) that lay outside the core syllabus. Even candidates for a Pass (or 'Common') degree had to show some measure of adventurousness by selecting at least one of the constituent elements of Mechanical Philosophy – mechanics, hydrostatics, pneumatics, sound, light, heat, and electricity – in which they would be examined, as well as in the principles of any two of the primary sciences. At least in principle, physics was thus given a prominence not accorded to the other sciences.

As both the champions of science and the sceptics knew, the acid test of the reform of 1850 lay in the take-up for the new school. And on this score, the NSS failed to live up to expectations. In a degree structure in which candidates for the B.A. could fulfil the requirements equally well by tackling, as well as the obligatory passage through Lit.Hum., what was perceived as the less-demanding material of the School of Law and Modern History or the more familiar syllabus in Mathematics, Natural Science immediately became the poor relation. The numbers attempting the School were small and they remained obdurately so, as the figures in Table 2.1 indicate. In the eight years from 1853, when the first candidates (two Passmen in Easter Term) presented themselves under the new regulations, to 1860, for example, only 71 took honours in the School, a figure that set the NSS numerically far behind Lit.Hum. (577), Mathematics (177), and its fellow newcomer, Law and Modern History (253). Numbers for the Pass degree were slightly higher, but the figure of 80 for the NSS in the same period was even more spectacularly eclipsed by those for Lit.Hum. (1567), Mathematics (758), and Law and Modern History (691).

As reader in experimental philosophy (latterly as professor, after the title of all readers was changed by statute in 1860), Walker worked tirelessly for the cause of science in the context of what he, like the other members of the Oxonian scientific lobby, recognized to have been a disappointing reform. His lectures, given (until 1860) in the Clarendon Building and treating the whole range of his discipline, retained their unusual drawing-power and regularly had audiences of over thirty, sometimes substantially more.[9] While these numbers reflected to some extent the demands of the new examination requirements, they clearly exceeded those of candidates attempting the NSS, as Walker's meticulously kept record of his lecture-fees

---

[9] The high attendances at Walker's lectures are confirmed by the signed certificates of attendance that have survived in OUA SP/85, 86, and 88.

Table 2.1. Honours and Pass Degrees, 1853–77

Term and Year	NSS		Lit.Hum.		Maths		Law/ Mod. Hist.		Theology		Mod. Hist.		Law	
	Hons	Pass	Hons	Pass	Hons	Pass	Hons	Pass	Hons	Pass	Hons	Pass	Hons	Pass
ET 1853	0	2	18	52	6	7	8	2						
MT 1853	1	0	44	59	10	28	13	19						
ET 1854	3	0	37	105	11	30	25	34						
MT 1854	0	2	29	72	15	35	22	42						
ET 1855	7	3	34	117	10	51	16	31						
MT 1855	5	4	44	89	8	48	20	60						
ET 1856	5	3	50	129	13	56	18	46						
MT 1856	4	1	38	117	10	60	16	69						
ET 1857	6	3	28	90	9	46	14	68						
MT 1857	10	4	36	108	9	45	19	48						
ET 1858	4	10	41	99	11	51	12	44						
MT 1858	7	10	48	78	14	57	15	53						
ET 1859	4	9	29	117	11	52	15	38						
MT 1859	5	11	33	111	11	52	13	50						
ET 1860	4	8	24	147	11	63	10	36						
MT 1860	6	10	44	77	18	77	17	51						
ET 1861	8	11	22	119	9	59	10	42						
MT 1861	5	21	34	120	12	69	23	32						
ET 1862	6	11	33	154	7	66	10	32						
MT 1862	6	18	23	125	14	87	21	35						
ET 1863	2	9	31	162	8	69	16	33						
MT 1863	6	5	28	96	16	60	19	59						
ET 1864	2	6	25	132	7	65	12	39						
MT 1864	8	3	50	74	12	58	16	31						
ET 1865	2	1	40	103	7	71	12	53						
MT 1865	8	1	57	76	3	72	17	51						
TT 1866	3	1	31	98	10	67	21	33						
MT 1866	4	0	53	76	8	69	28	29						
TT 1867	2	2	28	100	10	67	20	22						
MT 1867	7	2	57	110	4	71	43	29						
TT 1868	8 [?5]	2	25	178 [148?]	8	61	24	41						
MT 1868	8	0	61	?	8	?	29	?						
TT 1869	3	0	25	?	5	?	17	?						
MT 1869	12	1	61	76	9	79	42	29						
TT 1870	5	0	27	128	11	73	25	29	4	0				
MT 1870	10	0	56	93	9	74	35	41	5	0				
TT 1871	3	1	21	128	6	59	18	36	11	0				
MT 1871	10	0	69	73	17	64	34	30	14	0				
TT 1872	6	0	26	103	6	41	33	31	13	0				
MT 1872	7	0	85	71	12	86	22	28	35	0	12		15	
TT 1873	10	0	24	126	9	63	0	26	27	0	25		13	
MT 1873	7	0	64	84	10	62	0	27	30	0	35		18	
TT 1874	7	0	27	120	4	65		30	23	0	17		17	
MT 1874	17		60		10				26		47		18	
TT 1875	9		23		10				26		30		22	
MT 1875	11		77		13				29		34		24	
TT 1876	22		82		18				23		41		27	
MT 1876	9		24		8				23		18		15	
TT 1877	15		80		14				23		41		20	
MT 1877	15		14		9				14		22		17	

Sources: Complete lists of successful candidates, including those taking a Pass, appear in the *Oxford University Gazette* from its inauguration in January 1870. The sources for earlier years are the *Oxford University Calendar* and the full printed lists of successful candidates, at both Honours and Pass level, that the examiners published after each examination.

(a one-off payment of between £1 and three guineas, giving access to all subsequent lectures) confirms.[10] Yet Walker was manifestly dissatisfied and he remained vehement for change. In 1857, his dissatisfaction was directed at the first public examination – Moderations – that had been introduced in the reforms of 1850 as a hurdle to be passed between Responsions and the final examination. The examination had been conceived as a means of allowing basic, mainly linguistic knowledge to be tested at an intermediate stage, so as to leave more time for subsequent, more specialized study, whether of science, mathematics, philosophy, or history. In Walker's view, however, it had quickly become an intrusive diversion for candidates, many of whom devoted inordinate time and energy to securing high honours in what was intended to be no more than a qualifying test (Walker 1857: esp. 4). Walker was correct. But his protest went too brutally against the grain of rapidly established undergraduate and tutorial habits to have any hope of success, and in any case his energies had by now an alternative and more realizable target: that of securing premises for the teaching of science and the housing of the associated collections.

The campaign for the physical needs of the scientific professoriate had been revived as early as 1849, when Walker, Daubeny, and Acland were instrumental in calling a meeting of 21 members of Convocation in the lodgings of the Warden of New College, the Revd David Williams, and in the formation shortly afterwards of an Oxford Museum Committee that set the project for a worthy home for the sciences on its tortuous path to realization.[11] Reinvigorated in the spring of 1851 by the unheralded availability of £30 000 allocated for the purposes of a new museum from the profits of the University Press, and in the following year by the report of the Oxford University Commission, the champions of the museum project navigated resolutely and successfully through the choppy waters of conservative opposition and suspicion about the inexorably mounting costs of the building and its fitting out. On 20 June 1855, they were able to celebrate the laying of the foundation stone by the Chancellor, Lord Derby,[12] and five years later, with the accumulated investment spiralling towards the £87 000 that had been spent by 1867, they displayed the building to the scientific world during the meeting of the British Association for the Advancement of Science. It was there, on Saturday 30 June 1860, that over 700 people – dons, undergraduates, clergy, and members of the BAAS – had their first experience of the museum as they crowded into the still unfurnished library on the first floor to hear Thomas Henry Huxley engage in a

---

[10] A notebook recording the fees that Walker received term by term between 1849 and 1861, along with each term's printed announcement of his lectures, is in CLA, bookcase, shelf 3. See Figure 2.8.

[11] On the building of the University Museum and the debates and manoeuvres that led to and accompanied it see Fox (1997: esp. 645–65).

[12] The excitement of the ceremony is caught well in 'The University Museum, Oxford', *The Illustrated London News*, 26 (1855): 651–2, and in the report in *The Times*, 21 June 1855.

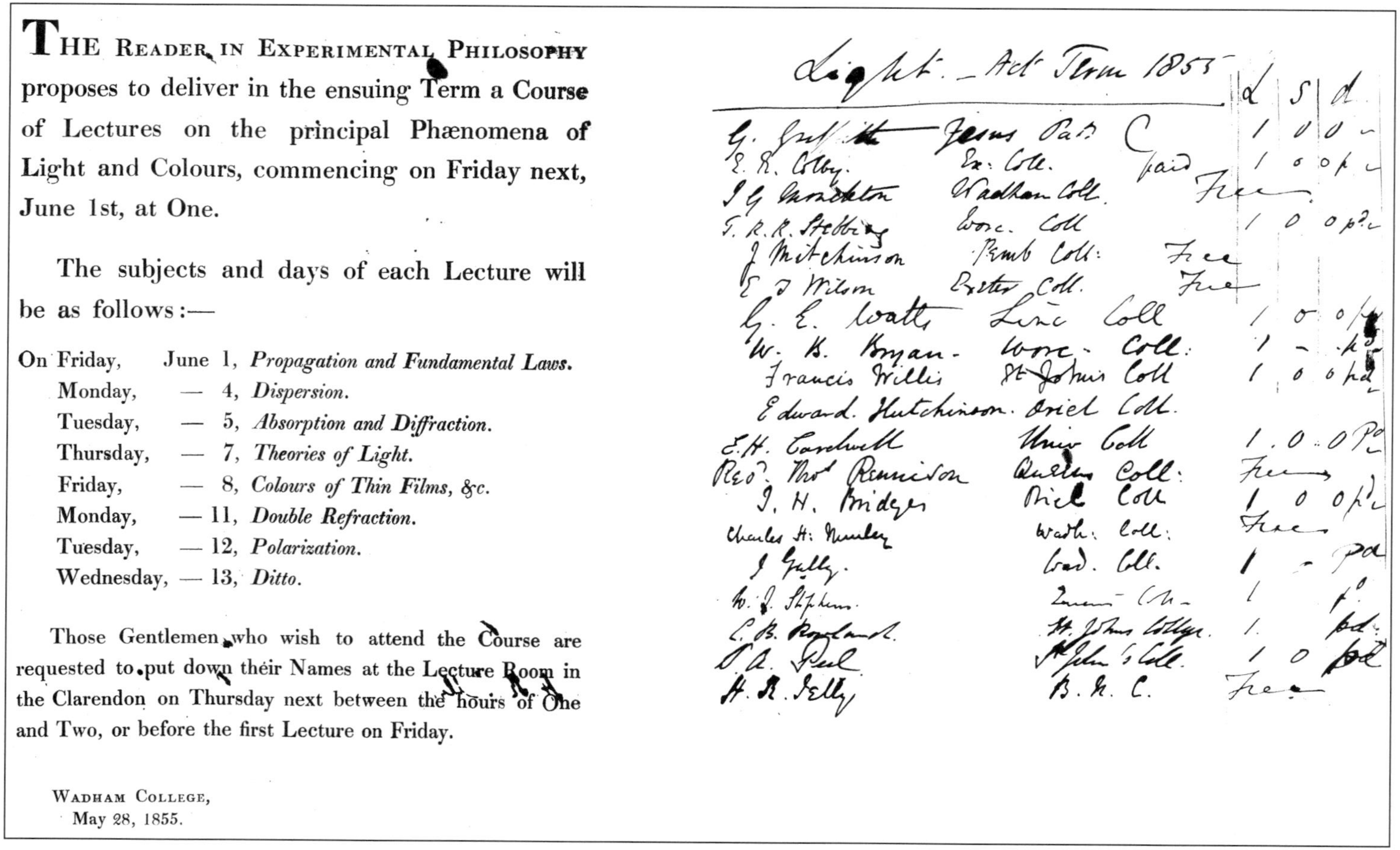

THE READER IN EXPERIMENTAL PHILOSOPHY proposes to deliver in the ensuing Term a Course of Lectures on the principal Phaenomena of Light and Colours, commencing on Friday next, June 1st, at One.

The subjects and days of each Lecture will be as follows:—

On Friday,   June 1, *Propagation and Fundamental Laws.*
Monday,   —   4, *Dispersion.*
Tuesday,   —   5, *Absorption and Diffraction.*
Thursday,   —   7, *Theories of Light.*
Friday,   —   8, *Colours of Thin Films, &c.*
Monday,   —   11, *Double Refraction.*
Tuesday,   —   12, *Polarization.*
Wednesday,   —   13, *Ditto.*

Those Gentlemen who wish to attend the Course are requested to put down their Names at the Lecture Room in the Clarendon on Thursday next between the hours of One and Two, or before the first Lecture on Friday.

WADHAM COLLEGE,
May 28, 1855.

Fig. 2.8 Facing pages from Walker's notebook (see note 10) recording those paying the fee for his lectures on 'the principal Phaenomena of Light and Colours' in Act (Trinity) Term 1855. The names include those of George Griffith, who undertook the teaching of physics during Walker's illness, 1862–5, and the future positivist John Henry Bridges, who had taken a third in Lit.Hum. at Wadham in Michaelmas Term 1854. Bridges's college affiliation appears as Oriel, where he was elected a fellow in 1855.

Fig. 2.9. The University Museum, inaugurated in 1860. From the Oxford University Almanack for that year.

tumultuous debate with the Bishop of Oxford, Samuel Wilberforce, about Charles Darwin's theory of evolution by natural selection.[13]

The opening of the University Museum, as the building was called from a very early stage in its planning, gave Oxford science the premises, under one roof, for which its advocates had striven for over a decade. As the collections were assembled from their disparate locations in the Old Ashmolean Building, the Clarendon Building, and Christ Church, the snags inherent in the structure – the difficulty of heating and lighting the huge central court, the leaks in the roof, and the impracticality of the lofty space of the chemistry laboratory – had yet to make their mark. In the eyes of most observers, Oxford had made an enviably lavish, even extravagant, provision for its men of science. And the professors who began to use the museum for work on and with their collections and apparatus – Walker in experimental philosophy, Benjamin Brodie in chemistry, William Donkin in astronomy, George Rolleston in the newly created Linacre chair of anatomy and physiology, Nevil Story-Maskelyne in mineralogy, Bartholomew Price in natural philosophy, John Phillips in geology, and John Obadiah Westwood, soon to become the first Hope professor of zoology – gave every impression of satisfaction.

In fact, the University Museum had been fashioned in such detail and for so long by those who now worked in it that serious discontent would have been

---

[13] In an abundant secondary literature on the debate see, for a recent account, Browne (2002): 114–25.

surprising. The disposition of the rooms allocated for the different disciplines, on two floors around the glass-covered central court, enshrined a belief in the underlying unity of the sciences and the need for easy communication between their premises that united the leading promoters of the project (see Figures 2.10 and 2.11). But while the allocation of space conveyed the priority of neighbourliness, it also reflected perceptions of the particular needs of each discipline. The chemical laboratory, for example, was housed in an oddly designed building, modelled on the abbot's kitchen at Glastonbury, that was set apart from the main structure, though connected to it by a corridor (Crosland 2003), while the rooms for anatomy were arranged round a small courtyard on the cooler north side of the building. The area given over to these disciplines, in each of which experimental work by students as well as by professors was envisaged, was more generous than that for experimental philosophy, mineralogy, and geology, all conceived as activities that had no need of significant facilities or laboratories for student exercises.

Experimental philosophy, like mineralogy and geology, was housed along the museum's south side. Provision for it consisted of a lecture-theatre, storage space for instruments (mainly in cupboards along the walls of the southern colonnade), and adjacent rooms described on the architectural plan as work-rooms, consisting of an upper room (the professor's private space) and a lower one, directly below it, designated as a laboratory. The lecture-theatre, which could accommodate audiences of about 150, was spacious. But practical teaching, conducted in the lecture-theatre and in one of the two adjacent rooms (each with a floor area of little more than 20 square metres) must have been difficult.[14] The contrast deserves comment. For the emphasis on lecturing and lecture demonstrations that this distribution of space represented was perfectly, and significantly, consistent with Walker's long-established pedagogical practices. There is no sense in which Walker had fought and then lost out in the planning and allocation of resources. He had evidently seen no need for large-scale laboratory instruction.

The removal to the Museum had virtually no effect on the numbers of undergraduates taking Honours in the NSS, which continued to hover at about ten a year (see Table 2.1). Numbers taking a Pass, after the still obligatory passage through Lit.Hum., on the other hand, did rise. But the boost was short-lived, and by the mid-1860s the Pass in Natural Science, which had 32 and 29 successful candidates in

---

[14] According to the plan reproduced in Figure 2.10, the pairs of rooms on either side of the lecture-theatre were given over to experimental philosophy. But a sketch by John Phillips indicating the proposed allocation of rooms in 1861 confirms that only the pair on the west side of the theatre was formally earmarked for the discipline; see minute-book of the Museum delegacy, 1858–66, OUA UM/M/1/2, tipped in at f.31, with minutes for 8 June 1861. On the early recognition of the inadequacy of this provision and the borrowing of one of the adjacent pairs of rooms to the east of the lecture-theatre see below, p. 47 and note 33.

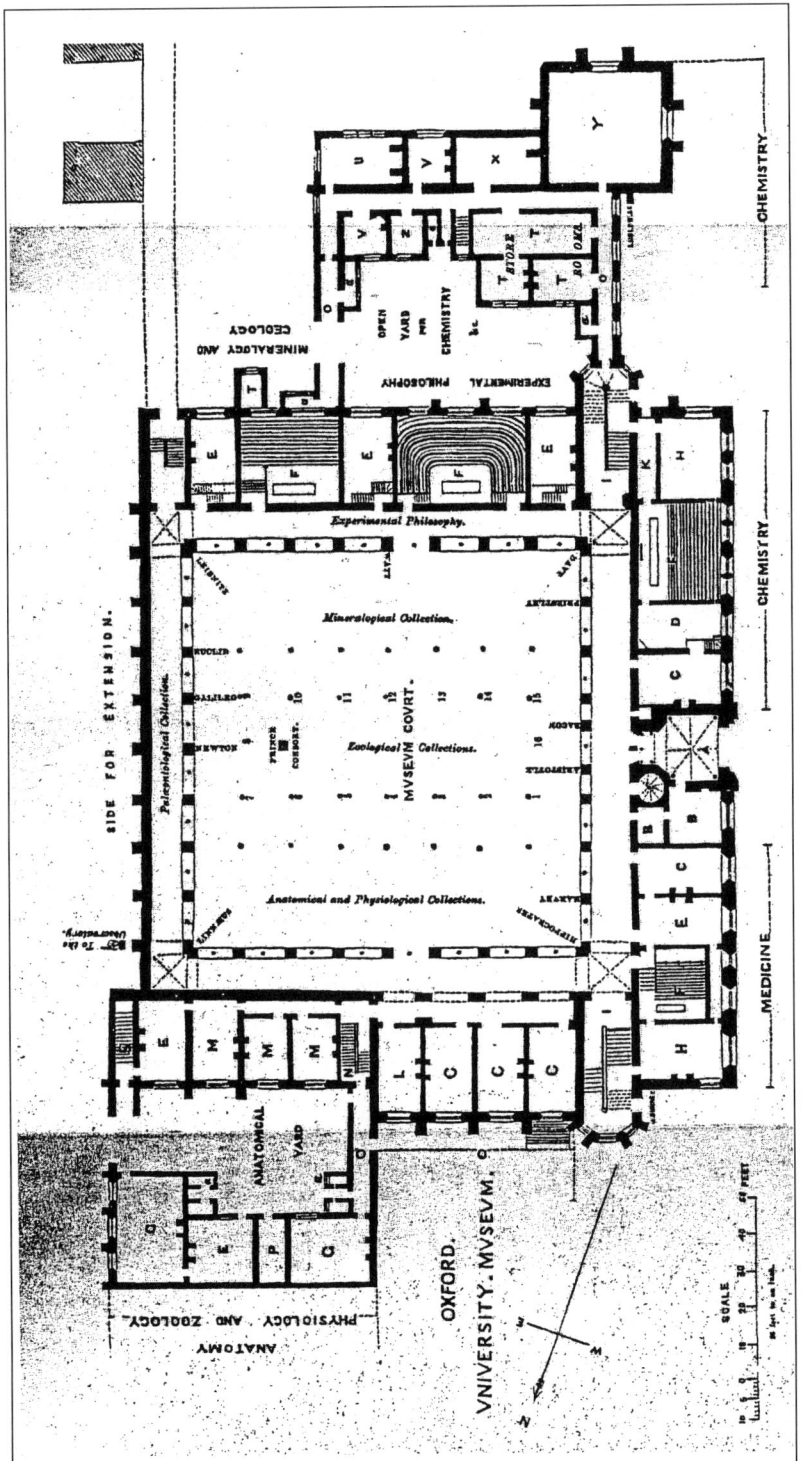

Fig. 2.10. Plan of the ground floor of the University Museum. From Henry Acland, *The Oxford Museum. The Substance of a Lecture*, 3rd edn (Oxford, 1866), frontispiece. The space originally allocated for experimental philosophy consisted of the lecture-room indicated on the southern side of the court and the two small rooms (one above the other) marked as E at the south-western corner. The rooms on the other side of the lecture-room, also marked E, were designated for mineralogy. However, in Nevil Story-Maskelyne's prolonged absence, they were used for experimental philosophy, first by George Griffith, then by Robert Clifton.

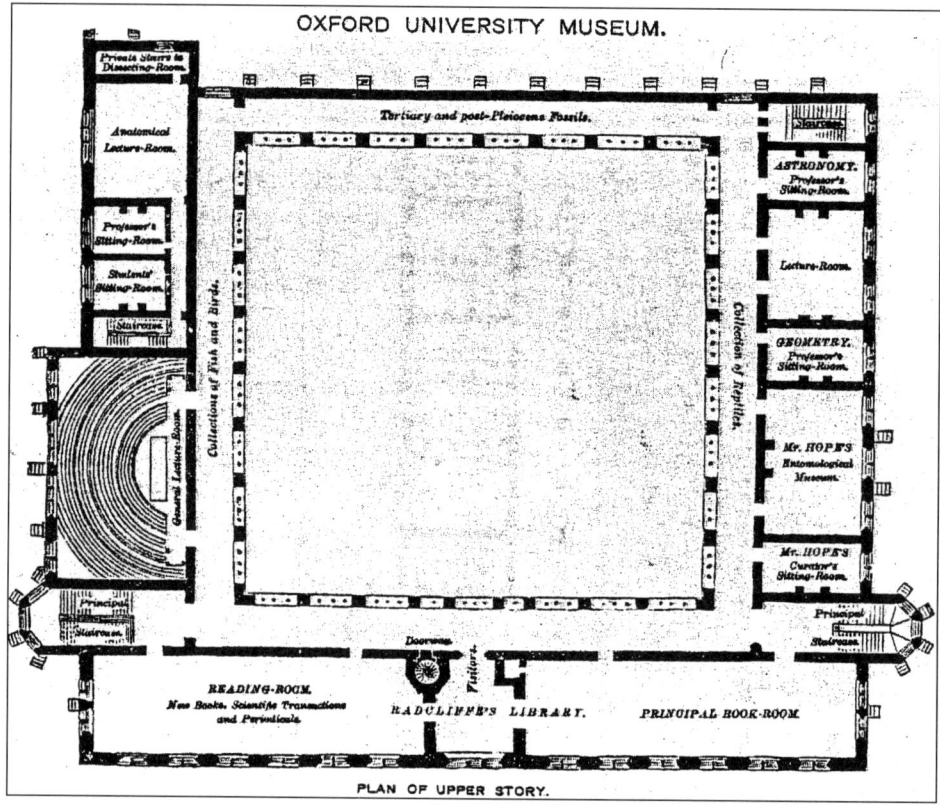

Fig. 2.11. Plan of the first floor of the University Museum. From Henry Acland, *The Oxford Museum. The Substance of a Lecture*, 3rd edn (Oxford, 1866), facing p. 63. Note the rooms allocated to the Savilian professors of geometry and astronomy and for the Hope entomological collection and its curator (Westwood).

1861 and 1862 respectively, was awarded to no more than two or three candidates a year. Behind such low figures there lurked the danger of the marginalization of science as a teaching subject. But the science professors had at least one straw at which they could clutch. This was the passing of the new examination statutes of 1864 in the face of vehement conservative opposition in Convocation. These statutes held out the prospect of a greater seriousness among the abler undergraduates working in the sciences, since it now became possible to graduate after taking at least third-class honours (or, from 1870, fourth-class honours) in any school, without going through the full Lit.Hum. mill: a Pass in Moderations was sufficient before moving on to whichever of the honour schools a candidate had in his sights. While the statutes were conceived as a modernizing measure that did something to favour the new schools, the reform had only a limited effect, in two respects. First, it did nothing to change the preference of the majority of undergraduates, who continued to favour the existing route of a Pass in Lit.Hum. and in one other school

(Mathematics and to a lesser extent Law and Modern History being by far the most popular choices). Secondly, and more importantly, it did not allow the complete avoidance of serious classical study; even a Pass in Moderations required a good command of Latin and Greek grammar and of Latin prose composition, as well as a knowledge of at least one Latin and one Greek author, the four gospels in Greek, and either logic or three books of Euclid and algebra 'as far as easy quadratics'.[15] The only consolation for the promoters of science was that it was now possible for undergraduates aspiring to specialize in natural science at honours level to begin doing so by the end of their second year.

Despite the limited nature of the reform, mechanical philosophy, as one of the primary sciences and hence a leading candidate for the advanced specialization that was required of all candidates for Honours, might have been expected to reap at least some benefit. But by 1864 the subject was in difficulties. Not long after Walker had completed the move from the Clarendon Building to the Museum, he showed signs of the failing health that culminated in his death in September 1865. The summer of 1860 was still a busy and productive one for him, as General Secretary of the BAAS and the deliverer of a major lecture on the sun during the Oxford meeting (Walker 1860). And the academic year 1860–1 passed well enough: in October 1860 and again in January 1861 he announced courses of lectures in the Museum.[16] But during the year 1861–2, responsibility for registration for Walker's lectures and almost certainly for the delivery of at least some of them passed gradually into the hands of George Griffith, an early graduate of the NSS who had taken first-class honours at Jesus in Michaelmas Term 1856 and had been appointed in the following year to one of the first college appointments in science, as lecturer (though never a fellow) in natural science. As Walker's capacities declined, Griffith filled the gap as best he could, offering both lectures and practical classes and even, from 1863, using the title of Deputy Professor of Experimental Philosophy. However, the arrangement was an unsatisfactory one that could only impede the development of the subject, and Walker's death was seen as the opportunity for a fresh start. Within a month, by October 1865, arrangements were in place for the appointment of a new professor at a salary of £500, the sum paid to Walker towards the end of his career.[17]

---

[15] The procedures and requirements for examinations under the various statutes were clearly presented year by year in the *OU Calendar*. For an excellent overview see also Curthoys (1997).

[16] The pattern of Walker's gradual withdrawal from teaching is conveyed in the termly announcements of professorial lectures and classes, a valuable collection of which is in Bodl. G.A. Oxon b.33. Also the few entries for Michaelmas Term 1861 and Lent Term 1862 in the notebook cited in note 10, above, appear to have been made by George Griffith.

[17] Walker's salary was raised from £230 to £300 in the early 1850s and to £500 following a recommendation of the University Commissioners in 1855. See *Oxford University Commission 1852: Report* 108, and *Oxford University Commission 1852: Appendix* 60; also *Evidence* 284. A summary of the changes in salary is in *OU Historical Register*: 63. The professorial salary of £500 that the University offered in 1865 was made up of £30 from Lord Crewe's benefaction, £270 from the University Chest, and £200 gained by commuting a

In the competition that followed, Griffith was the obvious internal candidate, and he entered the fray with gusto. The support he mustered, in a printed brochure of testimonials from a number of former pupils and, among Oxford scientists, from Phillips, Acland, Daubeny, Story-Maskelyne, the Balliol mathematician Henry Smith, and the Radcliffe Observer Robert Main, was plentiful (*Testimonials in favour of George Griffith* [1865]). In addition, his work as assistant general secretary of the British Association for the Advancement of Science won him a measure of outside support, including recommendations from Charles Lyell and Francis Galton. Main's affirmation, in his letter of support to Griffith, that 'in the whole circle of my scientific acquaintance, I do not know any one so well qualified as yourself to fill with honour and advantage to the University the vacant Chair' was a typically glowing endorsement (*Testimonials in favour of George Griffith* [1865]: 9). But Griffith's qualities were predominantly those of a reliable local man with broad scientific interests and a cultural range reflected in the seconds that he took in Mods in both classics and mathematics in 1855 (the year before he took his first in the NSS). He was seen first and foremost as, in Story-Maskelyne's words, 'a tried and successful teacher' (*Testimonials in favour of George Griffith* [1865]: 6), a reputation he had earned not only in the University but also at Winchester College, where he had taught for one day a week since 1863.[18] He was certainly not the distinguished external figure that the modernizing elements in the University, in particular on the Hebdomadal Council, seem to have been set on placing in the post.[19]

The suggestion (of uncertain origin) that an approach should be made to Hermann Helmholtz, at the time professor of physiology at Heidelberg, is a measure of the ambitiousness that was abroad as the prospect of an election to the chair began to materialize. In the event, however, nothing came of this idea: Max Müller, shortly to become Corpus professor of comparative philology, seems to have convinced the authorities that Helmholtz would not be interested, even if an enhanced salary of £700 were offered (Kurti 1984; Roche 1990). Whether

---

fellowship at Wadham College (a measure that did not preclude the professor's election to a non-stipendiary fellowship, if the college were so minded). In addition, the professor would be allowed to take a fee of £1 for each person attending his lectures and to apply for the purchase of apparatus from a sum of £3100 in consols administered for the purpose by the Vice-Chancellor and the Provost of Oriel from the bequest of Lord Leigh, an Oriel man. The provision for the new professor is summarized in 'A statement of the duties and emoluments of the professorship of experimental philosophy', a single printed sheet in Bodl. G.A. Oxon. c.81 (f. 290).

[18] On Griffith's work at Winchester and his views on the difficulty of advancing science in schools, see the evidence he gave to the Schools Inquiry Commission [Taunton] in *Schools Inquiry Commission. Vol. IV. Minutes of Evidence taken before the Commission* (London, 1868): 173–86. The Headmaster of Winchester, the Revd George Moberly, was one of several of Griffith's supporters who spoke warmly of his achievements at the school; see *Testimonials in favour of George Griffith* [1865]: 15.

[19] Acland's endorsements of Griffith (*Testimonials in favour of George Griffith* [1865]: 5) on the grounds of his familiarity with anatomy and physiology and hence of the appropriateness of his experience for the Oxford School of Natural Science reflects an admiration for breadth that sat uneasily with the emerging priority that many modernizers gave to specialization.

Helmholtz was ever approached (even informally by Müller) is unclear. But it seems likely that he only learned of the suggestion after the event, through a chance encounter with Henry Smith in Paris in the spring of 1866. A letter to his wife, written shortly after the meeting with Smith, indicates that he had no regrets:

It was said in course of conversation that there had been some notion of inviting me to go to Oxford as Professor of Physics. However, they could not offer more than £700 salary, which of course is more than we get in Heidelberg, but hardly enough to live comfortably in England . . . so I think Prof. Max Müller was right to say he could tell them decidedly that I should not accept it. (Koenigsberger 1906: 232)

With the diversion of Helmholtz's possible candidature having run its brief course, six other candidates (none of them identified) disappeared from view. This left the electors (a majority of them drawn, in accordance with an important recent reform, from the scientific professoriate[20]), with just one alternative to Griffith.

The alternative was Robert Bellamy Clifton, who offered the merits of a fine undergraduate record at Cambridge and five years' experience as the first professor of natural philosophy at Owens College, Manchester, where he had acquired the respect of his senior scientific colleague, the chemist Henry Roscoe, and a reputation as a popular lecturer.[21] Clifton's application also bore the aura of connexions far more extensive than Griffith's (*Application of Robert Bellamy Clifton* [1865]). His supporters included not only Roscoe but also J. P. Joule, William Thomson, J. C. Adams, G. G. Stokes, Isaac Todhunter (who, like Stokes, had taught him as an undergraduate), William Whewell, and, from Heidelberg, Robert Bunsen and

---

[20] By a reform characteristic of the modernizing trends of the mid-century, the practice of electing to the chair on the nomination of the Vice-Chancellor was abandoned in 1863. Following a debate whose keenness is reflected in the fly-sheets preserved in Bodl. G.A. Oxon c.79 (ff. 250–2), the electors were henceforth to be, in addition to the Vice-Chancellor and the Warden of Wadham College (the source of part of the salary; see note 17, above), the Savilian professor of astronomy, the Sedleian professor of natural philosophy, and the Aldrichian professor of chemistry. This was the only electoral board for a scientific chair (except for the recently created Linacre professorship of physiology) on which the professors of science held a majority. The nature of the opposition to the change is conveyed in critics' fears that the scientific professoriate might become a 'clique'. As the author of one fly-sheet put it, electors not holding appointments in the sciences would offer the merit of a special sensitivity to the bearing of scientific knowledge on the University 'as a religious place of education'.

[21] On Clifton see the obituary by Glazebrook (1921) and the unsigned notices in *Nature*, 107 (1921): 18–19, and *Monthly Notices of the Royal Astronomical Society*, 82 (1921–2): 248. More recent discussions of Clifton's work are to be found in Croft (1986: Chapter 9), and Gooday (1989: Chapter 6). A copy of Croft's study is available, with an almost complete set of the photographs accompanying the text, in the Clarendon Laboratory Archive. Another copy, without the photographs, is in the History of Science and Technology Seminar Room in the Modern History Faculty, Oxford, and an imperfect version, also lacking the illustrations, is in the Radcliffe Science Library (RR.x.384). Roscoe seems to have had a genuinely high regard for Clifton, whom he described as 'most popular' and whose lectures he judged 'admirable'; see Roscoe (1906: 109).

Georg Kirchhoff, who had met him on a visit to Roscoe's house in Manchester (Roscoe 1906: 73). If the lack of any previous contact with Oxford was a handicap (and the opposite was almost certainly the case), this did not weigh in the decision, and in November 1865 Clifton was appointed, leaving Griffith to become, soon afterwards, a full-time teacher at Harrow.[22] At the time, Clifton was still only 29, and his record of publication – a note on the early history of the signs '+' and '−' (Clifton 1866b)[23] and three modest papers on optics, one of them jointly authored with Roscoe (Clifton 1860; Clifton and Roscoe:1860–2; Clifton: 1866a) – was on the thin side. Nevertheless, on the basis of his success as an undergraduate and in teaching in Manchester (a burden that he insisted had seriously impeded his research), he arrived in Oxford with a reputation as one of the most promising physicists of his generation.

The son of Robert Clifton, a small landowner in the remote southern extremity of coastal Lincolnshire, and Frances Gibbons, Clifton had been brought up in the family home at Chapel Gate, Gedney.[24] After schooling in Peterborough and Brighton and a brief period at University College in London, where he studied under Augustus De Morgan and Richard Potter and won prizes for both mathematics and German,[25] he matriculated at St John's College, Cambridge, in March 1855. His undergraduate career (especially from his second year) was successful, although he found himself consistently pipped by his brilliant fellow-Johnian James Maurice Wilson, the future headmaster of Clifton College and later a prominent northern clergyman. In the twice-yearly college examinations that undergraduates at St John's were required to take for the first three of their four years in the college, Clifton appeared regularly in second place, behind Wilson. In the Mathematical Tripos, in 1859, he was Sixth Wrangler and second Smith's prizeman, beaten yet again by Wilson (who was both Smith's prizeman and Senior Wrangler). His performance was sufficient to secure his election as a fellow of his college and to lay the foundations for the appointment at Owens that he took up in 1860, when he was still only 24.

Clifton completed his lectures in Manchester by Easter 1866 and in the early spring took up residence in Oxford, establishing himself with his family and

---

[22] Griffith taught at Harrow, where he was the first (and, until 1876, only) science master. Regarded as increasingly ineffective in his later years, he retired in 1893 and died in 1902. See Stogdon (1937: vol 1, 602) and Tyerman (2000: 331–2, 382–3, 400).

[23] The note reflected his life-long interest in early mathematics, on which subject he accumulated a fine collection. See Gooday, Chapter 3, 90.

[24] On Clifton's family background and career in Cambridge, I have drawn on material in the archives of St John's College, Cambridge, especially the college examinations book (C15.8) and a small amount of associated material, including a copy of the entry for Clifton in the baptismal register for Gedney and a letter of 28 March 1855, by Isaac Todhunter, certifying that Clifton was 'duly qualified in point of manners & learning for admission as a member of St. John's College, Cambridge'.

[25] *The Times*, 2 July 1855: 9, where Clifton is recorded as having been runner-up for the college's senior mathematical prize and winner of the junior prize for German. I owe this reference to Stephanie Jenkins.

Fig. 2.12. Robert Bellamy Clifton, a photograph probably dating from the late 1870s, after just over a decade into his fifty-year tenure of the chair of experimental philosophy.

servants in a spacious Italianate residence in the recently built development of Park Town, within walking distance of the Museum.[26] As an ambitious outsider who had had no part in the earlier struggles to promote science in the University, he slipped easily into the role of a reformer keen to set his discipline on a less parochial and scientifically more distinguished path than Walker had ever thought possible or desirable. Although his experience in Cambridge had been primarily mathematical, he had attended Stokes's experimental lectures on optics, and now, as a physicist, he made experiment the core of his discipline. From the beginning of the Easter Term 1866 he settled into what was for many years to be his regular rhythm of lecturing twice a week (in a cycle, lasting three or four years, that covered the whole of

[26] Clifton had married Catherine Elizabeth Butler, daughter of the Revd Dr Butler of Reigate, in 1862. In the 1881 census return (in the transcription by the Church of Latter Day Saints) he is recorded as living at Portland Lodge (now 7 Park Town) with his wife and daughter, Catherine, aged 7. Three female domestic servants – a cook, a parlour maid, and a nurse – are also listed. Two of Clifton's three sons had already been born but are not mentioned; they may well have been at boarding school at the time. Clifton moved to a newly built house, 3 Bardwell Road, also in north Oxford, in 1893 and lived there until his death in 1921.

physics) and offering 'practical instruction', mainly to undergraduates aspiring to Honours in the School of Natural Science (Passmen having all but disappeared by the time of his arrival).

For the Hilary Term of 1867 (the only one for which detailed records are available), his lectures attracted three graduates, 24 undergraduates, and four advanced pupils from Magdalen College School, where one of the earliest school laboratories had been installed in 1863 (Stanier 1958: 158). Of those attending the lectures, six were also enrolled for his laboratory class (see Table 2.2). Such figures, though hard to reconcile with Clifton's assertion that eighty undergraduates had attended his first course of lectures, in 1866,[27] represented a brisk response and lent substance to his claim that both space and equipment, especially for laboratory teaching, were seriously inadequate. Annual sums of £150 a year from the University and about £90 from Lord Leigh's benefaction, and a generous special grant of £1000 that the University made in 1869, allowed some renewal of the now ageing apparatus that Walker had painstakingly accumulated until the collapse of his health in the early 1860s.[28] But support on this scale could not meet the needs of the practical teaching that Clifton regarded as an essential complement to his lecture demonstrations for undergraduates aspiring to Honours.[29] Quite apart from the requirement that candidates for Honours who chose mechanical philosophy as their special discipline in the NSS should receive laboratory training and undergo a practical examination,[30] it was such teaching, and such teaching alone, as Clifton insisted, that made it possible to exercise the student's all-important faculty of observation along with that of reasoning.

The keenness of Clifton's engagement and the air of modernity that fired his commitment to experimental practice won the minds, if perhaps not the hearts, of his immediate colleagues. By 1 November 1866, when he had been in the post for only a few months, a report by Phillips (in his capacity as Keeper of the Museum) to the Vice-Chancellor had endorsed unreservedly the case for providing the professor of experimental philosophy with additional laboratory space and space for workrooms and the storage of apparatus; the needs of zoology were said to be great, but Clifton's were the 'most pressing' of any of the science

[27] The figure of 80 appears on p. 3n of the printed report of 1 November 1866, cited in note 31, below.

[28] The £1000 almost certainly came from the fund for the purchase of apparatus referred to in note 17, above. In the years in which Griffith had replaced Walker, nothing had been spent on apparatus, apart from small sums necessary for repairs, as Henry Smith testified; see his letter in *Testimonials in favour of George Griffith* [1865]: 8.

[29] Clifton expounded his pedagogical principles with particular clarity in his evidence to the Devonshire Commission in 1870. See *Devonshire Commission. First, Supplementary, and Second Reports* 1872: *Minutes of Evidence* 186.

[30] The requirement that part of the examination for such candidates (whichever science they chose) should be practical was introduced in 1862.

professors.³¹ Within days, the report was on its way to Hebdomadal Council, with the full support of the Museum delegates. A second report followed in February 1867; even more forceful than the first, it was based again on a letter from Phillips to the Vice-Chancellor and destined for transmission, via the delegates, to Council. Now, the constraints that Clifton suffered were elaborated in even starker detail, partly by Phillips and partly by Clifton himself in the personal statement that he added to the report.³² Even with access to additional space borrowed from mineralogy (space of which Story-Maskelyne, with his time largely devoted to his post of keeper of mineralogy at the British Museum, was making little use), the provision for experimental philosophy fell far short of Clifton's expectations. Practical instruction, which Clifton seems to have conducted in the borrowed space, in the lower of the pair of small rooms belonging to mineralogy, could still only be offered to a handful of pupils (see Figure 2.10).³³ The conditions were so unsatisfactory that, according to Phillips (echoing Clifton), those seeking the highest honours had no choice but to leave Oxford and work in one of the new generation of physics laboratories that were beginning to be established or refitted in continental and even some British universities.

Clifton's requirement, as he elaborated it in his personal statement, was for six rooms that would serve as student laboratories, in addition to battery rooms, workshops, and storerooms, a 'physical Cabinet' for the secure housing of apparatus, and a private laboratory for the professor's personal research. At this stage, Clifton envisaged an extension of the southern front of the Museum in an easterly direction (an expedient suitable at once for optical experiments and for the preservation

---

[31] The report, in the form of a letter from Phillips to the Vice Chancellor, dated 1 November 1866, was printed under the heading 'For the use of the Museum delegacy only' and then reissued, after approval by the Delegacy on 10 November 1866, for submission to Hebdomadal Council. Copies are in the minute-book of the delegacy, 1858–66 (OUA UM/M/1/2, f. 86), the (unpaginated) minute-book for 1867–75 (OUA UM/M/1/3), and an important collection of correspondence and printed papers concerning the funding and building of the Clarendon Laboratory (OUA UC 10/33). In a brief introductory paragraph, the Delegacy urged Council to commend the proposed expenditure, as well as expenditure for zoology, to the University. Clifton was evidently dissatisfied with the lecture-theatre as well, since he did not lecture in Michaelmas Term 1866, while refurbishment (in particular for lighting and ventilation) was in progress; see Clifton to Phillips, 8 March 1867, OUA UM/F/4/3.

[32] Copies of this second report are in the minute-book of the Delegacy for 1867–75 (OUA UM/M/1/3) and in OUA UC 10/33. The core of the report was Phillips's letter of 1 February 1867 (pp. 1–4), followed by Clifton's personal statement (pp. 4–8). After discussion by the delegacy on 2 March 1867, the report was passed on for consideration by Hebdomadal Council.

[33] In his letter of 16 September 1867 to Sir William Thomson (see Appendix III), Clifton refers to the space allocated to him, consisting of his lecture-room and the two adjacent rooms: a 'small office' on the upper level and a laboratory underneath, used for preparing lecture-demonstrations. Such practical instruction as he could offer was conducted in the room borrowed from mineralogy. The latter room, which exists virtually unchanged today, could scarcely have accommodated more than half a dozen pupils at any one time. Henry Smith's testimonial confirms that Griffith, while deputizing for Walker, held practical classes in the same cramped space. In Smith's words, the room was 'quite unfit for such a purpose; it being impossible without further outlay to perform in it exact experiments on frictional electricity, or to observe with accuracy the solar spectrum'; see *Testimonials in favour of George Griffith* [1865]: 8.

Table 2.2. Those attending Robert Clifton's classes, Hilary Term 1867

Name	College	Year of BA	Honours in Natural Science/Mathematics	Mods and Other Schools and Distinctions	Career, etc.
Allen, Francis Hordern	Pembroke	1869	Pass in Natural Science (TT 1869)	Pass in Lit. Hum. (TT 1869)	Vicar of All Saints, Moss, Yorks (1875)
Beecroft, George Audus Beaumont	Christ Church	1868 (B.Mus. 1867)	2 Natural Science (MT 1868)	2 Classical Mods (1866)	Died in 1873
Bennett, L.	Pembroke				Probably George Spencer Leigh Bennett, who took his B.A. in 1867 and was vicar of Long Sutton, Lincolnshire, in 1886
Billing	Magdalen College School				Probably Campbell Pymar Billing, who matriculated at Magdalen in 1867 but took no degree
Bird, Thomas	Oriel	1868	Pass in Natural Science (MT 1867)	Pass in Lit. Hum. (MT 1868)	
Boswell, Robert Bruce	Lincoln	1870	2 Natural Science (MT 1869)		
Champneys, Francis Henry	Brasenose	1870 (B.Med. 1875)	1 Natural Science (MT 1870)		St Bartholomew's Hospital (1870); Radcliffe travelling fellow (1872); and a medical career
Conroy, John [Sir John Conroy, Bart.]	Christ Church	1869	1 Natural Science (MT 1868)		Lecturer in chemistry, Keble (1881–5), then fellow of Balliol; see Simcock, Chapter 4, 134–7
Coxe, Hilgrove	Corpus Christi	1868	3 Mathematics (TT 1868)		Rector of Albury, Oxon. (1870); vicar of Hinckey, Berks (1879–80), and of Pyrton, Oxon (1880)
Donkin, William Frederick[†]	Magdalen	1868	2 Natural Science (MT 1868)		Lecturer in natural science, Keble (1875–7), and tutor (1877–80)
Dyer, William Turner Thiselton	Christ Church	1867	2 Mathematics (TT 1867) 1 Natural Science (MT 1867)	2 Maths Mods (MT 1865)	Assistant director (1875–85), then director (1885), Kew Gardens

Name	College	Year	Honours	Career
Forbes, L. [Arthur Litton Armitage Forbes?]	Wadham			Arthur Litton Armitage Forbes matriculated at Wadham in 1864. He appears to have left without taking a degree but went on to a medical career, practising as a surgeon in London
Glanville, John Usher	Exeter	1868	3 Natural Science (MT 1868)	Various curacies held from 1869
Goolden, Walter T.	Magdalen College School			Walter T. Goolden matriculated at Magdalen in 1867 and, after taking a first in the NSS in MT 1871, graduated in 1872
Gwyn-Jeffreys, Howell†	Balliol	1867	1 Mathematics (TT 1867)	Barrister, Lincoln's Inn (1870); J.P. Glamorgan and Hereford
Haslam, John Bailey*†	Christ Church (incorporated)	[BA Cantab, 35th Wrangler and 4th Classic 1866]	2 Maths Mods (ET 1865) 3 Classical Mods (ET 1865) Johnson Memorial Prize (1867)	Fellow of St John's, Cambridge (1867–73); science master, Clifton College (1867); teaching career (classics) followed by service as Government Inspector of Schools (1874–94)
Heath, Christopher Henry Edmund*†	Pembroke	1866	1 Maths Mods (MT 1865), 3 Classical Mods (MT 1865), Senior mathematical scholar (1868)	Barrister, Lincoln's Inn (1871)
Hodgson, Richard Greaves	Christ Church	1867	1 Mathematics (MT 1867)	Assistant master (1868–70), master (1871–9), then headmaster (1879), King's School, Canterbury
Lankester, Edwin Ray	Christ Church	1868	1 Natural Science (TT 1868)	Burdett–Coutts scholar (1869)	Radcliffe travelling fellow (1870); fellow, Exeter (1872); professor of zoology and comparative anatomy, University College, London (1874); professor of natural history, Edinburgh (1881)

Table 2.2. (Continued)

Name	College	Year of BA	Honours in Natural Science/Mathematics	Mods and Other Schools and Distinctions	Career, etc.
Meredith, Thomas	Exeter	1868	3 Natural Science (TT 1868)		Archdeacon of Singapore (1882)
Morrell, George Herbert	Exeter		2 Natural Science (MT 1867)	1 Maths Mods (TT 1865), Junior mathematical Scholar (1865)	The name on Clifton's list is 'Mull' but the pupil was undoubtedly G.H. Morrell, later M.P. and a leading member of the Oxford brewing family
Moseley, Henry Nottage	Exeter	1868	1 Natural Science (TT 1868)		Radcliffe Travelling Fellow (1869); Fellow of Exeter (1876–81); Linacre professor of human & comparative anatomy and fellow of Merton (1881)
Pye, A.	Magdalen College School				
Reeves, Thomas James	Exeter	1868	2 Natural Science (TT 1868)		
Reinold, Arthur William*†	Brasenose (1866), Merton (1867)	1866	1 Mathematics (MT 1866); 1 Natural Science (MT 1867)	1 Maths Mods (TT 1865); Junior mathematical scholar (1865)	Fellow of Merton (1866–9); Dr Lee's reader and senior student of Christ Church (1869–73); professor, Royal Naval College, Greenwich, 1873–1908
Richardson, William Moore	Merton	1869	2 Mathematics (MT 1868)	1 Maths Mods (TT 1866)	Vicar of Wolvercote, Oxon (1883)
Robinson, Ellis Ashton†	Balliol	1868	3 Natural Science (MT 1867)		Private tutor to Carew family; Headmaster, Marylebone Grammar School; proprietor of preparatory school, Dulwich
Smart, Edward Rowland	Jesus	1868	1 Natural Science (TT 1868)		Rector of St Mary's, Tobago and canon of St George's Cathedral, St Vincent (1878)

Name	College	Year	Course	Notes
Squire, Lovell	St Mary Hall	1867	2 Natural Science (MT 1867)	Assistant master at Bootham School, York; subsequently kept a Quaker school in Falmouth and died in San Francisco (1885)
Stephenson, Henry Stillington Grey	Queen's	1869	2 Natural Science (MT 1868)	
Yule, Charles John Frances	Magdalen College School		[1st in Natural Sciences Tripos, Cambridge, 1872]	Tutor in natural science, Magdalen (1873–84); Fellow (1873–1905); Holy Orders and a clerical career (1885–1900). Yule matriculated at Balliol in January 1868 but migrated to St John's College, Cambridge in May 1869

* Indicates a graduate

† Indicates participation in Clifton's laboratory class, in addition to lecture attendance.

The only non-collegiate affiliation given in the second column is Magdalen College School. The names are presumably those of pupils, not masters, in all cases.

Sources: The table summarizes information in a list of students, in Clifton's hand, in the University Archives (UM/F/4/3). A covering letter from Clifton to John Phillips is dated 8 March 1867. I am grateful to Mark Curthoys for drawing my attention to this source, which forms part of material that Phillips was preparing for a report to the Vice-Chancellor on attendances in the Museum, and for allowing me to draw on his first draft of the table, and to Stephanie Jenkins for the leads she has given me on certain of the students mentioned.

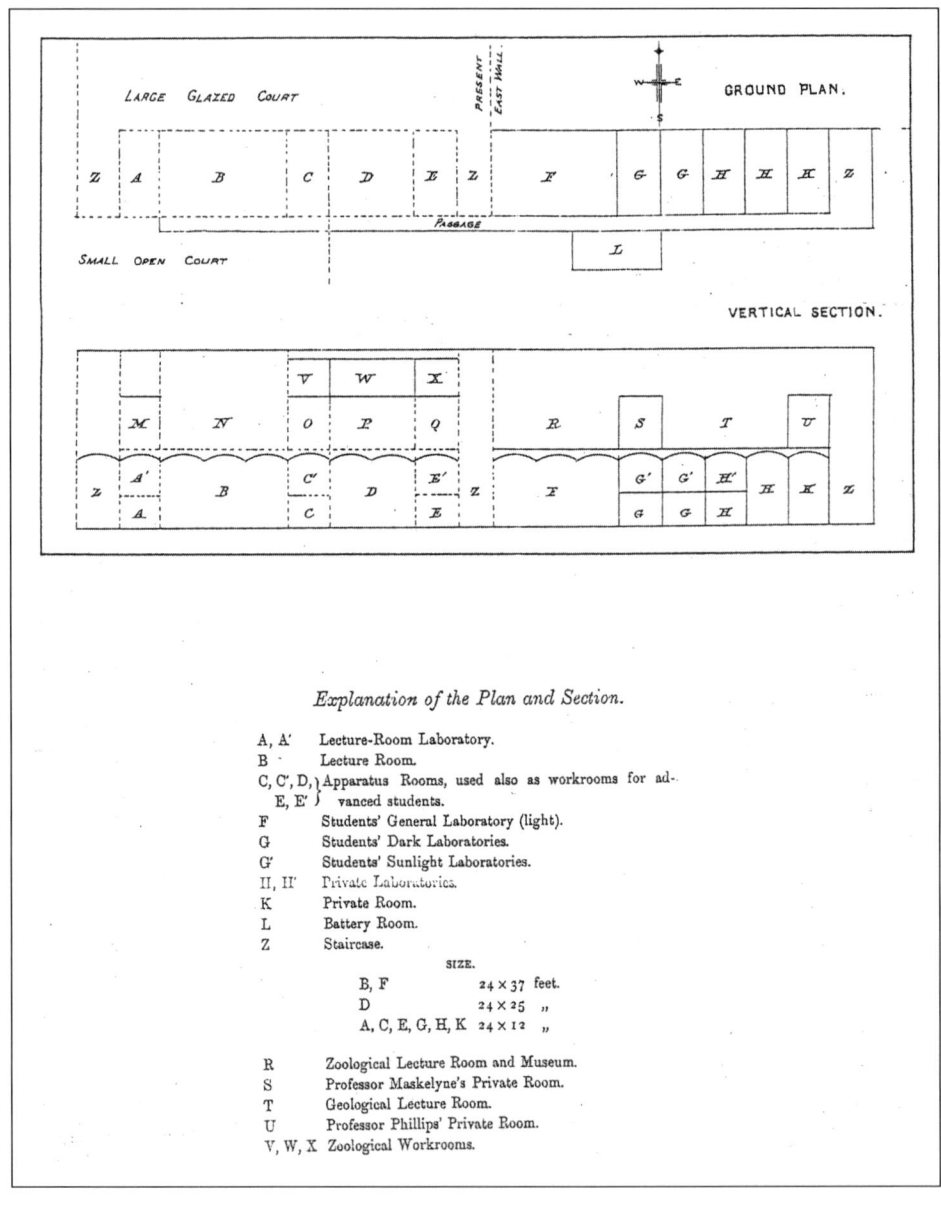

Fig. 2.13. Ground plan and elevation for an alternative scheme for extending the south front of the University Museum eastwards. This scheme was proposed by Clifton in February 1867 in a statement appended to the second of two printed reports prepared by John Phillips for transmission by the delegacy of the University Museum to Hebdomadal Council; see this chapter, note 32. The scheme would have entailed the reallocation of the mineralogy lecture-room (D) and the small adjacent rooms C,C[1] and E,E[1] as work-rooms for advanced students. In addition, the whole of the ground floor of the extension would have been allocated to experimental philosophy, with zoology, geology, and mineralogy occupying the full length of the south front at first-floor level. The scheme was abandoned once the funds for a purpose-built Clarendon Laboratory became available.

of instruments from the harmful effects of rooms facing 'the cold quarters of the sky'); see Figure 2.13. Although modest by comparison with the free-standing laboratory that was eventually built, the proposed structure reflected the grandeur of Clifton's ambition. On his plan, not only the easterly extension but also the space on the southern side of the Museum court currently allocated to geology and mineralogy would become the domain of experimental philosophy: geology and mineralogy, along with zoology, would move to the first floor of the extension, while zoology would be found additional space on the second floor of the existing south front.[34] Even some of those who sympathized with Clifton found the scheme audacious, as the startled admiration that the shy, retiring Donkin expressed in a letter to Phillips in March 1867 makes clear. Clifton's strategy put Donkin in mind of 'the farmer who wanted a new gate, and asked the landlord to build him a new house'. 'However', he added, referring to the extra rooms for which Clifton was campaigning, 'I sincerely hope he will get them'.[35]

By the time Donkin made his comment, developments on a front initially unrelated to Clifton's aspiration were opening the door to the financial windfall that was to transform the face of Oxford physics. For almost a decade, the trustees of the Clarendon Trust, established by Henry Lord Hyde, the great grandson of the Earl of Clarendon, had been in intermittent correspondence with the Vice-Chancellor, with a view to making the accumulated funds available to the University. By a codicil to his will, dated 10 August 1751, Hyde had bequeathed, for the benefit of the University, the profits arising from his grandfather's *History of the Rebellion* and other writings.[36] Hyde's stated intention was that the money should be applied as a beginning for a fund for supporting 'a Manège or Academy for Riding and other useful exercises in Oxford'. Alternatively, if that proposal was unacceptable, it should be used for such other objects as the trustees would 'judge to be most for the honour and benefit of the University, and most conducive to publick utility'.[37] By November 1858, the assets of the Trust stood at £8803 10s. 4d. in 3 per cent consols, in addition to a 'small balance' in the hands of its bankers. In the trustees' view, the sum was too great to be allowed to lie idle and unappropriated, and the Duke of Newcastle, as senior trustee, duly made what proved to be a momentous, apparently unsolicited first approach to the Vice-Chancellor.[38]

---

[34] Clifton's personal statement set out this plan, along with a sketch, on the proposed extension. See Figure 2.13.   [35] Donkin to Phillips, 15 March 1867, OUA UM/F/4/3.
[36] On the terms of the bequest and an incomplete outline of the exchanges between the trustees and the University in the person of the Vice-Chancellor, see the printed announcement, dated 9 December 1867, of the trustees' offer. Copies are in the Register of Convocation, 1854–71 (OUA NEP SUBTUS Reg Bu), between ff. 433 and 434; also in Bodl. G.A. Oxon c.83 (f. 332). Much of the original correspondence is in OUA UC 10/33.
[37] The words are those of the codicil to the will, authenticated as correct in a letter from the Duke of Newcastle to Francis Jeune as Vice-Chancellor, 2 November 1858, in OUA UC 10/33. The printed codicil is also in OUA UC 10/33.   [38] Duke of Newcastle to Vice-Chancellor, 2 November 1858, OUA UC 10/33.

In the exchanges that followed in the late 1850s and 1860s the University's prevarication must have tried the trustees' patience. The Duke of Newcastle's request for suggestions of ways in which the funds might be disbursed for the benefit of the University came to nothing, despite his insistence that the trustees would be willing to consider proposals for expenditure other than on a riding academy. It seems that the Vice-Chancellor expressed an interest in the possible deposit of the Trust's papers in the Bodleian, but he evidently did not respond to the Duke's core request.[39] After some years of silence, the trustees' further request, in May 1863, for ideas about the use of the funds fell foul, first, of the Vice-Chancellor's insistence that Hebdomadal Council would need to be consulted before a formal approach could be made to the trustees,[40] and then of the Duke of Newcastle's sudden death in October 1864. In fact, the mixture of the cumbersome procedures for decision-making and the Duke's death was providential, as was the emergence of William Ewart Gladstone as a resolute pacemaker among the trustees (now reduced to Gladstone, Sir William Heathcote, and the ailing Marquess of Lothian). Further delay followed Convocation's rejection of the proposal for a riding school on 13 December 1864 and the trustees' rejection, on 23 June 1865, of the alternative suggestion that the funds, now amounting to £10 697 3s 10d in consols, should be transferred to the University and the interest used to lay out and maintain the Parks 'as a place of recreation and exercise'.[41]

The saga of false starts and prevarication meant that the matter was still in abeyance in the autumn of 1866, when Clifton's agitation edged the plan for a physics laboratory onto the list of candidates for expenditure. It was a candidate that squared well with the trustees' principle that whatever scheme was to benefit should be as 'definite, tangible, & complete in itself as may be'.[42] But the University had competing priorities. In April 1867, the Vice-Chancellor reported the view of Hebdomadal Council that the most urgent need was for new Examination Schools, a measure, long advocated, that would ease the mounting pressure on space in the Bodleian Library.[43] Only if that proposal were unacceptable to the trustees would

---

[39] Duke of Newcastle to the Vice-Chancellor, 25 November 1858, OUA UC 10/33, refers to a letter of 15 November 1858, now lost, in which the Vice-Chancellor had evidently made his suggestion about the papers. The trustees took the view that they could not part with the papers permanently. See Gladstone's somewhat pained letter to the Vice-Chancellor, now J. P. Lightfoot (Rector of Exeter College), 19 November 1864, OUA UC 10/33.

[40] Vice-Chancellor, J. P. Lightfoot, to the Duke of Newcastle, 20 May 1863, in reply to a letter of 11 May 1863 from Henry Boodle, solicitor and secretary to the trustees; both letters in OUA UC 10/33.

[41] Gladstone's perception of the need to disburse the funds as quickly as possible is conveyed in a series of letters to the Vice-Chancellor – of 19 November 1864, 26 November 1864, 25 February 1865, and 23 June 1865 – all in OUA UC 10/33. The figure of £10 697 3s 10d is given in the letter of 26 November 1864. The draft of a reply (probably to the letter of 23 June 1865), in which the Vice-Chancellor, who was still Lightfoot, expressed his inability to go back yet again to Convocation, exuded a lethargy that Gladstone must have found exasperating. The proposal concerning the Parks seems to have been made in a letter of 20 February 1865 from the Vice-Chancellor to Gladstone. The letter is missing but is referred to in Gladstone's correspondence.   [42] Gladstone to Vice-Chancellor, 23 June 1865, OUA UC 10/33.

[43] Vice-Chancellor to Gladstone, 5 April 1867, OUA UC 10/33.

Council wish to propose the alternative of the construction and fitting-out of laboratories and other facilities for experimental philosophy. The latter option was 'rendered desirable, if not necessary, by the rapid development of this branch of science since the Foundation of the Museum'. But it remained very plainly Council's second choice.

On 3 May 1867, less than a month after receiving the Vice-Chancellor's response, Gladstone notified him of the trustees' momentous decision to reverse Council's priorities.[44] The proposal for a laboratory was deemed preferable, and barely a month later the elderly Dublin architect Thomas Deane was engaged on preliminary drawings for the building that was eventually constructed, in an appropriately gothic style, at the north-west corner of the Museum site (see Figures 2.15 and 9.1, and dustjacket).[45] Even at this stage opposition rumbled on. In September, Clifton felt he had to solicit Sir William Thomson's written endorsement both of the plans and of the necessity for students of physics to receive practical instruction.[46] And in November, Sir William Heathcote's confirmation of the trustees' approval for the scheme and for Deane's plans was tinged with anxiety about the firmness of Convocation's decision.[47] As late as 4 February 1868, when Convocation finally approved the design, it did so in the face of a rival claim on behalf of the Taylor Institution for modern languages, an attempt to resurrect a case in favour of the construction of a university swimming bath that Congregation had submitted to Council a year earlier, and a conservative rearguard action led by Charles Neate of Oriel (long known as an opponent of what he saw as profligate expenditure on the University Museum) and J. R. Magrath, later to become Provost of Queen's and Vice-Chancellor.[48] The Christ Church mathematician Charles Dodgson also weighed in, though in a tone of amused rather than belligerent scepticism. His

---

[44] Gladstone to Vice-Chancellor, 3 May 1867, OUA UC 10/33.

[45] Deane to Vice-Chancellor, 11 June 1867, OUA UC 10/33. Deane had practised on his own in Dublin since the death of his partner and the designer of the University Museum, Benjamin Woodward. It is not clear whether by this stage the plan for a free-standing building had replaced the original idea for an extension on the southern side of the Museum.

[46] See Clifton's letter of 16 September 1867 to Thomson, reproduced in Appendix III.

[47] Sir William Heathcote to Vice-Chancellor, 22 November 1867, OUA UC 10/33.

[48] Register of Convocation, 1854–71, 434, and *Oxford University Herald*, 8 February 1868: 8. See also Clifton's four-page preparatory memorandum of 25 January 1868, 'The offer of the Clarendon trustees', in OUA UC 10/33. Another copy is in Bodl. G.A. Oxon. b.32 (f. 24), and the text is reproduced in the *Oxford University Herald*, 1 February 1868: 8–9. The proposal for a swimming bath was resurrected in a one-page fly-sheet, dated January 1868 and signed 'W.F.S. and J.R.M.', identifiable as the Revd Walter Short, fellow and junior dean of New College, and Magrath. The fly-sheet reminded readers of a 'memorial' in support of the project which upwards of 120 members of Congregation had submitted to the Hebdomadal Council in February 1867 but which Council had peremptorily rejected on 3 May 1867. Copies of the fly-sheet, entitled 'The Clarendon trustees, the Hebdomadal Council, and the swimming bath memorialists', are in OUA UC 10/33 and Bodl. G.A. Oxon c.84 (f. 409); the text also appeared in the *Oxford University Herald*, 1 February 1868: 8. The case on behalf of the Taylor Institution was made in an unsigned fly-sheet 'To the members of Convocation', dated 30 January 1868, in Bodl. G.A. Oxon b.32 (f. 24). The fly-sheet called for the support of members of Convocation 'in the hope that, at least, some portion of the Fund' should be allocated for the benefit of the Taylor Institution.

satirical squib, extravagantly outlining the need for rooms and outdoor spaces in which various mathematical procedures could be conducted ('a piece of open ground for keeping Roots and practising their extraction', and a long, narrow strip of ground for 'investigating the properties of asymptotes'), was certainly intended to draw attention to Clifton's grandiose aspirations.[49] But the fact that Dodgson published his squib two days after the vote in Convocation suggests that no serious harm was intended.

And indeed no harm was done. Convocation remained firm in its conviction when, on 20 June 1868 (by which time Clifton had been persuaded to trim his original specification to the tune of about ten per cent), it took the definitive step of accepting the Clarendon trustees' offer of a maximum of £10 300 for the cost of buildings and fittings (though not the apparatus).[50] On 10 July, the eight tenders for the work were opened in the London house of Sir William Heathcote, and that of the prominent Oxford builder, Symm, was chosen (Croft 1986: 39–40; Law 1998: 54–6). Exchanges between Lord Caernarvon and Clifton continued into the early months of 1871. But from January 1869 the funds were fed in regular instalments through the London and Westminster Bank, allowing building to proceed without evident hitch. By October 1870 Clifton, aided by a demonstrator, A. W. Reinold (see below, p. 64), and a newly appointed assistant, Henry Walter, was using the laboratory for practical instruction, though not initially for lecturing, which only began there at the very end of the month.[51] A year later, with his former premises now reallocated and refitted for chemistry, he was fully established in the laboratory, offering his twice-weekly lectures there, as well as laboratory supervision.[52]

Despite Convocation's decision of June 1868, the generosity of the provision in the new building continued to irk conservative critics anxious that the growth of a costly non-collegiate provision for science risked upsetting the traditional order of Oxford. When Clifton's case for financial support to meet the enhanced cost of running the Clarendon came before Convocation two years later, it encountered

---

[49] Dodgson's two-page squib, 'The offer of the Clarendon trustees', was dated 6 February 1868 and addressed to 'My dear Sandford'. A copy is in Bodl. G.A. Oxon b.32 (f. 24), along with other fly-sheets that circulated at about the time of the vote; it is reproduced in Hannabuss (2000a: 197).

[50] Clifton had been persuaded to reduce the estimated cost from £11 436 to £10 285, according to a letter of 24 June 1868 from the Vice-Chancellor to Lord Caernarvon (who had now taken over correspondence with the University on behalf of the trustees, following Heathcote's serious illness); OUA UC 10/33. The printed formal proposal for acceptance of the trustees' offer was circulated by the Vice-Chancellor on 15 June 1868 before its approval *nem. com.* in Convocation five days later; see Bodl. G.A. Oxon c.84 (f. 248) and the Register of Convocation, 1854–71 (see note 36), f. 449.

[51] *Oxford University Herald*, 29 October 1870, 8. Nothing is known of Walter, except that his appointment, at the age of 27 and with a wife and child, was made from four candidates at a meeting of the Museum delegacy on 20 October 1870 (UM/M/1/3).

[52] Clifton's announcements in the *Oxford University Gazette*, from its foundation in 1870, show that by October 1870 laboratory instruction was offered on Tuesdays, Thursdays, and Saturdays from 10 a.m. to 4 p.m. By January 1871, the laboratory was open daily, the fee for pupils wishing to work for three days a week being £3 per term.

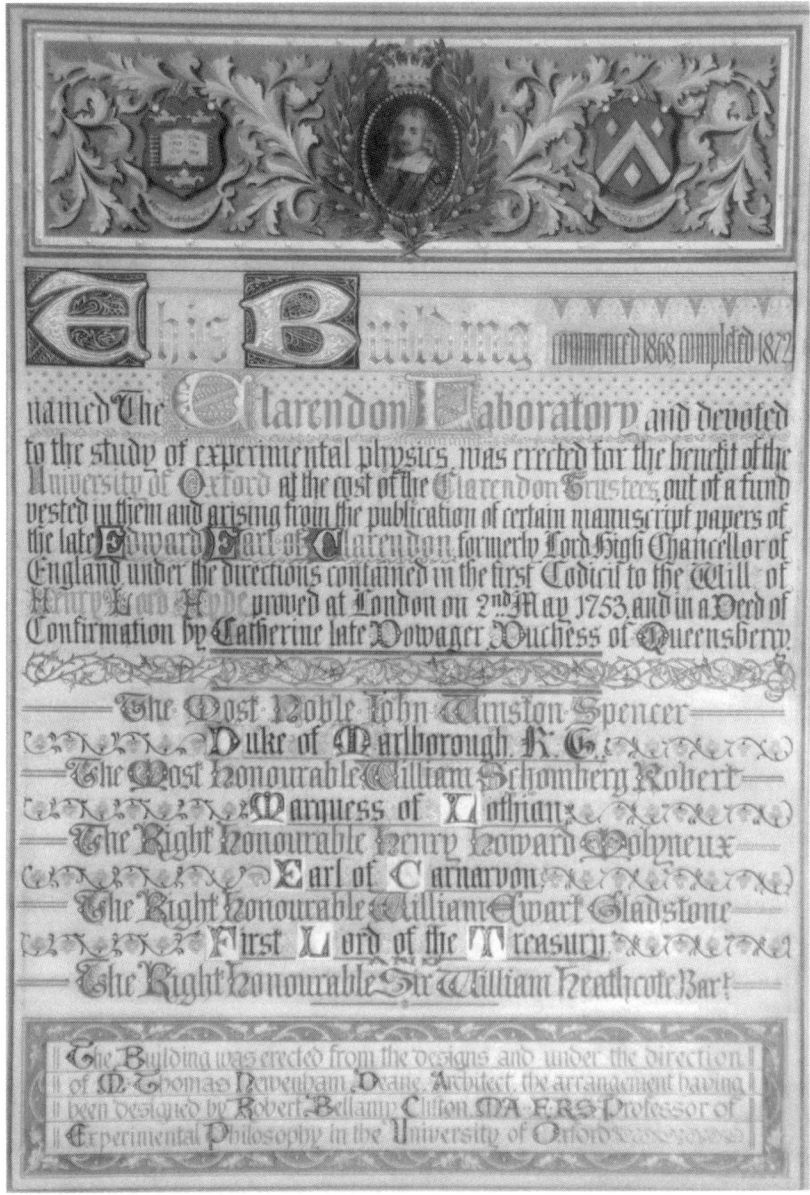

Fig. 2.14. Illuminated parchment commemorating the building of the first Clarendon Laboratory. Now displayed on the first floor of the Clarendon Laboratory's Lindemann Building, this once hung over the fireplace in Robert Clifton's room in the original laboratory; see Figure 2.19. Photograph by Chris Stammers, Photographic Unit, Department of Physics.

opposition. His request for an increase in the cost of assistance from £110 to £165 (to cover Walter's wages and those of a servant, a man to attend to the furnaces, and a female cleaner) was approved without dissent, but significant minority votes opposed the allocation of £150 for the demonstrator's salary and £200 for general running expenses.[53] Clifton was probably oblivious to the significance of the shot that had been fired across his bows. He seems only to have felt satisfaction that, despite his status as a young newcomer to Oxford, he had carried progressive opinion with him to win a victory for the University's move to reform and modernization. And he set to with a will.

Clifton's meticulousness was evident in every facet of the new laboratory, as it was in the instruments, including many selected from the finest pieces displayed at the Universal Exhibition of 1867 in Paris. The guiding principle of the design was that the delicate apparatus on which students worked should be disturbed as little as possible; hence it was the students, rather than the instruments, who should move about the building.[54] To this end, most of the space on both the ground and first floors was divided into a number of small rooms, each devoted to a particular activity.[55] On the ground floor, in addition to the professor's two private laboratories, there were rooms for weighing and measuring, spectrum analysis, heat, radiant heat, and static electricity, while rooms on the first floor were designated for current electricity, acoustics, and optics, a small mathematical lecture-room, a library and student common room, and a retiring room for the professor. In addition, the roof-space along whole of the 91-foot long western side served as a gallery for optical experiments, and the space above the large lecture-theatre (designed for 150 students and occupying roughly the south-eastern quarter of the building) was fitted out for photography. Central to the whole structure was a large court-area, extending from the ground to the roof, around which the rooms were arranged and in which experiments necessitating the use of the full height of the building could be conducted. And beneath it all was a basement room for the study of magnetism, store-rooms, and battery-rooms, the space where C. V. Boys was to conduct his determination of the gravitational constant in 1894, as a visitor from the Royal College of Science in London.[56]

---

[53] The report on the Convocation of 3 June 1870, in *OUG*, 1, no. 18 (7 June 1870): 1–2, records votes of 67 *placets* and 16 *non-placets* on the proposal for the demonstrator's salary, and 60 *placets* against 15 *non-placets* on the request for running expenses. Clifton's brief paper justifying the expenditure appeared in *OUG*, 1, no. 16 (24 May 1870): 2–3. By May 1871, the laboratory's budget had been increased by an additional £50 for 'current expenses'; see *OUG*, 2 (1871): 175 (20 May 1871).

[54] Clifton's letter to Stokes, 12 March 1870, bears further witness to his concern for the sensitivity of the instruments under his care. In the letter (CUL Stokes Collection, Add. MS 7656, C 714), Clifton solicited Stokes's support in his protest against a proposal to open Parks Road to traffic that would pass within fifty feet of the laboratory.

[55] The most detailed source for the disposition of the rooms and the specifications of the building and fittings is the set of nineteen architect's drawings, each signed by Thomas Deane and the builder, Joshua Symm, in OUA UD 36/5/1-19. I am grateful to the Keeper of the University Archives, Simon Bailey, for bringing these drawings to my attention and hastening the process of conservation so that they could be consulted. Other useful sources are the unsigned article, with a picture and floor-plan (see Figure 2.15), in *The Builder*, 27 (1869): 367–9, and the sketches accompanying Clifton's reply to James Clerk Maxwell's request for information about the Clarendon Laboratory, referred to on p. 60; see Figure 2.16. For an informed early account of the building see also Moore (1878: 281–2).

[56] See Gooday, Chapter 3, 110–14, and Figures 3.10–3.13.

Fig. 2.15. The Clarendon Laboratory, from *The Builder*, 27 (8 May 1869): 367. The illustration appears in an unsigned article on the proposed laboratory, which was still eighteen months away from completion.

In its structure, fittings, and apparatus, the Clarendon Laboratory, as the new building was called after a brief period in which it was referred to as the Hyde Institute or simply the Physical Laboratory, was a showpiece redolent with significance. For Clifton, it represented both a personal triumph and a crucial step towards a provision that would allow him to realize his ambitions for his discipline. With room for forty students working simultaneously (Moore 1878: 282) and more than 750 square metres of laboratory space available to him on the two main floors (in addition to the lecture theatre and the areas in the basement and under the roof), his working conditions had been transformed. More generally, the laboratory conveyed the University's readiness to embrace the modern world. And it encapsulated the success with which the science lobby and its leading sympathizers, mainly though not only on the Hebdomadal Council, had advanced their case for over two decades. Not everyone regarded the new building as an embellishment to the architectural treasures of Oxford. An unsigned article in *Jackson's Oxford Journal*, for example, commented on its heavy, rather gloomy character and its tendency to overshadow the red brick of Keble College on the other side of Parks Road:

The effect of the building in connection with the New Museum is not satisfactory. It does not look well as an adjunct to the Museum, and its proximity to the College only adds to the darkness of the locality.[57]

---

[57] *Jackson's Oxford Journal*, 16 October 1869: 6. I am grateful to Stephanie Jenkins for drawing this source to my attention.

But, for all but the most diehard critics of science, it remained a matter of pride that Oxford now found itself with one of the first two purpose-built physics laboratories in Britain, the other being Thomson's in Glasgow.[58] In the eyes of most Oxonian observers, it was a cause of particular satisfaction that the University not only possessed what J. P. Earwaker, a Merton undergraduate, judged to be 'the most perfect physical laboratory in the world' (Earwaker 1870–1: 171) but that it had also maintained its lead over Cambridge in the material provision for science. That lead had been established early. For although the first examinations for the new Natural Sciences Tripos had been held in Cambridge in 1851, two years before those for the NSS in Oxford, the museums and lecture-rooms on the Free School Lane site had not been inaugurated until 1865, five years after the University Museum, and only then after more than a decade of struggle against a scepticism distinctly more resolute than had been encountered in Oxford (Willis and Clark 1886: vol. 3, 145–90; Shipley 1913: 298–43; and Searby 1992: 203–33). In terms of scale, too, Oxford could preen itself on having consistently led its arch-rival. Cambridge's new buildings for zoology, comparative anatomy, mineralogy, and botany, and the refitting of the thirty-year-old accommodation for anatomy and chemistry had cost something of the order of £30 000, barely a third of what had been spent on the University Museum, and it had left physics and laboratory teaching in chemistry virtually unprovided for.[59] It was only in 1868, when new examination statutes brought heat, electricity, and magnetism into the syllabus for honours in the Mathematical Tripos, that such neglect came to be seen as intolerable (Warwick 2003: 264–7 and 291–8), and only then that the events leading to the opening of the Cavendish Laboratory between the autumn of 1873 and spring of 1874 were put in train.[60]

It is a measure of the Clarendon's standing that when James Clerk Maxwell, newly appointed to the Cavendish chair of experimental physics in 1871, began planning his laboratory, he approached Clifton (as well as Thomson in Glasgow and P. G. Tait in Edinburgh) for advice and received, in return, the floor-plans reproduced in Figure 2.16.[61] Clifton, as Maxwell observed in a postcard to Tait, had done an excellent job:

Clifton has had terrible work and has done it well. Now he is a Plumber, now a Scene shifter and Property man, now a Bricklayer &c through all the trades mentioned by the learned Martinus Scriblerus (Harman 1995: 636).

[58] For a description of Thomson's laboratory, which was opened in 1870 as part of the University of Glasgow's new buildings to the west of the city, see Gray (1896–7).

[59] I take the figure of £30 000, a sum that seems to have included both construction and fittings, from *Devonshire Commission. Third Report* (1873): xlii. Willis and Clark give a figure of just under £25 000.

[60] The growth of science as an undergraduate study in Cambridge showed many parallels with that in Oxford and was not conspicuously precocious. For example, the requirement that candidates for the Natural Sciences Tripos must first pass in either the Mathematical or the Classical Tripos was only abandoned in 1861. Even when this constraint was removed, the number of candidates for the NST only began to grow significantly in the 1870s and more particularly the 1880s. See Wilson (1981–2) and MacLeod and Moseley (1982).

[61] CUL Maxwell Collection, Box 5, Add. MS 7655/Vj/1 (ii) and (iii). I am grateful to Dr Joanie Kennedy for drawing my attention to this material, which is also cited in Gooday (1990: 6–46).

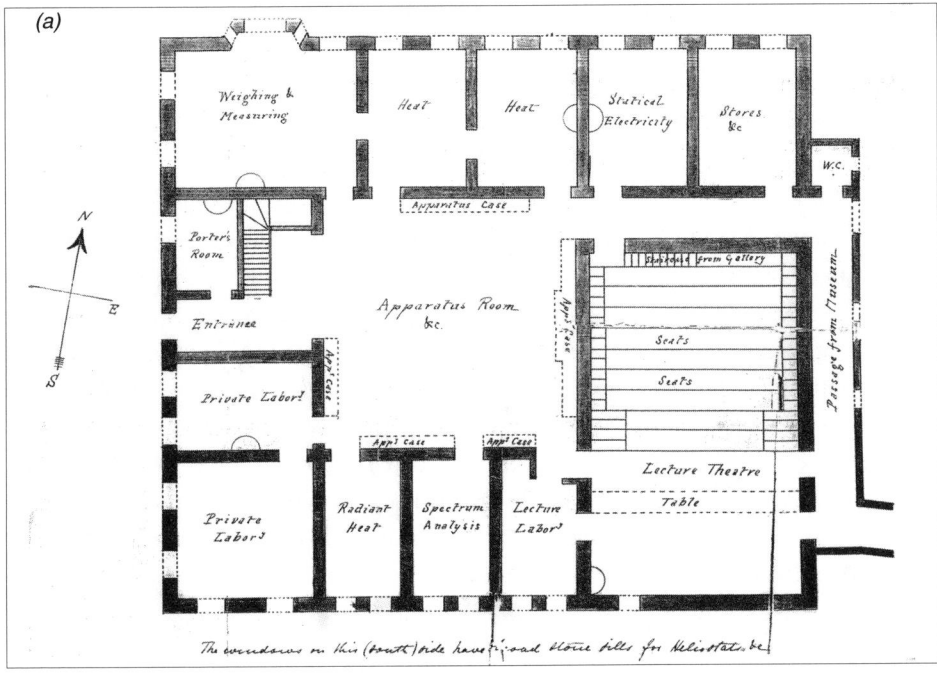

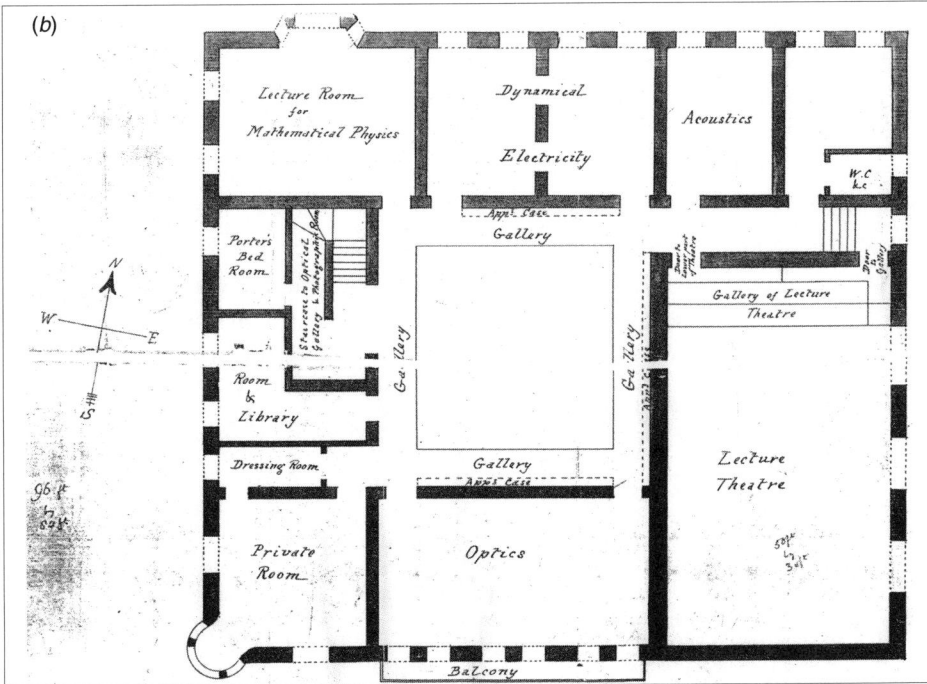

Fig. 2.16. Architectural plans of (*a*) the ground floor, and (*b*) the first floor of the Clarendon Laboratory, sent by Clifton to Maxwell, c.1871–2, when Maxwell was seeking advice on the design of the Cavendish Laboratory. As in the Clarendon, each of the rooms of the Cavendish was allocated to a particular area of physics. From Cambridge University Library, Maxwell Collection, Box 5, Add. MS 7655/Vj/1 (ii) and (iii).

How far Maxwell's Cavendish was influenced by the Clarendon is unclear. But the result was a laboratory that at this stage, despite its later fame, in no way eclipsed its Oxford counterpart. Certainly, the building, for which the initial estimate was £8450, cost rather less than the Clarendon (Willis and Clark 1886: vol. 3, 183).

Such comparisons must have heartened Clifton, and both he and the science lobby of which he was now an essential, though seemingly never dominant, member entered the 1870s with high expectations. Clifton's own confidence in the possibility of further expansion is reflected in the ambitiousness of his response to a letter of 10 May 1873 in which the Vice-Chancellor, following the report of a specially appointed committee of the Hebdomadal Council, invited professors in all disciplines, scientific and non-scientific, to outline the needs of the various areas of professorial teaching.[62] As Clifton saw it, the overwhelming need was for manpower, and he argued for the creation of three additional professors of physics and one of experimental mechanics, each with an associated demonstratorship, and for the eventual establishment of a chair and demonstratorship in civil engineering. While he conceded that it might be possible to create only one of the chairs immediately, he insisted that this one chair, in physics, was an urgent necessity if he was to be able to concentrate on the parts of the discipline in which his main strengths lay – optics and acoustics – and leave other branches, in particular heat and electricity, to the new professor.

The case for the new posts rested on an elaboration of the evidence that Clifton had given in 1870 before the Royal Commission on Scientific Instruction and the Advancement of Science, chaired by the Duke of Devonshire.[63] Then too, he had stressed the difficulty that one person had in covering the whole range of physics. The point was a fair one, although neither in 1870 nor now in 1873 was Clifton the only representative of his discipline in Oxford. The potentially most valuable alliance would have been with his professorial colleague the Revd Bartholomew Price, a prominent Oxford man and Sedleian professor of natural philosophy since 1853, whose lectures, including regular series on optics, mechanics, and other areas of applied mathematics, led candidates for the Honour School of Mathematics far into the realm of physics.[64] Relations between Price and Clifton appear to have

---

[62] Clifton's response, dated 21 February 1874, is on pp. 34–6 of a seventy-page printed document containing the replies of all the professors. Copies are in the Museum of the History of Science (MS Gunther 65) and Bodl. G. A. Oxon 8° 124 (24).

[63] *Devonshire Commission. First, Supplementary, and Second Reports* 1872: Minutes of Evidence 186–94. See also *Devonshire Commission. Third Report* 1873: xxiii, where the commissioners supported the case for two new professorships of physics (in addition to two new chairs in chemistry and one in pure mathematics), a chair of mathematical physics, and one in applied mathematics and engineering.

[64] Price had come through the Oxford Mathematical School, taking a first at Pembroke in 1840. His teaching as Sedleian professor was directed at candidates for Honours in the same school, in which at this time there were optional papers (taken by the most ambitious students) in Mechanics (including the dynamics of material systems), Optics, Hydromechanics, and Astronomy. For an accessible example of papers set on these subjects, see *OUG*, 1, supplement to issue no. 2 (1 February 1870): 2–8.

been perfectly cordial. But there are no signs that the two men ever worked closely together in a way that might have allowed a broadening of Clifton's preoccupation with fine experimenting or a qualification of his repeated assertion that while mathematics was an essential tool of the physicist, his own teaching did not require any significant preparation in the subject.

Beyond the professoriate, a number of college-based initiatives also offered the possibility of disciplinary collaboration (Simcock, Chapter 4). At Trinity, a succession of able Millard lecturers in physics offered instruction not only to Trinity men but also to undergraduates from across the University: Archibald Simon Lang Mac Donald, Lazarus Fletcher, Alexander Macdonell, and Harold Bailey Dixon, all of them distinguished young products of the NSS, held the post successively between 1873 and the mid-1880s.[65] The initiative at Trinity, in fact, formed part of an intercollegiate provision in which Magdalen and Merton (always an ambitious science college) also played a leading part. At Magdalen, as tutor in natural science from 1869, Edward Chapman (first in Natural Science in MT 1864) acted as the hub of the arrangement, which gave undergraduates access to instruction not only in physics (by the Millard lecturer) but also in chemistry (by Thomas Wyndham, until his sudden death in 1876, and then by William Herbert Pike) and in biology (by Chapman), augmented by practical classes in physiology and physiological chemistry (by C. J. F. Yule).[66]

The collaboration in which Chapman and his colleagues engaged was a successful one. Between 1869 and 1877, a total of 44 Magdalen men worked in their college laboratory, with more than seventy from other colleges, and eight candidates from Magdalen took firsts in the NSS.[67] But college appointments lay outside the professor's domain and, offering as they did the scientific equivalent of tutorials or

---

[65] It was as the holder of another Millard lectureship, in experimental mechanics and engineering, that the Revd Frederick John Smith (in his later years Jervis-Smith) laid the early foundations of engineering in Oxford from 1885. See Simcock, Chapter 5.

[66] On Chapman and the instruction that was offered in the laboratory, lecture-theatre, and other rooms in the Daubeny building at Magdalen see the evidence he gave, with Yule, before the University of Oxford Commission on 31 October 1877 and the printed statement of his evidence dated 29 October 1877, both in *University of Oxford Commission 1881: Minutes of Evidence* 219–24; a copy of the printed statement, slightly modified, is also in Bodl. G. A. Oxon 8° 1079(5). A vivid insight into Chapman's increasing involvement in the Great Central Railway Company (of which he was for some time vice-chairman) and in social and (conservative) political life emerges from his collection of invitations, menus, and other ephemera, most dating from about the turn of the century, in Bodl. 3995 c. 1. Wyndham, a fellow of Merton, had taken a first in the NSS in MT 1865. Pike, who held a Ph.D., left Oxford in 1879 and went on to become the first professor of chemistry in the faculty of medicine at the University of Toronto. On Yule, a fellow of Magdalen from 1873 to 1905 and a tutor from 1873 to 1884, see Table 2.2, Gunther (1924: 408–14), and the brief biographical entry in Foster (1893: 311). Despite Chapman's long association with Magdalen, marriage prevented his being elected to a fellowship until 1885 (following a relaxation of the rule on celibacy that affected appointments across the University from 1882).

[67] *University of Oxford Commission 1881: Minutes of Evidence* 221. As Chapman observed, his figure of seventy did not include students using the Magdalen laboratory who were pupils of the Millard lecturer.

Fig. 2.17. Lazarus Fletcher. During the 1870s, Fletcher served as demonstrator in physics, under Robert Clifton, and Millard lecturer in physics at Trinity College. After marriage (in 1880) and his consequent resignation from his fellowship at University College, he served as keeper of the mineralogy department at the British Museum (Natural History) from 1880 to 1909.

intercollegiate catechetical lectures in the humanities or mathematics, they contributed little to Clifton's dream of a provision that would make Oxford a leading centre of European physics. The one, partial exception to this generalization was the Dr Lee's readership in physics that Christ Church created, with a salary of £350, in 1869. The college appointed to the post A.W. Reinold, a pupil of Clifton's who had taken firsts in Mathematics and Natural Science in, respectively, 1866 and 1867 (Stocker 1921). Reinold, who also achieved the distinction, rare for a scientist, of election to a fellowship of Merton in 1866 (which he retained until his marriage in 1869), was quickly set on a trajectory that could have made him a leading figure in Oxford physics. For, in addition to his base in the Christ Church Laboratory (converted from its former use as the anatomical Laboratory once anatomy had been relocated in the University Museum), he also served as Clifton's first demonstrator in the Clarendon, a position that raised his combined annual income to £500, the equal of Clifton's. But in 1873, after almost four years in which he lectured three times a week, giving a complete coverage of physics, including the areas of electricity, magnetism, and heat in which Clifton felt less at home, he left for the chair of physics at the Royal Naval College, Greenwich.

The momentum that Reinold engendered was never quite regained. His successor in the Dr Lee's readership, Robert Edward Baynes, had an academic record comparable with Reinold's (having taken firsts in Mathematics and Natural Science in MT 1871and MT 1872, respectively) and he was an outstanding tutor. Moreover, his textbook *Lessons on Thermodynamics* (1878a), based on his lectures in the Clarendon in 1876, displayed a command of thermodynamics that embraced with equal authority both the mathematical and the experimental facets of the subject; it was clear, rigorous, and informed by the latest continental as well as British contributions of Rankine, Tait, Maxwell, Zeuner, Verdet, and Saint-Robert. But thereafter, Baynes published only a handful of very short papers, an introductory textbook on heat (Baynes 1878b), a revision of Stewart's *Elementary Treatise on Heat* (1895), and a translation of Oskar Emil Meyer's *Die kinetische Theorie der Gase* (1899), and stayed for almost fifty years in his post without ever emerging as a significant physicist (Baker 1921–2). His wholehearted dedication to college duties (crucially, he did not take up the demonstratorship vacated by Reinold), his activities in the higher reaches of local Freemasonry, and a lack of self-confidence seem to have extracted a heavy toll on his intellectual life.

If, in retrospect, 1873 cannot be said to have been a watershed, it did mark an inflexion in the fortunes of physics in the University. Arthur Rücker, who took firsts in Mathematical Mods and then, in the Final Honour Schools of both Mathematics (MT 1870) and Natural Science (MT 1871), was a worthy successor to Reinold in the demonstratorship at the Clarendon. But he stayed for barely a year, and his departure in 1874 for the chair of physics in the new Yorkshire College in Leeds deprived Oxford of yet another potentially leading figure (Thorpe 1916). To replace him, in October 1874, Clifton again appointed a high-flying Oxford graduate, William Nelson Stocker (firsts in Mathematics and in the NSS in TT 1873 and TT 1874, respectively). Stocker set down deeper roots in Oxford than Reinold and Rücker had done, becoming a fellow of Brasenose in 1877 and staying in his post at the laboratory until 1883, when he was appointed professor of physics at the Royal Indian Engineering College at Cooper's Hill, near Egham. In the Clarendon, he began strongly, and Clifton was evidently conveying genuine admiration when he described him, in a letter to Stokes in 1876, as 'a most excellent demonstrator' and a man who had both the 'power' and the 'will' to engage in original work, even if in this respect he judged him to be not quite the equal of Rücker.[68] But Stocker's nine years as demonstrator were marked by his own and the laboratory's loss of intellectual momentum. By the time of his letter to Stokes, Clifton observed that Stocker was becoming swamped not only by his responsibilities in the Clarendon but also by the reading ('wearisome and destructive

---

[68] Clifton to Stokes, 21 August 1876, in CUL Stokes Collection, Add. MS 7656, C717. A passage from the letter is quoted in Gooday, Chapter 3, 92.

Fig. 2.18. This image of a youthful Arthur Rücker is taken from a *carte de visite* photograph produced in London, presumably during the latter part of his Oxford period c.1873–4, just before taking up his position as professor of physics and mathematics at the new Yorkshire College of Science, founded in Leeds in 1874.

to real progress') that the serious pursuit of a college fellowship required. The truth may also be that Stocker simply allowed himself to become too easily overwhelmed and that he lacked the drive and creative spark of his predecessors in the demonstratorship. At all events, his dedication to teaching (which included the supervision of laboratory classes, sole charge of the teaching of heat and of weighing and measuring, and the giving of occasional lecture-courses in place of Clifton) was never translated into an active engagement in research or in the discipline beyond the laboratory and his college.[69]

If Clifton's statements are to be believed – and there seems no reason to doubt them – his first years in the Clarendon were busy. It is not clear whether at this stage

[69] Like Baynes, but unlike Reinold and Rücker, Stocker never became a Fellow of the Royal Society. The brief biographical information on him in Foster 1893: 349 and *Who's Who in Oxfordshire* (1936): 347, is interestingly complemented by the printed application and testimonials that he submitted, unsuccessfully, for the professorships of mathematics and physics at University College, Nottingham in 1881 and of experimental physics at University College, Liverpool. Stocker, who remained at Cooper's Hill until 1901, spent his later years back in Oxford, helping out in the Clarendon during the First World War and pursuing his passion for strenuous walking. On the circumstances of his resignation from Cooper's Hill see Cameron (1960: 17–18). J. G. Griffith, the son of I. O. Griffith (a demonstrator in the Clarendon between the wars and a fellow of Brasenose from 1920 to 1941), gives an interesting personal reminiscence of Stocker in an interview recorded in 1983. The recording, with others assembled by A. J. Croft, is in the Clarendon Laboratory Archive.

Fig. 2.19. Clifton's private room on the first floor of the Clarendon Laboratory. From an album of photographs of the laboratory, the University Museum, and the Pitt Rivers Museum. The album, assembled by Clifton's son, Walter Bellamy Clifton, in the 1890s, is in the Clarendon Laboratory Archive. It is cited hereafter as 'Clifton album'.

he had already adopted the curious practice, recalled many years later by Richard Glazebrook, of working through the night, once his family had retired, and continuing until 7.30 or 8.00 a.m., at which time he slept for two hours or so before appearing in the laboratory at about 11.00 a.m. (Glazebrook 1921: ix). But, according to the evidence he gave before the University of Oxford Commission in 1877, he worked in the laboratory for a minimum of 76 hours a week in term-time and for much longer than that – up to 17 hours a day – at times of exceptional pressure.[70] The vacations, during which Clifton turned to the general managerial work of the laboratory, the maintenance of the apparatus, and the ordering of new equipment, were scarcely less full. It was a schedule that since 1875 had left Clifton, on his own estimate, with no more than a fortnight a year free from university work and hence with virtually no opportunity for independent research and little time for even keeping up with the subject.

The burden of running a laboratory and mounting a programme of lectures and practical instruction extending far beyond the areas of optics and sound, in

[70] *University of Oxford Commission 1881: Minutes of Evidence* 21–9 (24).

Fig. 2.20. Hall in the Clarendon Laboratory, showing the cupboards used for storing apparatus. Clifton album.

which he felt most at home, would have weighed heavily on any professor. But Clifton's extreme fastidiousness made him his own worst enemy and contributed to his feeling of being perpetually overworked.[71] To the Commission of 1877, he stated that the preparation of each lecture took at least twenty hours, and the work was of such complexity that, in his view, it could not be delegated: 'it would take as long to tell the lecturer's assistant what to do as to do it'.[72] In addition, on each of the three days that he devoted every week during term to laboratory teaching he would spend eight hours instructing and keeping an eye on students scattered over all three floors of the extensive building. How many students had to be supervised in this painstaking way and how intrusive they really were, or need have been, is not clear. What is evident, however, is that the burden did not lie in any marked expansion of advanced teaching. For, although the numbers of Honours candidates in the NSS rose through the 1870s from 15 in the two

---

[71] His letters of the mid and later 1870s to Stokes (CUL Stokes Collection, Add. MS 7656) are revealing in this respect. They show Clifton repeatedly apologizing for the lateness of the referee's reports that Stokes had asked him to prepare for publications of the Royal Society.

[72] *University of Oxford Commission* 1881: *Minutes of Evidence* 24.

Fig. 2.21. Gallery on the first floor of the Clarendon Laboratory, also used for storing apparatus. Clifton album.

examinations in 1870 to 30 seven years later (see Table 2.1), only a minority of these candidates, perhaps four or five a year, were seriously engaged in experimental philosophy.

A measure that did change the profile of physics teaching, on the other hand, was the introduction of the Preliminary Honour examination in the early 1870s. The new examination, in chemistry and what was designated as 'mechanics and physics', had to be taken before candidates could go on from Moderations, in which they would already have had to achieve at least a Pass, to the Final Honour School.[73] Its aim was in part the raising of standards in the School. But, more importantly, the examination served to maintain the breadth of an NSS syllabus that now gave candidates for Honours who so wished the option of being examined in the Final Examination in only one of the three primary sciences, as opposed to all three of them, as was previously required.[74] From the time the first

---

[73] The regulations for the Preliminary examination were approved, as part of the new examination statutes, in Congregation on 1 June 1871 and in Convocation a week later. See OUG, 2 (1871): 212–13, 223–4, and 235 (2 June, 6 June, and 13 June 1871).

[74] The revised statutes for the Honour examination in Natural Science were passed in Congregation on 18 and 22 March 1870; see OUG, 1, no. 10 (29 March 1870): 9. The permitted degree of specialization was limited by the requirement that candidates for Honours in Biology must satisfy the examiners that they possessed 'a competent knowledge of Physics and Chemistry', while those seeking Honours in either Physics or Chemistry had to satisfy the examiners that they had 'a competent knowledge of the other of these two subjects'.

examinations were held in Easter Term 1872, Clifton perceived the influx of students who had no intention of taking physics beyond Preliminary Honour level as a threat to his vision for his discipline. Certainly, the numbers were substantial; a total of 40 candidates entered for the paper in mechanics and physics in the two sittings of the examination that took place in 1876–7, for example.[75] But Clifton coped, characteristically, by distancing himself, so far as he could, from direct involvement with such students. As the announcements in the *Oxford University Gazette* make clear, much of the lecturing and virtually all of the examining and supervision of laboratory classes fell to his demonstrators and the college lecturers, leaving Clifton with the task of general oversight and the largely self-imposed stress of protecting delicate apparatus from the ravages of clumsy undergraduate hands.

Elementary teaching was also kept within bounds by the continued scarcity of candidates for the Pass degree attempting papers in physics. New statutes for this degree, approved in February 1872 and first implemented for those sitting the examination in Michaelmas Term 1874, obliged candidates for a Pass to choose from a list of twelve subjects distributed between three groups: Group A, essentially classical literature and Greek and Roman history; Group B, which gathered what were broadly regarded as modern disciplines in the humanities, such as English and European history, modern languages, geography, political economy, and law; and Group C, in which there were four subjects in the area of mathematics, physics, and chemistry.[76] To achieve a Pass, candidates had to be examined in three of the twelve subjects, no more than two of the subjects being drawn from the same group. Predictably, the majority of candidates chose subjects in Group A (nearly all of them opting to be examined in two classical texts), with significantly fewer venturing into Group B and even fewer into Group C. Within Group C, moreover, physics was the least popular choice; in the examinations on 1875 and 1876, a total of only two candidates were examined in the subject.[77]

For someone as exigent and fastidious as Clifton, teaching to weaker students would have been soul-destroying, and in any case the smaller the numbers, the more

[75] 25 candidates entered for one or both of the two subjects for the examination in MT 1876, 41 in ET 1877, including 17 and 23 candidates, respectively, for the mechanics and physics paper; see *OUG*, 7 (1876–7): 84 and 389 (14 November 1876 and 14 May 1877). With a pass rate of 85 per cent, the examination was not an unduly formidable obstacle.

[76] *OUG*, 3 (1872): 44 (6 February 1872). Congregation approved the measure on 1 February 1872 after protracted discussion during the previous autumn and only by a majority of 67 to 43 The disciplines in Group C were defined, in the words of the statutes, as: 1) The elements of geometry, including geometrical trigonometry, 2) The elements of mechanics, solid and fluid, treated mathematically, 3) The elements of chemistry, with an elementary practical examination, 4) The elements of physics, not necessarily treated mathematically.

[77] As the pass lists in the *Oxford University Gazette* indicate, the great majority of candidates selecting a discipline from Group C chose the first option, on the elements of geometry.

time he had for the task of running the laboratory. Typical questions in the Pass and, from the early 1870s, Preliminary Honour papers called for definitions of basic physical concepts, descriptions of simple experimental procedures, and only the most elementary numerical calculations. One of the fifteen questions on the Pass paper of Trinity Term 1870 in Mechanical Philosophy, for example, read:

Describe a method for observing and measuring the linear dilatation of substances by heat.

The coefficient of linear expansion of iron for one degree centigrade being .000012, what will be the length at 20° centigrade of iron rails which are a thousand metres long at zero?[78]

In the Preliminary Honour examination, the standard was marginally higher, although the predominantly descriptive, qualitative style of the questions was similar. Typical questions among the twelve set in Physics in Trinity Term 1872 were:

Explain the terms – angle of incidence, critical angle, refractive index.

What is the critical angle for the substance whose refractive index is $\sqrt{2}$?

and

What is an induced current of electricity? How are induced currents produced and what are their properties?[79]

In the Mechanics paper, which also contained twelve questions, candidates were asked:

Describe and explain the principle of the ordinary mercury barometer. If the height of the barometer be 760m.m., and the specific gravity of mercury be 13.596, what is the atmospheric pressure in kilograms per square metre?

In the Final Honour examination, by contrast, Clifton expected candidates to attain a considerably higher standard, especially in the papers on the individual specialities within physics (usually limited in the 1870s to Light and Heat). In Physics in Trinity Term 1874, one of the ten questions read:

Explain Jamin's method of determining the compressibility of water. What is the influence of temperature on the compressibility of water, alcohol, and ether?[80]

---

[78] *Oxford University Examination Papers. Second Public Examination. School of Natural Science. Trinity Term, 1870* (Oxford and London: James Parker, n.d.), in Bodl. 2626 e.17.

[79] These questions and the one that follows from the Mechanics paper are from a bound set of examination papers for the Second Public Examination in Natural Science, 1872–7 (Bodl. 2626 e.59).

[80] This and the following question are from the set of papers cited in the previous note. Among the many sets of papers in the Bodleian, a particularly valuable one (Bodl. 2626 d.35) is that assembled by Joseph Frank Payne (first in Natural Science, 1862). The papers, which include those set for certain college fellowships and university scholarships as well as for university examinations, date from between 1856 and the later 1860s. The case of Payne illustrates well how a science-based career could be built on success in the NSS. Within a year of taking his first, he had won the Burdett–Coutts scholarship for advanced study in science (with special reference to geology) and had been elected a fellow of Magdalen. From 1865 to 1867, he held a Radcliffe travelling fellowship, which allowed him to train in medicine and to embark on a distinguished medical career, while retaining his fellowship (as he did until 1883).

In the paper on Light, the eight questions were tougher. The following question makes the point:

For what sections of the wave surface, whose equation is:

$$\frac{a^2 x^2}{a^2 - r^2} + \frac{b^2 y^2}{b^2 - r^2} + \frac{c^2 z^2}{c^2 - r^2} = 0,$$

does the double refraction partially follow the law of a negative and of a positive uniaxal crystal? What is the defect from complete coincidence?[81]

Clifton's intention was clearly that the strongest undergraduates should achieve a standard quite comparable with that required in physics for the Mathematical or Natural Sciences Triposes in Cambridge, and he chose his textbooks accordingly. The books he recommended were the latest in the field and, in many cases, far from easy. As his advanced teaching got under way in the 1860s, he prescribed not only the second edition of Edmund Atkinson's translation of Adolphe Ganot's comprehensive but untaxing *Traité élémentaire de physique* (1866) and the mathematically undemanding treatises by Walker (1851–2) and Charles Brooke (1867) but also the Müller-Pouillet three-volume *Lehrbuch der Physik und Meteorologie* (in German) and the three volumes of Jules Jamin's *Cours de physique de l'Ecole Polytechnique* (in French).[82] For his specialized lectures, even more demanding texts were proposed. On acoustics, for example, he recommended John Tyndall's *Sound* and Helmholtz's *Die Lehre von Tonempfindungen* (in German), while on heat the second volume of Jamin's *Cours* was to be supplemented by Balfour Stewart's substantial *An Elementary Treatise on Heat* (1866).

Clifton's inclusion in his recommendations of works in foreign languages reflected at once the high standards he expected of his pupils and the gaps that existed among up-to-date textbooks in English. The obvious remedy to the gaps – of encouraging the writing of new works – seems initially to have interested him. In December 1866 what was referred to as 'Mr Clifton's series' was considered by the Oxford University Press's School Book Committee.[83] In reality, the books in

---

[81] This question is typical of the increasingly taxing questions that came to be set during the mid and late 1870s, when Clifton's ambition was at its height.

[82] Jamin's *Cours*, published in three volumes between 1858 and 1866, gives a measure of the high standards that Clifton sought to impose. The first volume was designed for candidates for entry to the Ecole Polytechnique, while the second volume (mainly on heat, sound, and the properties of gases) and the third (devoted to advanced electricity and magnetism) were pitched at the level of courses within the Ecole. In addition to the advice on textbooks that appeared in the *University Calendar*, more detailed information was given (for all disciplines) in the termly published sheet advertising 'Lectures of professors'. Several issues of the sheets for this period are in Bodl. G. A. Oxon c.84 and 85.

[83] Minute-book of the School Book Committee, 42–3 (meeting of 4 December 1866) and 44 (14 December 1866). The volumes considered were by Clifton (optics), Price (mechanics and hydrostatics), Donkin (acoustics), Balfour Stewart (magnetism), Esson (electricity), Millar (crystallography), and Bence Jones (physiological optics). I am grateful to Martin Maw for facilitating my access to the minute-book, now in his care in the archives of the Oxford University Press. The alternative solution, of commissioning

question were intended for the recently launched Clarendon Press series of elementary but authoritative texts in both the sciences and the humanities for student use (Sutcliffe 1978: 19–24). The series had been agreed upon in 1864, and with the strong backing of Bartholomew Price (first on the School Book Committee and then, from 1869, as secretary of the delegates to the Press) it got off to a brisk start. On the science side, Alexander W. Williamson's *Chemistry for Students* (1865), Stewart's *An Elementary Treatise on Heat* (see above), and A. G. Vernon Harcourt's *Exercises in Practical Chemistry* (1869) were all substantial contributions. But most of the possible volumes that were discussed in 1866 and those that the delegates formally approved for publication at meetings in 1868 and 1869 failed to materialize.[84]

Clifton's own promised volume, an *Elementary Treatise on Optics*, was one of those that never appeared.[85] And nothing more was heard of other volumes with similar titles – by Price (on mechanics and hydrostatics), William Esson (on electricity), Story-Maskelyne (on mineralogy and crystallography), and George Griffith (on physiological optics) – despite their being formally commissioned in 1869. Donkin, however, was one author who went some way towards fulfilling his contract; at least the first part of his *Acoustics* was published (posthumously and after an heroic battle against illness) in 1870 (Donkin 1870). George Rolleston's *Forms of Animal Life* also appeared in 1870, and eight years later Robert Baynes published his *Lessons on Thermodynamics* (see above) in the series. Despite these volumes by Oxford authors, it is hard to see such a modest level of production as anything but a missed opportunity for exploiting a shop-window for the University's science. The patchy level of commitment on the part of Oxonians contrasted tellingly with the more high-profile contributions to the series by Tait and, more spectacularly, Maxwell.[86]

An ordering of priorities that placed so little emphasis on low-level teaching meant that Clifton's criteria for success or failure centred almost exclusively on the all too few Honours candidates who chose to specialize in physics. The evidence he gave before the Commission of 1877 conveyed the keenness of his hopes and frustrations with regard to this type of pupil (*University of Oxford Commission* 1881: *Minutes of Evidence* 22–3). The conditions in which he was trying to advance his discipline not only impeded his own engagement in advanced work but also inhibited

---

translations of foreign texts, was considered, but it did not find favour. The Press consulted Tait about the idea of translating Helmholtz's *Die Lehre von Tonempfindungen* and his *Physiologisches Optik* (published as part of Gustav Karsten's *Allgemeine Encyclopädie der Physik* and soon to be reissued as *Handbuch der physiologischen Optik*), for example. But the proposal was rejected; see minute-book of the School Book Committee, 53 (7 June 1867) and 54 (3 July 1867).

[84] The titles that were approved for publication appear in the minutes of the meeting of the delegates on 29 October 1869, in 'Orders of the delegates of the Clarendon Press 1853–1881 and accounts 1854–1867', 161, in the archives of the Oxford University Press.

[85] Clifton's volume, with some of the others that did not appear, was frequently advertised, for example in *The Student's Handbook to the University and Colleges of Oxford* (Oxford: Clarendon Press, 1873).

[86] On the inclusion in the series of Maxwell's far from elementary *Treatise on Electricity and Magnetism*, see note 96, below.

the work of his ablest students. College fellowships for which a physicist could realistically compete were far scarcer than those available in the classical subjects or even mathematics, so that the traditional Oxonian incentive to excel in the Schools was limited. One consequence was that gifted young men, especially those who might be deterred by the requirement of celibacy attached to most fellowships, were commonly lured into the booming profession of school-teaching, where salaries were higher than those that Clifton offered his demonstrators (£150 p.a.) and where some undergraduates were even offered comfortable posts without sitting their final examination.[87] In his complaint on this score, as in all his evidence to the Commission, Clifton presented himself as the champion of specialization and excellence rather than of breadth and large numbers. It was a high intellectual ideal and one that distanced him from Acland's more capacious, multidisciplinary vision of a proper scientific education. It was also one that the conditions of the mid and late 1870s within Oxford were making unrealistic.

Among the circumstances that constrained Clifton's ambitions after the initial euphoria of his occupation of the Clarendon was the gathering tide of demands for the material support of sciences other than experimental philosophy. One conspicuous new focus for expenditure was astronomy, now under the leadership of the Revd Charles Pritchard, Fourth Wrangler in the Cambridge Mathematical Tripos in 1830, who arrived in 1870 to occupy the Savilian chair, vacated in the previous year by the death of Donkin. By December 1872, despite his age – he was now well into his early sixties – and an experience of teaching limited to almost thirty years as a schoolmaster, Pritchard had fashioned a plan for a 'school of astronomical physics' in which the priority would be research rather than the instruction of undergraduates.[88] For this purpose, the tiny free-standing observatory that had housed the modest equipment he had inherited from Donkin, and in which he had installed, unsatisfactorily, his own 10-inch equatorially mounted reflecting telescope, was seriously inadequate. But, in the face of this potentially fatal impediment, Pritchard immediately found an ally and mentor in the ways of the University in the person of the Keeper of the Museum, John Phillips. It was Phillips, ageing himself and anxious to ensure that his own astronomical observations (mainly conducted in a private observatory erected in the garden of the Keeper's house) would be continued, who fired Pritchard's zeal for a new building and equipment.

The guidance that Phillips gave Pritchard was crucial for the passage through Convocation of the proposal for the replacement of the old observatory with a far

[87] In a letter of 21 August 1876 to Stokes (CUL Stokes Collection, Add. MS 7656, C717), Clifton expressed a disappointment that clearly weighed with him, following the departure of William B. Croft to a post at Winchester College immediately after taking a first in the NSS in Trinity Term 1875.

[88] Pritchard's plans for astronomy in Oxford appear in a printed appeal to Convocation, dated December 1872, in the minute-book of the Museum delegacy, 1867–75, OUA UM/M/1/3. The text is reprinted, with a short additional paragraph on the proposed location for the observatory, in *OUG*, 4 (1873): 43 (11 February 1873).

larger one situated on the edge of the Parks 200 yards to the east of the Museum (Hutchins 1992: 54–64; 1994: esp. 219–37). The generosity of the University's new provision for astronomy, consisting of a grant of £2500 for the purchase of a 12¼-inch equatorial refracting telescope by Sir Howard Grubb of Dublin and £1500 for the building, was immediately matched by the gift of a 13-inch reflecting telescope, donated by the wealthy amateur astronomer Warren De la Rue.[89] By the time he died in 1893, Pritchard had not created a 'school' on the scale that he had planned. But, equipped with the new observatory (built to Charles Barry's design and inaugurated in 1875), a lecture-theatre (added in 1877–8), and telescopes, especially the De la Rue instrument, of outstanding research quality, he had certainly placed Oxford among the leaders of world astronomy by the early 1880s.

These were good years too for chemistry. Again the catalyst was an appointment, in this case of William Odling (London-trained in medicine and so, like Clifton and Pritchard, another outsider to Oxford) as Benjamin Brodie's successor in the Waynflete chair of chemistry in 1872. Throughout his forty years in the chair, Odling showed no personal interest in experimental research. But he managed his discipline effectively enough. A two-storey extension to the laboratory, built between 1877 and 1879, allowed chemistry to continue the expansion from the cramped quarters of the Glastonbury (or abbot's) kitchen that it had begun when it moved into the premises vacated following experimental philosophy's departure from the Museum in 1871.[90] The amount allocated for the construction and fittings – of up to £7000 – stood as another demonstration of the University's continued readiness to promote science. But in years in which agricultural depression was beginning to 'squeeze' Oxford's resources, to use John Dunbabin's term (Dunbabin 1997: 405), any generosity towards another science necessarily diminished Clifton's prospects of securing further expansion for physics. On the one hand, therefore, Convocation's allocation of a second £1000 for the purchase and maintenance of equipment in February 1877 gives the lie to any suggestion that physics was neglected.[91] This was small beer, however, by comparison with the funds that Hebdomadal Council, ever since the internal inquiry of 1873 (see above, p. 62), had

---

[89] *OUG*, 4 (1873): 42–3 and 73 (11 February and 4 March 1873).

[90] Even the refitting of the space in the Museum had left chemistry with overcrowded quarters inadequate for the more than fifty students wishing to undertake practical work. For Odling's insistence on the need to double the laboratory space and provide more staff to assist with the teaching see the unpaginated pamphlet, dated 22 April 1874, that he addressed to the Museum delegacy; a copy is in the Museum of the History of Science, MS Gunther 65. Later pleas are Odling's 'Statement as to chemical laboratories in the University Museum', *OUG*, 7 (1876–7): 126–7 (5 December 1876), and his letter of 11 April 1876 to the Dean of Christ Church, H. G. Liddell, ibid., 360 (28 April 1877). Liddell had been Vice-Chancellor at the time of the inquiry of 1873 (see p. 62) and was now an influential Pro-Vice-Chancellor. The votes in Convocation on 30 January 1877 and 5 June 1877 are reported in *OUG*, 7 (1876–7): 219 and 434 (30 January 1877 and 5 June 1877).

[91] For Clifton's request to the Vice-Chancellor (27 January 1877) and Convocation's approval of the grant (13 February 1877) see *OUG*, 7 (1876–7): 232–3 and 239 (6 February 1877 and 13 February 1877). By now, the residual scepticism towards physics, reflected in 7 *non-placets* (against 48 *placets*), was somewhat less strong

hoped to make available for improving the University's provision for teaching, especially non-collegiate professorial teaching.

In a restatement of the needs of the various boards of studies in 1877, Council was unequivocal in its recognition of the importance, for physics, of creating a second professorship in the subject and a professorship in mechanics and engineering.[92] By now, the ideal solution that Clifton had articulated in 1874 – of creating three professorships of physics as well as one in 'experimental mechanics' – was stated by him to be 'impossible'.[93] But even Council's more modest aspiration, which reflected Clifton's new realism, was to remain a dead letter for over two decades. Despite the undimmed zeal for research and advanced teaching that Clifton expressed to the University Commission in 1877, his optimism must soon have been tried by the inability of Council to deliver on its plans for improving the conditions for the work of professors and departments and to persuade colleges to assume a greater share of the cost. Through his own family too he knew at first hand the emerging difficulties of an economy dependent, as the University's to a large extent was, on agriculture. In 1873, on the death of his father, Clifton had become responsible for the family estate in a far from prosperous corner of Lincolnshire, and he took his responsibilities to his tenants there with characteristic seriousness, visiting Gedney frequently and engaging with the consequences of rural depression. It was the type of burden that weighed heavily with him.

Despite Clifton's gathering recognition of the limitations of what he might achieve, the fact remains that by 1877 he had established Oxford as one of the leading centres for physics in the British university system. It was part and parcel of that achievement that he himself had accumulated a full quiver of the accolades that went with his status as an Oxford professor. The virtually inevitable fellowship of the Royal Society had come his way in 1868 (following his much earlier election as a fellow of the Royal Astronomical Society in 1860), and in Oxford he had secured a potentially useful base within the college system through his election as a fellow of Merton in 1869 and an honorary fellow of Wadham shortly afterwards. Within the Royal Society and in his relations with the wider world of science and in public life,

---

than it had been in 1870; cf. note 53. More striking was the opposition in Convocation on 5 June 1877 to the allocation of up to £7000 for the extension to the chemistry laboratory (22 *non-placets* against 62 *placets*) and of up to £2400 for the additions, especially the lecture theatre, to the University Observatory (27 *non-placets* against 46 *placets*); see OUG, 7 (1876–7): 434 (5 June 1877).

[92] Council's sympathetic view of the needs of the Boards of Studies is abundantly clear in its decision to publish in full the replies it had received from the interested parties between 1873 and 1877. It did so in the 'Statement of the requirements of the University adopted by the Hebdomadal Council on the 19th of March, 1877, with the papers upon which it was founded', published as a Supplement to issue no. 254 (28 April 1877) of the OUG, 7 (1876–7): 333–64. While stating the need for a second professorship of physics and professorship of mechanics and engineering, Council urged that, as an immediate measure, a second demonstrator in physics should be appointed, followed (after some extension of the laboratory) by a third; ibid., 335. [93] Clifton to the Dean of Christ Church, 2 May 1876, ibid., 360.

he remained a prominent and respected figure in such tasks as examining and refereeing papers, and in these contexts he continued until his later years to assume his share of responsibility. On home-ground, on the other hand, he seems to have been less sure-footed. Even before 1877, as I have argued, there had been signs of the suspicion that his ambitions persistently raised in more conservative circles, and in the securing of alliances beyond the immediate confines of the scientific professoriate he never displayed the tactical skills of an Acland or a Phillips. This is not to say that he was damagingly eccentric or wholly isolated. Plainly he was not. Indeed, in the priority he gave to Honours students and his insistence on high standards, he can be seen as both following and fostering the reforming tide that, by the 1860s, was leading the University to greater seriousness and specialization in undergraduate studies. In this, he might have found a natural ally in Mark Pattison, then in his prime as Rector of Lincoln College and a champion of the professors and the ideals of disinterested learning that they were intended to promote (Pattison 1868: esp. 231–70, and Sparrow 1967: 105–49). But there is no evidence that the two men ever sought to make common cause.

While he fought doughty personal battles, therefore, Clifton failed to forge the alliances that would have helped his discipline. In no context, for example, does he seem to have associated himself publicly with the broader currents of change within the University. His timorous, ungregarious personality, manifested in a social life focused on the receiving of students for Sunday lunch or tea in his home and in the recurringly self-deprecating tone of his correspondence, was almost certainly at the heart of the rather solitary nature of his engagement.[94] Almost effortlessly, physics in Oxford had done well in the late 1860s and early 1870s; at that stage, it had not mattered that the head of the discipline had sought, in his dealings with his colleagues as in physics, 'to eliminate or to minimize friction' by avoiding unnecessary confrontation and going 'quietly on his way' (in the words of the *Oxford Magazine*'s revealing comment on him on his retirement in 1915[95]). But consolidation in the more difficult circumstances that had come to prevail by the mid-1870s required unprecedented levels of tenacity and visibility. In this new world, Clifton was not the man for the job, and his own slight record in research and that of his laboratory could not prevent critical eyes from making comparisons with what was afoot elsewhere. In Glasgow, William Thomson and his purpose-built laboratory (inaugurated in 1870) were going from strength to strength (Gray 1896–7), while in Cambridge Maxwell was beginning to make waves, not least through the

---

[94] The unsigned obituary in *Nature* (Anon 1921a: 18) probably catches Clifton's personality accurately when it describes him as 'a man of great learning, but also of great deliberation'. An example of Clifton's self-deprecation appears in his letter of 16 April 1879 to Stokes (CUL Stokes Collection, Add. MS 7656, RS 1403), in which he apologizes for the 'great length' of his report on a paper by Gordon: 'I am a very feeble creature with a pen and the result is that my performances with that implement are desperately prolix.'

[95] *Oxford Magazine*, 34 (1915–16): 18.

publication of his *Treatise on Electricity and Magnetism* (Maxwell 1873),[96] and to propel the Cavendish towards the eminence that fed the increase in the numbers of undergraduates studying physics in either the Mathematical or the Natural Sciences Tripos from the 1880s, an increase that Oxford conspicuously failed to match (Wilson 1981–2: 335–66; MacLeod and Moseley 1982: 202–9). Perceptive observers, who certainly included Clifton himself, also saw that a laboratory, such as the Clarendon, that had appeared as a showpiece in 1870 risked, seven years on, being eclipsed by those of the biggest continental and American universities. Between 1873 and 1878, for example, the University of Berlin spent £77 000 (over six times as much as had been spent on the Clarendon) on its new physics institute, and in 1882 August Kundt's chair at the University of Strasbourg was endowed with a smaller but finely conceived and handsomely equipped laboratory costing £29 000 (Cahan 1985: 15–37).

Whether a professor more forceful and adroit in his personal relations than Clifton or one better able to bring his early aspirations for research to material fruition would have allowed Oxford physics to cope better with the challenges that were accumulating by 1877 must remain an open question. Yet such failings as might be laid at Clifton's door do not constitute anything like the whole story. Likewise, the idea of a *genius loci*, hostile to science, is hard to reconcile with the evidence of a university in which powerful forces, notably but not exclusively on Hebdomadal Council, were intent on modernization and change.[97] A major obstacle to the realization of Clifton's early ambitions, on the other hand, did lie in the small number of undergraduates wishing to study science, especially physics, at an advanced level.[98] On this score, the examination statutes told significantly against Oxford. For even after the reforms of 1870, the still considerable hurdle of Moderations, predominantly classical and occupying most of the first two years of the majority of undergraduate careers, delayed an engagement with science for what appears, in retrospect, a damagingly long period of time. The contrast with Cambridge on this score is striking. There, undergraduates were able to begin serious mathematical study immediately on arrival at the University; at no stage did they experience the same pervasive dominance of the classical tradition that so coloured the experiences and choices of undergraduates in Oxford.

One conclusion that emerges from any account of the fortunes of physics in Oxford before 1877 is that an analysis cast only in terms of failure or missed opportunities is inappropriate. It is inappropriate too to seek explanations for what

---

[96] Although Maxwell's work responded most immediately to the integration of electricity and magnetism in the syllabus for the Cambridge Mathematical Tripos, it was published as a title in the Clarendon Press Series of textbooks in 1873, after being commissioned in 1868. On the circumstances of the publication see Achard (1998). [97] The point is well made by Howarth (1987: 335–6 and 349–53).
[98] Again, the point is well made by Howarth (1987: 358–60 and 366; 2000: 488–92).

occurred solely in the shortcomings of Robert Clifton, who has tended to bear the brunt of criticism, essentially for not laying the kind of foundations on which, over the next two decades, the Cavendish Laboratory rose to international eminence. Clifton, as I have argued, may not have been quite the man for the job of navigating amid the complex interactions between disciplinary lobbies, reforming and conservative factions, and the diverse expectations of undergraduates in a period of accelerating academic reform tempered by the beginnings of economic depression. The fact remains, however, that in the period discussed in this chapter things generally went well for Clifton's discipline. That said, it is clear that by the later 1870s the context of Oxford physics was changing rapidly, and it was by its response to those changed circumstances that the subject was about to be tested.

3

Robert Bellamy Clifton and the 'Depressing Inheritance' of the Clarendon Laboratory, 1877–1919

Graeme Gooday

> All the available space in the Laboratory is now devoted to the instruction of students preparing for the examination in the School of Natural Science, and it will in future be impossible to offer facilities to advanced students wishing to engage in Research.
>
> (Clifton 1910–11: 959).

> ... a delightful old gentleman and kindness itself. But for him physics had reached its end in the 1890s.
>
> O. F. Brown, Clarendon demonstrator, 1910–13, reminiscing on the septuagenarian R. B. Clifton in 1968.[1]

The account of the early Clarendon laboratory offered by Fox in the preceding chapter shows that Robert Clifton's early career in Oxford up to 1877 was relatively dynamic, and marked by optimism for future development of University physics. And yet Clifton's subsequent endeavours in Oxford up to his retirement in 1915 have been widely stigmatized for the absence of internationally significant accomplishments by either the professor or his students. This chapter will not, however, indulge in counterfactual speculation on what Robert Clifton might have done. Instead, I investigate what Clifton actually did in the period 1877–1915, whom he helped in their research, and what he tried to do to develop Oxford physics, often without success in an increasingly unsympathetic and financially inclement environment.

This chapter draws upon and extends Chapter 6 of my unpublished Ph.D. thesis, 'Precision measurement and the genesis of physics teaching laboratories in Victorian Britain', University of Kent at Canterbury, 1989. I would like to thank Robert Fox, Jack Morrell, and Tony Simcock for the generous and helpful advice they provided on earlier versions of this chapter.

[1] Letter from O. F. Brown to A. J. Croft, 18 September 1968, CLA.

In seeking to explain Clifton's activities, I resist the peculiarly twentieth-century assumption that the scholar's duty to publish research necessarily takes priority over the commitment to high-quality teaching. Given that the whole question of whether universities should primarily be devoted to teaching or research was moot in later Victorian Oxford, this is a topic on which historians ought perhaps to maintain a strategic agnosticism (Engel 1983). Although after 1878 Clifton only refereed for scholarly journals rather than contributing articles to them, his creativity was expressed in developing new lecture apparatus, investigating the construction of safety lamps for coal-miners in the 1880s, and supporting the research of others in his laboratory, notably Charles Vernon Boys in the 1890s.[2] I suggest Clifton generally subordinated research to the higher imperative of training students to the most rigorous standards of experimental physics. That is why he was devoted to the upkeep of 'his' laboratory and the design, construction, and preservation of its instruments, and not least to developing the good 'character' of his students and their future prospects – rather than arrogating the resources of the Clarendon to research activity. In that regard, we can understand perhaps Clifton's later management of the Clarendon Laboratory as a correlate of his gentlemanly responsibilities regarding the Lincolnshire estates he took over upon his father's death in 1873 (Howarth 1987: 37).

One important point that I am at pains to emphasize is that Clifton was by no means opposed to research *per se* throughout his career in Oxford. After all, the principal evidence of any such opposition comes from the quotation above, dating from 1911, and it was not disapproval of research but *overcrowding* that Clifton cited (not entirely ingenuously) as his specific reason for formal abandonment of new investigations. This is the strongest evidence that the pre-war Clarendon was not as moribund as F. A. Lindemann later claimed (see below). Although we might give some credence to the suggestion, made over a half century later by his demonstrator in 1911, O. F. Brown, that Clifton considered innovation in physics to have ended in the 1890s, I show that optical research did go on in the Clarendon until at least 1907. It is more helpful to consider Clifton's later career as marked by a long-running frustration at the University's refusal to grant him resources for teaching and research proposed by the University of Oxford Commission of 1877.[3] After all, such resources were granted to J. S. E. Townsend when he took up the Wykeham chair in 1900 to implement an extension of the subject that Clifton had himself proposed 23 years before. In a period when the financially straitened University unsurprisingly favoured the projects of young bloods like Townsend rather than those of antiquated sexagenarians, Clifton lacked the relevant political skills and tact to secure the funding for 'experimental philosophy' that he achieved in his first decade at Oxford. Indeed, the death of his friend and powerful University ally Henry J. S. Smith in 1883 was probably a significant factor in Clifton's subsequent

---

[2] In these regards, my account of Clifton differs from that given in Howarth (2000: 459–61).
[3] See Fox, Chapter 2, 76.

failure ever again to win the support of Convocation for his plans. Clifton's undiplomatic petulance in later years can be seen in his persistent (and contentious) claim that a Clarendon Laboratory filled almost to capacity by undergraduates could not make equipment available for research.

Picking up on themes in Fox's chapter, I show that Clifton's disappointments arose both from his idiosyncratic conduct and the ways in which local political contingencies thwarted any ambitions he had to maintain the Clarendon as a significant centre for anything but teaching in selected areas of physics. While Clifton maintained a profile outside Oxford in the Royal Commission on Accidents in Mines (1879–86) and as President of the Physical Society of London (1882–4), the critical reception of his 1877 paper on electrical potentials led him, I suggest, to lose his nerve for research publication. So, after failing to persuade the University to grant him a new electrical laboratory in 1887, Clifton devoted his time and laboratory space to the new University-imposed duty of teaching large numbers of 'Prelim' students. From 1900 we see the melancholy spectacle of Clifton's eclipse by the younger and more dynamic Townsend and his retreat into curmudgeonliness, before retiring at the age of seventy eight, having held his chair for fifty years.

THE RECEIVED VIEW OF CLIFTON AND THE CLARENDON LABORATORY

Most critical accounts of Clifton's career at Oxford appear to have drawn rather heavily on the claims made by his professorial successor F. A. Lindemann. This is despite the obvious fact that Clifton had left Oxford four years before Lindemann was appointed in 1919 and that Lindemann seems neither to have met Clifton nor seen the Clarendon in operation during the latter's tenure. Indeed, Lindemann's views on Clifton imbibed a great deal from the rather unsympathetic testimony of Townsend and especially Tizard, who was Townsend's laboratory demonstrator and Lindemann's early close friend; hence they should not simply be taken at face value (Birkenhead 1967: 56; Clark 1965; Fort 2003). Such evidential problems did not inhibit the Earl of Birkenhead in quoting directly from Lindemann's own testimony in his biography *The Prof in Two Worlds*. Widely read and cited by historians of Oxford physics, Birkenhead followed Lindemann's views that the 'moribund' state of the Clarendon in 1919 reflected 'great discredit' upon Clifton. Birkenhead interpreted Lindemann's 'depressing inheritance' of 'scarcely any apparatus' suitable for carrying out research, no facilities for electric power, and, above all, 'no staff of research physicists' as a firm indication that Clifton had been 'entirely' opposed to research (Birkenhead 1961: 90). Yet the testimony on which Birkenhead relied had been constructed by Lindemann over the years to show a special need for the raising of research funds for the Clarendon Laboratory. In 1939, for example, Lindemann had sought the support of the Hebdomadal Council for an unusually large investment in his laboratory research regime. To maximize the impact of his appeal, he denigrated the state of physics due to both Clifton (by then long dead) and Townsend in the period before his appointment twenty years earlier.

Lindemann told Hebdomadal Council that to explain why a great increase in the grant to the Clarendon Laboratory was required, it was 'necessary to go back some way into the history of that institution'. His short history displays such a degree of historical inaccuracy and chauvinism that there can be little doubt as to his rather *expedient* reliance on Clarendon mythology to win finance for research:

Up to the year 1870 Oxford and Cambridge were of approximately equal standing in the Physical Sciences. In 1868[sic] Professor Clifton was appointed to the Chair of Experimental Philosophy and the Clarendon Laboratory was built 4 years later [sic][4]. At about the same time[sic][5] the Cavendish Laboratory was built at Cambridge. In the Cavendish laboratory a large amount of fundamental research was carried out and by 1918 the Cambridge Physics School was the best and largest in England. My predecessor, Professor Clifton, did not encourage research, the Physics school dwindled, despite the erection the Electrical Laboratory at the beginning of this century, and in 1918 the reputation of the Department had sunk almost to zero [sic][6]. . . .

In my predecessor's time the laboratory had been closed three days a week, no electricity was laid on, there was one assistant with a footlathe and one part-time demonstrator. So far as I can ascertain no original work was published in the 10 years preceding my appointment. Teaching, no doubt adequate in its way, could scarcely arouse interest, still less enthusiasm among the undergraduates in such circumstances.[7]

The latter paragraph relates mostly to what Lindemann found in the Clarendon only *after* it had been running at minimal level during the Great War and without Clifton's professorial guidance for four years. Saying nothing about the University's persistent refusal from 1887 to 1915 to grant Clifton the extra resources he requested, Lindemann asserted of his own early years in the chair that:

In view of the complete lack or ordinary facilities and *a fortiori* of modern apparatus, it was necessary to spend the greater part of the finances generously allocated to the Department by the University in equipping the Laboratory, developing a workshop, and in meeting the running expenses of a research department.[8]

---

[4] Clifton was in fact appointed to the chair in 1865, and the Clarendon began to be used in October 1870.
[5] After the Chancellor of Cambridge University, the seventh Duke of Devonshire, reached an agreement with the University to fund a chair of experimental physics and the relevant buildings in 1870–1, the Cavendish Laboratory was constructed between 1872 and 1874. See [Anon.] 1910: 4–7 and Harman (1995: 33–37, 615–700, 840–76; 2002: 3–13).
[6] For a contrasting view of Townsend's researches in the Electrical Laboratory see Lelong, Chapter 6.
[7] HCP, 1939, OUA HC1/1/173: 227–9. I thank Mark Curthoys for drawing this document to my attention. [8] Ibid.

In this pragmatically formulated account, it is striking that Clifton is made the scapegoat for the lack of research facilities in the Clarendon in 1919, and the University is cast as the benign benefactor. As successful as this was in getting Lindemann what the Clarendon needed in 1939, the irony in his interpretation will become apparent when we see how the University refused similar levels of benefaction in the second half of Clifton's Oxford tenure.

Other historiographical themes that have often been borrowed from Birkenhead's biography of Lindemann include the myth that Clifton had originally, and lamentably, been appointed in preference to the eminent German physicist and physiologist Hermann von Helmholtz. A passing reference to this in Birkenhead's book in relation to Lindemann's own early ambitions for the chair in late 1913[9] is what subsequently inspired the late Nicholas Kurti's analysis. Kurti claims that Helmholtz's friend Max Müller, soon to be elected Corpus professor of comparative philology, was responsible for dissuading the University from approaching Helmholtz on the ostensible grounds that Oxford could not afford a suitable salary for him (Kurti 1984, 1985; Fox, Chapter 2, 42–3). Insofar as this led to the appointment of Clifton rather than Helmholtz, Kurti argues that Müller's action was the root cause that allowed the putative fifty years of 'neglect' of physics under Clifton (Kurti 1984: 314).[10]

Such speculation about a possible alternative history of the Clarendon under Helmholtz is best informed by a comparison with Clifton's professorial contemporaries at the Cavendish Laboratory opened in Cambridge in 1874: James Clerk Maxwell (in the Cambridge chair from 1871 to 1879), Hon. John W. Strutt, Lord Rayleigh (1880–4), and Joseph John Thomson (1884–1918). Kurti argues that under their aegis, Cambridge 'justly earned the reputation of being the centre of excellence for physics' by virtue of the 'discoveries' made by Cavendish and ex-Cavendish researchers. One might note, however, that the great proportion of Maxwell's innovative research was published before he became busy in the Cavendish Laboratory, notably his famous *Treatise on Electricity and Magnetism* of 1873 (Maxwell 1873). Soon thereafter Maxwell became occupied in the time-consuming duties of being both a university professor and leading a busy academic physics laboratory, rather analogously in fact to Clifton's situation in Oxford. As a gesture to the Cavendish's benefactor, the seventh Duke of Devonshire, Maxwell managed to replicate Henry Cavendish's experiments on electrical discharge, editing and publishing the 1770s manuscript

---

[9] 'The Professorship in question is for physics, the Clarendon Lab. The present "encumbrance" is Clifton. He has been there since about 1870. They chose him in preference to Helmholtz who wanted to come. Tizard thinks I might get it, I can hardly believe it. Townsend who is the other physicist there is very nice. I lunched with him today. He is one of Carson's Ulstermen. Clifton says he is going to retire next September'; F. A. Lindemann to his father A. F. Lindemann, November 1913, cited in Birkenhead (1961: 56).

[10] Kurti (1984: 314) suggests that Müller probably also feared that Helmholtz might 'outshine' his reputation as the most famous scholar in Oxford.

shortly before his untimely death in October 1879. Yet, as Peter Harman notes, by 1877 Maxwell was concerned at how much he had had to give up in undertaking his public office; only in his last two years in the chair could he develop substantially new theoretical work in statistical mechanics and thermodynamics (Harman 2002: 2). Of course, as Kurti implicitly recognizes, it was the 'Cavendish men', rather than Maxwell alone, who were responsible for creative research in Cambridge during the 1870s and after. Yet Kurti does not consider the extent to which Clifton's students and demonstrators at the Clarendon in the 1870s – R. E. Baynes, Arnold Reinold, Arthur Rücker, and Lazarus Fletcher, for example – were arguably as disposed to research and publication as their Cambridge counterparts (Rücker 1873–4; Thorpe 1916: xxii)

In pursuing Kurti's Oxford–Cambridge comparison in this new direction, it is enlightening to consider the parallels between Clifton and Maxwell's successor, Lord Rayleigh, who was a member of the landed gentry, albeit of a somewhat higher order. Rayleigh took up the Cavendish chair in pursuit of an enhanced income in the period 1879–84 during a severe agricultural depression. While in Cambridge, he engaged in both undergraduate teaching and especially research, and it was only due to the latter that the Cavendish began to acquire real national and international fame. At the Cavendish, Rayleigh undertook, with his sister-in-law Eleanor Sidgwick, important work with precision measurement techniques on electrical standards (Rayleigh and Sidgwick 1884). It is notable, however, that once the agricultural depression was over, Rayleigh returned to his estates, those clearly being his highest priority.[11] The parallel with Clifton here is enlightening since it draws our attention to the way that a landed gentleman might place a higher value on the proper management of his estates than on large-scale research in experimental physics. In contrast with Rayleigh, though, Clifton did not treat the role of professor of experimental physics as a secondary career to support his family during times of agricultural difficulty.[12] Rather it was just such agricultural difficulties in the early 1880s that harmed Oxford's finances to an extent that made it easy for Clifton's opponents to defeat his ambitions for Clarendon expansion (Engel 1983).

A third historiographical theme to which Birkenhead's biography lends support concerns Clifton's attitude towards research. Birkenhead's claim, following Lindemann, is that Clifton was 'entirely opposed to research' (Birkenhead 1961: 90). This is a curious claim, one that evidently stems from Lindemann's acquaintance with only the decaying and 'depressing' post-war inheritance of his long-departed

---

[11] The third Baron Rayleigh wrote to his mother on November 17 1879 of how he and his wife responded to a campaign in Cambridge for him to take up the chair vacated by Maxwell: 'Neither of us like the idea of living at Cambridge, but perhaps I ought to take it for 3 or 4 years, if they can get no one else fit for the post. It would fit pretty well with the agricultural depression'. The five years that Rayleigh spent in the post were productive, but the strain severely tested his health (Strutt 1924: 99–100, 147–8).

[12] For Rayleigh's career at the Cavendish see Strutt (1924: 99–100, 147–8), and Glazebrook (1910).

predecessor. Certainly, as Bleaney and Fox have noted, the view that 'the wish to do research betrays a certain restlessness of mind' has often been *attributed* to Clifton (Bleaney 1988: 285; Fox 1997: 686). The absence of any identifiable primary source for this alleged remark, and especially the lack of date or context for it, naturally lead Bleaney and Fox to treat this attribution with some caution. I show at the end of this chapter that the only direct evidence of Clifton's associating research with a 'restlessness' of mind comes from O. F. Brown's 1968 reminiscences of a comment he heard as a Clarendon demonstrator more than a half century earlier, in 1910–13. As the timing of this matches Clifton's June 1911 pronouncement in the *Oxford University Gazette* that the Clarendon had become too overcrowded for research (see epigraph above), we might instead infer that Clifton's hostility to research was somewhat circumstantial and perhaps limited only to the last five of his fifty years in the chair (Clifton 1910–11). That Brown lived on until at least 1968 with such a vivid memory of Clifton's curmudgeonly twilight years helps to explain why that particular geriatric image of him has so long prevailed among physicists in Oxford and elsewhere. By contrast, there was no surviving oral tradition to carry down to posterity any tales of the relatively dynamic Professor Clifton of the 1870s.

Fox observes in his chapter how easily contemporaries might have inferred from Clifton's palpable immersion in the time-consuming demands of Clarendon teaching in the 1870s that, even quite early in his career, he attached a low priority to research. Yet Clifton's dedication to high-quality teaching by no means allows us to conclude that he disapproved of research *per se*. Indeed, if this were true, we cannot arrive at sensible answers for the following important questions: Why did he publish a research paper on electrical contact potentials in 1877? Why did he agree to be President of the Physical Society of London in the years 1882–4, making oral presentations to the Society on some of his investigations on lenses in 1882–3? Why did he undertake extensive research for the Royal Commission on Accidents in Mines between 1879 and 1886 in an attempt to develop a new safety-lamp for miners? Why did he so generously support C. V. Boys of the Royal College of Science in London when Boys used the cellars of the Clarendon Laboratory for four years of delicate experiments to redetermine the Newtonian constant of gravity G from 1890? Why, when his laboratory became crowded with Preliminary students in the early 1890s, did he publicly construe this as an impediment to the conduct of research? Why did he act as a referee for research papers published in the *Philosophical Transactions of the Royal Society* as late as the mid-1890s? And why indeed was optical research going on in the Clarendon Laboratory until at least 1907?

None of these questions can easily be answered if it is assumed that throughout his career Clifton consistently dismissed research as something to be undertaken only by the 'restless'. Rather, we might consider Clifton's view of research as evolving over time in response to local contingencies and becoming more indifferent as his conservative ambitions were repeatedly frustrated from the 1880s.

CLIFTON AND THE CLARENDON'S HEYDAY

As Fox has shown in the preceding chapter, the liveliest period of the Clarendon under Clifton's tenure was evidently the decade or so after it opened in late October 1870. Its centrality in the development of the new subprofession of experimental physics in Britain can be seen in the way that, in architectural terms at least, the Clarendon soon came to be seen as a model design for an academic physics laboratory. As Clifton's former pupil and demonstrator Arthur Rücker said in his Presidential Address to members of Section A of the BAAS gathered in the Clarendon for the Oxford meeting of 1894: '... it has served as a type. Clerk Maxwell visited it while planning the Cavendish Laboratory, and traces of it can be found in several of our university colleges' (Rücker 1894). Indeed, in drafting his design of the Cavendish, Maxwell worked from plans of the Clarendon given to him by Clifton in the early 1870s (Schaffer 1992; Fox, Chapter 2, 60–2 and Figure 2.16). The Cavendish director sought to glean more than just architectural advice from Oxford: contrary to the myths of Cambridge-focused historiography, the Cavendish Laboratory was hardly a vigorously flourishing centre of publication on innovative experimental physics during the six years after it opened in 1874. For that innovative phase one must look to at least the tenure of Lord Rayleigh from 1879 to 1884, and more plausibly to J. J. Thomson's succession, notably the late 1890s, when his studies on cathode rays began to attract a significant community of graduate researchers (Gooday 2001).

As a site of high-quality undergraduate teaching in experimental physics the Clarendon had few serious rivals in England until the Cavendish Laboratory under Rayleigh began systematically to accept undergraduates studying for the Natural Sciences Tripos in 1880. It is interesting to note further that Clifton's lecturing was conspicuously more popular than Maxwell's.[13] John Ambrose Fleming recorded only one fellow-student joining him at Maxwell's last two years of lectures (Fleming 1934: 64). By contrast, in 1870 Clifton claimed to attract audiences of up to forty

---

[13] It has been argued that the termination of Maxwell's contract with King's College London in 1865 was precipitated by his ineffectiveness in teaching undergraduates, especially as he was rapidly replaced by William Grylls Adams; see Gooday 1989: Chapter 5. Although Harman emphasizes that Maxwell resigned rather than being sacked, he notes that Maxwell did so in the middle of the session and continued to live in London. It would have been more conventional for Maxwell to continue to the end of the academic year. Maxwell offered to do so, but King's evidently declined his offer (Harman 1995: 3–4).

students preparing for their finals in the School of Natural Science.[14] Whereas Maxwell's teaching and writing explored the more technical mathematics of electromagnetic theory (Warwick 2003: 287), Clifton took extraordinarily time-consuming measures to prepare his lectures and demonstrations in order to make them as clear and effective as possible to undergraduates. One obituarist, almost certainly a former pupil, described him as an 'excellent and inspiring lecturer' (Anon. 1921). The point was to show prospective young science masters at public schools how the job of teaching experimental physics should be done, Clifton having borrowed much from the model of G. G. Stokes's lectures on optics that he attended in Cambridge in the late 1850s:

> [Clifton's] great idea was that his lectures would give his class instruction in giving experimental lectures; in the summer term he showed his men[15] the solar spectrum in great detail with a grating spectroscope. He also lectured on the optical properties of crystals and the phenomena of polarisation [sic], doing everything with minute care and great deliberation (Glazebrook 1921: vii).

In contrast to Maxwell, Clifton devoted his time and energies to the expert cultivation of instrumental practice in lectures. Clifton apparently spent an 'enormous amount of time' in designing and fitting up apparatus for lecture purposes, so that his lectures were often 'more of the nature of laboratory demonstrations' (Anon. 1921a: 19). Such was Clifton's obsession with the precise details of his demonstration experiments that subjects were taught in depth rather than in breadth: in the lecture theatre, time was 'no consideration' for Clifton (Anon. 1921a: 19). Being, however, a busy manager of his family estate, and an unusually *large* man not apprenticed in the artisanal arts,[16] Clifton's apparatus was actually constructed – and often reconstructed – in the laboratory workshops by the Clarendon's caretaker and instrument-maker. One such person, John Hornsey, was employed in this dual capacity for almost the whole of Clifton's time in the laboratory, from 1874 to 1912.[17] In fact, given the time and care invested in the more delicate and precise items used for lecture demonstration purposes, such as Rowland gratings and

---

[14] Clifton reported to the Devonshire Commission on 12 July 1870 that over the preceding three years the average number of students in his lectures had been 40 (excluding female students and non-matriculated attendees); *Devonshire Commission. First, Supplementary, and Second Reports* 1872: 187 . Howarth (1987: 370) notes that in 1886, Clifton had 20–25 students in his class.

[15] There were occasional female students in the Laboratory, but these 'ladies' were always distinguished from 'members of the University'. See Table 3.2.

[16] In his letter of 18 September 1968 to A. J. Croft (see note 1), O. F. Brown wrote of Clifton that he was a very large man with 'perhaps the biggest hands and feet I have ever seen, and the nails of a chinese mandarin of the old school'.

[17] Of Hornsey's retirement, Clifton said in his annual report in 1913 for 1912: 'At the end of the year, in consequence of failing strength, at the age of nearly 77, Mr John Hornsey retired from the service of the University, having for 37½ years faithfully discharged the duties of caretaker of the Laboratory and instrument-maker to the Department' (Clifton 1912–13).

Fig. 3.1. Workshops in the Clarendon Laboratory, regarded by Clifton as an essential facility in any physics laboratory. Clifton album.

Fig. 3.2. Instrument-shop in the Clarendon Laboratory, also visible in the distance in Fig. 3.1. Clifton album.

Michelson interferometers, Clifton seems to have been unenthusiastic about lending them to undergraduates with ambitions for research. Regarding such equipment, Richard Tetley Glazebrook reported in his obituary of Clifton:

> Much of this was designed and redesigned by [Clifton] until perfection, or something approaching it, was reached, and so much loving care had been spent on an instruments that it needed to be kept jealously under lock and key, taken out from time to time to be dusted and cleaned, possibly to be used in a lecture, but entrusted never to the careless handling of a student of physics (Glazebrook 1921: vii).

As one informed observer put it:

> The laboratory having been built, it had to be equipped with apparatus which was a labour of love to Clifton, for he was a born instrument-maker. Much of the apparatus is of his own designing, with the result sometimes that when an instrument had been brought to perfection it had become too sacred to be entrusted to the common herd (Anon. 1921a: 19).[18]

Insofar as Clifton only used some apparatus once a year and otherwise secured it in the Clarendon's display cabinets, he appears to have been indulging in a form of gentlemanly connoisseurship. The same practice marks the way in which he brought together his remarkable collection of antique mathematics books, later stored in a specially built library, the 'Folly,' at his home.[19]

Outside the lecture-theatre in the laboratory's experimental rooms, Clifton was initially quite actively involved in practical instruction of the students. This was a remarkably intensive regime of precision experimentation with the types of apparatus that his students actually were allowed to touch, and consisted almost entirely of repetitions of 'known' experiments carried out 'with as much accuracy as possible.' (Anon. 1921a: 19). Indeed, the laboratory books in which Clifton or his demonstrators used to record the results of students experimentation showed that students were expected to work to produce results reliable to at least three and usually four significant figures.[20] One who many years later had evidently not forgotten the

---

[18] This author comments that 'research in the modern sense was not welcomed with open arms; the apparatus was too jealously guarded'. Anon (1921a: 19)

[19] Glazebrook's obituary of Clifton recorded an episode indicative of his dedication to rare books even as a Cambridge student: 'From early days he collected mathematical books of historical interest ... He had a large collection of old Euclids, arithmetics, etc. and knew more about their contents than bibliophiles usually do. There is a story that, while still an undergraduate, Whewell sent for him and asked him his price for a copy of Calendri's 'De Mathematica Opusculum' [sic] (Florence, 1491), which he had bought for 2s. 6d. Clifton declined to part with it, and Whewell, who wanted the book for a friend said, 'Well, I have done my duty by my friend, but now – you keep that book!' (Glazebrook 1921: viii)

[20] The Clarendon Laboratory archives hold two books recording the measurement results of both male and female students from 1886 to c.1905: 'Mechanics & Physics Results Hilary Term 1886–Easter Term 1894' and 'Mechanics & Physics Results Michaelmas Term 1894–[1905]', CLA, bookcase, shelf 3. These unpaginated volumes systematically recorded the dates on which each student undertook a particular measurement in a cycle of a dozen or so experiments; the handwriting is not Clifton's so is likely to be that of one of Clifton's demonstrators. Results were recorded to between three and five significant figures without any estimate of error, as was the custom in contemporary British physics; see Gooday (2004: 76–8).

personal ennui of such a training experience, contended that this 'sound grounding' in 'accurate' experimental work 'no doubt bore good fruit later in many cases' (Anon. 1921a: 19). Whatever tedium was experienced in the laboratory, however, was given compensation by the Professor's invitations to them to have Sunday lunch with him and Mrs Catherine Clifton (whom he had married in 1862), who apparently kept a 'most hospitable house'.[21] Indeed, Clifton's students and demonstrators do seem to have been very fond of the gentlemanly professor for the care he took in their futures: 'He devoted himself to his pupils, both in Oxford and afterwards in obtaining posts for them' (Anon. 1921a: 19).

Clifton himself seems to have paid less direct attention to laboratory teaching as the 1870s progressed, especially as his own former students – generally those who had taken Schools in both Mathematics and then Physics – proved to be effective demonstrators in their own right. Notable among these were Arnold Reinold

Fig. 3.3. John Viriamu Jones, 1890. From Edward Bagnall Poulton, *John Viriamu Jones and other Oxford Memories* (London, 1911), frontispiece. After serving as demonstrator in the Clarendon Laboratory under Robert Clifton, Jones went on to become the first principal of Firth College, Sheffield in 1881 and, two years later, the first principal of the University College of South Wales at Cardiff.

---

[21] The lunches continued regularly until Catherine became seriously ill in 1893. But, according to Glazebrook (1921: ix), they were revived before her death in 1917.

(1870–3), Arthur Rücker (1871–4), Lazarus Fletcher (1875–7), and John Viriamu Jones (1879–81) whom Clifton assisted in securing some prestigious positions.[22] Indeed such was the academic reputation of the Clarendon's demonstrators that in 1876 Clifton was approached by his friend and former mentor G. G. Stokes, Lucasian professor at Cambridge, to advise upon which of his workers in the Clarendon would be suitable for employment at the Cavendish Laboratory. Clearly Stokes would not have made such enquiries unless the Clarendon had acquired a reputation for training first-rate physicists under Clifton's guidance; moreover, Clifton was able to suggest six possible Oxford candidates for the Cavendish posts.[23] From this it could be inferred that in the early years of the Cavendish, its Oxford counterpart by no means lay in a subordinate position.

Ironically, however, the burden of teaching at the Clarendon had grown to such an extent with the ever-increasing demand for the training of science teachers that even his laboratory staff found it very difficult to carry out much research. As he wrote to Stokes in 1876 of his current and 'most excellent' demonstrator, W. N. Stocker, appointed two years earlier:

His work with me has been so heavy that he has had no opportunity of doing original work, but I think it is not for the want of power or will. He has taken entire charge of the teaching [in] the department of weighing, measuring and heat, and has helped me greatly in optics. I never have occasion to interfere in his department as I always find everything going well when I do look in . . . [I] feel sure that Stocker has originality from the care with which he gets over difficulties in the laboratory work.[24]

Whilst the Clarendon thus gained recognition outside its walls, it continued to receive the general internal support of the University as a centre of teaching in the precise techniques of physical measurement. When Clifton appealed to the Vice-Chancellor to grant £1000 for 'elaborate measuring instruments' in February 1877, he met very little opposition. Included in his list of required measurement apparatus were balances, diffraction benches, quadrant electrometers, magnetometers, a theodolite, a torsion balance and so forth. Clifton emphasized that these were now needed in multiple quantities since 'the increasing number of students and the requirements of practical examinations render it necessary to have more than one instrument for the same purpose'. Convocation gave Clifton a clear indication of confidence in his work by voting 48 to 7 in favour of this move.[25]

The equipment that Clifton thereby acquired was very useful in enabling him to continue a scheme of research into the electrical behaviour of metals and liquids

---

[22] For more details see Fox, Gooday, and Simcock, Chapter 1, 17–18; on Reinold see Anon. (1921b).
[23] Clifton to Stokes, 11 March 1876, CUL Stokes Collection, Add. MS 7656, C 716.
[24] Ibid.
[25] *OUG*, 7 (1876–7): 232–3 (6 February 1877). This includes a letter from Clifton to Vice-Chancellor, 29 January 1877, making the case for further expenditure on apparatus.

Fig. 3.4. Engine-room in the Clarendon Laboratory, with an unidentified figure (possibly the laboratory's caretaker and instrument-maker John Hornsey or Clifton's assistant Henry Walter). The alternating-current Wilde dynamo (Fig. 3.7) is in the foreground (*left*) Clifton album.

with the assistance of his two 'very efficient' laboratory demonstrators. His paper was intended as the start of a research programme involving precise measurements of the 'difference of potential produced by contact with different substances'. In May 1877 he sent an incomplete preliminary investigation to *the Proceedings of the Royal Society*, where it was published a few months later. Although he stated his hope that 'at some future time to be able to communicate a more complete investigation with reliable quantitative determinations' (Clifton 1877: 299), this wish was never fulfilled, for reasons that will be discussed in the next two sections. As he told the investigative government-appointed University of Oxford Commission in October of the same year, although he had been able to use the results of his research in his lectures, he was sure that the research 'had seriously interfered' with his attentions to the laboratory students; hence he would 'hardly venture to undertake such work again' (*University of Oxford Commission* 1881: 22).[26] Nevertheless, Clifton's relative confidence in the bright future of the Clarendon – if only sufficient resources could be made available for its operations – is clear from his testimony to the Commission.

[26] For further discussion of Clifton's evidence to the Commission see Fox, Chapter 2, 73–7.

Fig. 3.5. Switchboards in the Clarendon Laboratory. These were presumably managed by Clifton's assistants to control the electrical supply to different parts of the building. As there was no mains electricity in the laboratory until after the First World War, the supply was almost certainly taken from the Manchester and Wilde machines illustrated in Figures 3.6 and 3.7. Clifton album.

Fig. 3.6. The Clarendon Laboratory's Manchester dynamo. This was a standard model manufactured by Mather and Platt in the mid-1880s to the designs of John Hopkinson, F.R.S., who worked as an electrical consultant in collaboration with his brother Edward Hopkinson. Like Clifton, John Hopkinson had been a high flier in the Mathematical Tripos at Cambridge. Clifton album.

Fig. 3.7. Alternating-current dynamo designed by Henry Wilde, F.R.S. Wilde is generally agreed to have been co-inventor of the self-exciting dynamo in 1866–7, along with Charles Wheatstone and Werner von Siemens, although in vigorous litigation he regularly asserted his personal claim to priority. Since Wilde was a resident of Manchester, Clifton might well have met him there during his period as professor at Owens College between 1860 and 1865. Clifton album.

CLIFTON'S TESTIMONY TO THE 1877 OXFORD COMMISSION

In the year of the University of Oxford Commission, Clifton's teaching had been busier than ever, with the laboratory accepting its maximum of 16 students. Even with his two demonstrators to assist him, he now claimed to be spending a minimum of five and a half hours per day with his general class. With his most advanced students alone he sometimes spent this much time, often extending such personal tuition to a period of several days. The laboratory was this full because the national demand for teachers of physics was, as Clifton put it, 'greatly in excess of supply'. More students than ever were drawn to seek a laboratory training in order to qualify as science teachers – particularly students who had been sent to Oxford by physics masters who were former pupils of his (*University of Oxford Commission* 1881: 23). Apart from straining his teaching resources to the maximum whilst he had been attempting to carry out research, Clifton found that the level of this demand had another deleterious consequence upon the prospects for research at the Clarendon:

> As soon as the best students have passed the University examinations, and often before they have taken the University examinations . . . they are bought off from continuing their studies by the larger schools . . . if they can say that they have had a year's training in the laboratory, very frequently the certificate of having passed the University examinations in physics is not required. At the present moment I have lost two of my best pupils, neither of them having been through the schools, both of them being taken away to fill appointments as school teachers. It follows that the men who are most fitted to help in research are fallen away before they can be of any use (*University of Oxford Commission* 1881: 22–3).

In these circumstances, Clifton's reputation as a trainer of high-quality physics teachers actually worked to his disadvantage in another sense too, for his laboratory demonstrators were also attracted to teaching in the wealthy public schools:

> It has frequently happened lately that I have been consulted about appointments for teachers which have been vacant, and I have been quite unable to find any pupil of mine who was not already engaged. During the present period I am unable to find a pupil disengaged who is qualified to take the [vacant] office of second demonstrator in our laboratory and the funds at my disposal are not sufficient to enable me to induce a man to leave schoolwork (*University of Oxford Commission* 1881: 23).

Such apparently was the ironic result of Clifton's success as a trainer of physics teachers. Without a second demonstrator and with the ever-increasing content of contemporary physics brought about by the continuing growth of the subject, Clifton's strategy to maintain the highest standards of teaching was to restrict the scope of his laboratory instruction. Whilst Clifton taught optics, he assigned the duties of instruction in heat and weighing and measuring to his sole remaining demonstrator, effectively abandoning the subjects of electromagnetism and acoustics (*University of Oxford Commission* 1881: 23). Although in his lectures he still maintained a fairly broad coverage of the subject, he considered this narrowing of his laboratory curriculum vital in order for it to be possible to pursue the 'minuter details' of each subject. As he agreed with Commission's chairman, Lord Selborne, the progress of physics chiefly depended upon the minutiae of precision physics: 'at the present time especially, the progress of physics seems to me to depend on the progress of methods of exact measurement' (*University of Oxford Commission* 1881: 24). Hence, even if he were not able to carry out research himself, through his teaching of laboratory techniques of exact measurement Clifton considered that he trained his students in the research skills they needed in order to contribute to the advance of physics.

After hearing such evidence from Clifton and the rest of the Oxford professoriate, the Commissioners framed recommendations in May 1878 that were sympathetic to Clifton's position. That is to say that they acknowledged research as a legitimate function of the University yet at the same time subordinated it to the professorial duties of excellence in teaching (Engel 1983: 189). To assist Clifton in carrying out teaching and research in this order of priority, they followed the course initiated by Hebdomadal Council, which in 1876 recommended the foundation of a chair of physics by New College, along with two associated demonstratorships to share the burden of running the Clarendon (Anon. 1876: 24). Since there was considerable optimism in early 1878 that the income of the University would increase throughout the following decades, both from agricultural improvements and college tenancies, it was assumed by all concerned that the financial requirements of the science chairs would easily be met (Engel 1983: 201). This alas proved not to be the case, and the resulting financial stringencies entailed that the expansion in physics supported by Clifton was not enacted until 1900 – and then ironically in ways that marginalized his status in a manner that he could hardly have predicted in 1877.

Long before then, however, Clifton's nerve in promoting research cannot have been helped by the critical reception accorded to his 1877 paper on contact potentials, to which the next section is devoted.

CLIFTON'S 1877 RESEARCH PAPER ON VOLTAIC ACTION

The question of how to account for the action of voltaic cells, and in particular the source of the voltaic effect, was a topic intermittently debated for over a hundred years after Volta's initial work was published in 1799–1800. As Sungook Hong has pointed out, even after three decades of intense battle between the rival camps affiliated to William Thomson and James Clerk Maxwell at the end of the nineteenth century, the dispute was not fully resolved (Hong 1994). That debate was launched in 1862, when William Thomson revived a version of Volta's original theory that proposed that the source of the electromotive force in a cell lay at the surface of 'contact' between two conductors – metals or electrolytes. During the 1870s, Maxwell and some of his allies had argued, on the basis of his theory of potential, that no such potential difference could *persist* at the meeting point of two conductors. Rather, the electromotive force was chemically engendered and was physically to be detected only in the difference of potential between the conductors and the air next to them.[27] Several experimenters attempted to resolve this conflict in the mid-1870s, especially allies of William Thomson such as William Ayrton and

---

[27] I am grateful to Professor Hong for allowing me several opportunities to discuss this subject with him.

John Perry, then professors at the Imperial College of Engineering in Japan. Their work has featured more prominently in the history of this debate than has Clifton's, and this section of the chapter explains why this was so.

In the winter of 1875–6, Ayrton and Perry undertook some experiments to establish support for Thomson's view. They investigated the potential difference between the successive metallic and liquid elements in a compound voltaic cell to see whether these 'internal' potential differences due to heterogeneous 'contact' actually did change the potential difference of the cell. According to them, the result of their experiments showing that such a change was actually observable militated decisively in favour of the Thomsonian view. Evidently delighted with such results, on 6 May 1876 Ayrton and Perry posted off their draft paper 'The contact theory of voltaic action no. 1' from Tokyo to their erstwhile mentor at the University of Glasgow. The paper arrived by ship a few months later in time to be read at the 1876 BAAS meeting in Glasgow by Sir William Thomson himself. Ayrton and Perry thereby hoped to receive public recognition of priority for what they clearly conceived to be a decisive experiment in favour of the Thomsonian theory. This, however, was obviously one of many British Association meetings not attended by Robert Clifton. News of the Ayrton and Perry paper seems to have reached neither Clifton at Oxford nor the referees who subsequently supported publication of his paper on the same subject in the *Proceedings of the Royal Society* in 1877.

Clifton's experiments on the topic of 'voltaic contact' were undertaken in two phases between term-time teaching during the Christmas vacation in December 1876 and January 1877 and then during the Easter vacation in March 1877.

Notwithstanding his warm rapport with William Thomson, his interest in the voltaic controversy seems not to have been stimulated by any partisan affiliation to the disputing parties. Rather it seems to be linked to his interest in instruments, especially Thomson's quadrant electrometer, and in the virtuosic exercise of precision measurement with such devices. Indeed, apart from reference to Thomson's experimental technique, citations in the published version of his paper are exclusively to continental researchers, viz. Becquerel, Péclet, Pfaff, Buff, and Kohlrausch. His aim was to follow their line of attack, especially Kohlrausch's, in investigating the 'difference of potential arising when a metal and liquid are brought into contact' at all the heterogeneous points of contact within the cell. His aim was to 'determine the origin of the difference of electrical potential that is exhibited at the terminals of a voltaic element' (Clifton 1877: 299). Although he adopted a practical method not unlike Ayrton and Perry's, he did not frame his experiments to produce the kind of definitive conclusions to which they had laid claim. Clifton concluded his account rather with the suggestion that his data yielded values of potential difference that were 'much less' than previous investigations had led him to expect; these were measured to three and four significant figures, with information on errors left conventionally implicit – see Table 3.1. Clifton encountered problems with his apparatus that he did not fully resolve, such as the persistent diminution of the potential of his standard Clark cell, and vowed to rectify these at some future point. Certain other unspecified circumstances – perhaps teaching duties – had also 'prevented' him from making much progress with quantitative measurements. So Clifton sent his results to the Royal Society on 22 May 1877 merely as an interim account, hoping that 'at some future time' he would be able to communicate a more complete investigation with 'reliable' quantitative determinations. (Clifton 1877: 312).

Table 3.1. 'Table showing the difference of potential of copper terminals of voltaic elements in which no current has circulated' (Clifton, 1877: 313–14)

Composition of elements [Details of Clifton's specification of electrolyte concentrations have been omitted for the sake of clarity, except where essential to explain difference in data]	Difference of potential in terms of	
	Standard Daniell	Volt
Tin in dilute sulfuric acid . . . and Tin in solution of caustic potash	0.532	0.574
Tin in distilled water and Tin in solution of caustic potash	0.713	0.769
Copper and zinc (both well cleaned) in distilled water		
When metals are first immersed	0.760	0.820
After immersion for 1.5 h	0.821	0.886
After immersion for 3 h	0.838	0.905
Copper and amalgamated zinc in dilute sulfuric acid	0.856	0.924
Copper in distilled water and copper in solution of potassium cyanide	0.923	0.996
Daniell: – Copper in saturated solution of copper sulfate and amalgamated zinc in dilute sulfuric acid (1 volume commercial sulfuric acid to 10 volumes of distilled water)	0.958	1.034
Daniell: – Copper in saturated solution of copper sulfate and amalgamated zinc in dilute sulfuric acid (1 volume commercial sulfuric acid to 8 volumes of distilled water)	0.982	1.059
Carbon and zinc in a saturated aqueous solution of potassium biochromate	0.996	1.074
Copper in saturated solution of copper sulfate and copper in solution of potassium cyanide	1.102	1.189
Smee: – Platinized silver and amalgamated zinc in dilute sulfuric acid	1.193	1.288
Leclanché: – Solution of ammonium chloride in distilled water	1.268	1.369
Grove: – Platinum in commercial nitric acid	1.504	1.622
Platinum in acid solution of potassium bichromate and amalgamated zinc in dilute sulfuric acid	1.678	1.811
Carbon and zinc solution of potassium bichromate	1.701	1.835

Despite these caveats, and lacking any reference to Ayrton and Perry's prior research on the subject, Clifton's paper was immediately published in the Royal Society's *Proceedings* after it was read at a Society meeting on 14 June 1877. Given the sailing time between the two countries, the relevant issue of the *Proceedings* only reached Ayrton and Perry in Japan in December 1877 – where it was greeted with considerable outrage by the brash young pair. They seem to have (mis)interpreted Clifton's rather tentative paper as if he had actually claimed priority over them in deciding experimentally between the rival theories of voltaic action. Why did they interpret this letter in such an extreme and extraordinary manner? Ayrton himself was certainly a highly combative individual, having been brought up in a family of lawyers and was known for being somewhat prickly on matters of priority, especially as he and Perry were young scholar-engineers still attempting to establish their international reputations (Gooday and Low 1998–9). The rather more telling issue here is that Perry had a particular vendetta to settle with the Oxford professor, relating to an incident in the summer of 1873 in which Perry had applied without success for a fellowship at Clifton's own college, Merton. On finding subsequently that his M.A from Queen's College, Belfast was not sufficient to render him eligible for the fellowship, Perry wrote to *Nature* in October that year publicly accusing Clifton and the college in question of having misled him in failing to point out his disqualification at the time of his application. Perry moreover attacked Clifton for not offering him access to the apparatus of the Clarendon in order to prepare for the fellowship examination. Clifton's crushingly patronizing reply published on 23 October 1873, and the petulance of Perry's rejoinder a week later cannot have left the two men with much good will for future encounters.[28]

The letter that Ayrton and Perry wrote to the *Telegraphic Journal* on 14 December 1877 (published 1 March 1878) condemning Clifton's paper, and explaining their chronological priority over Clifton's allegedly similar experiments, was remarkably splenetic even for this 'prickly pair':

If the investigation in question had been of merely ordinary importance we should not have deemed it necessary to point out the priority of our experiments to those of Professor Clifton's; but when the fact is remembered, a fact not very evident from Professor Clifton's paper, that a series of experiments such as we performed [in 1875–6], clears up the long-standing discrepancies between the chemical and contact explanations of voltaic phenomena, and so is of extremely great importance in the science of energy, we trust we may be pardoned for claiming the priority is due to us (Ayrton and Perry 1878a: 99–100).

---

[28] Clifton to the editor of *Nature*, 18 October 1873, 'Oxford science fellowships', *Nature*, 8 (1873): 528; John Perry to the editor of *Nature*, 28 October 1873, 'The Oxford science fellowships,' ibid.: 549–50.

They did not stop short at claiming 'priority' and showing irritation at Clifton's failure to cite their work. Ayrton and Perry devoted much of their letter to ridiculing his allegedly similar experiments as being fundamentally incompetent – and thus ironically highlighting the way in which Clifton had not offered results remotely like theirs in the first place. Their letter lists the following criticisms of Clifton's experiments:

(i) these were undertaken with apparatus that was less delicate 'by far' than theirs – implicitly remarkable given how much nearer he was to expert instrument makers than they had been out in Japan;

(ii) that where Clifton only offered qualitative indications of the *sign* of a potential difference between two substances in contact, Ayrton and Perry offered full numerical values as well;

(iii) that Clifton's excessive use of six insulating stems indicated his poor comprehension of the risk of losing electrical charge from the apparatus;

(iv) Clifton, unlike Ayrton and Perry, was unable to measure the potential difference between two liquids;

(v) he claimed that no current had circulated in the copper terminals in open circuit after liquid immersion had taken place, whereas Ayrton and Perry claimed to have shown that there was most definitely a transient current due to the 'soaking in' and 'soaking out' of the dielectrics involved;

(vi) Clifton failed to realize that Clark cells do not generally offer an immutable potential difference, nor are mutually calibrated with other Clark cells, and indeed they add witheringly that Clifton's data to prove this point would only be 'of value to practical men' viz., engineers not trained at college (Ayrton and Perry 1878a: 99–100).

In effect, Ayrton and Perry's attack on Clifton's ineptitude in citing fellow researchers was extended to a personalized attack on his inability to achieve competence in even rudimentary tasks in experimental physics. To rub salt into Clifton's wounds, when Ayrton and Perry published the completed version of their papers on the 'contact' theory voltage in the *Proceedings of the Royal Society*, they returned Clifton's perceived snub by making no reference to his earlier paper.[29] Although Clifton's precise response to this attack and then slighting is not known, it is not hard to see in the light of the above why he never quite found the nerve to publish a research paper again. Indeed, as Peter Harman notes, even a request by

---

[29] Ayrton and Perry (1878b). For further discussion of Ayrton and Perry's papers and the longer-term response to them, including Maxwell's famous critique of 1879 in which he disagrees with their interpretation but jests that their 'scientific energy' was 'threatening to displace the centre of electrical development and to carry it out of Europe and America to a point much nearer to Japan', see Hong (1994: 244–7 and 262–6).

the Royal Society to act as referee for a paper on an electrical topic in July that year left him in a state of 'fear and trembling'.[30]

His later curmudgeonly response towards aspiring researchers in the period 1911–15 might have owed something to resentment at the mauling his 1877 paper received at the hands of Ayrton and Perry. Even so, Clifton did not – contrary to myth – cease to undertake research. He moved rather to redouble his efforts in work on instrumentation and technology, but did not publicize the detailed results of his labours through learned journals. He found other public forums in which he could discreetly offer his expertise to the world.

THE PHYSICAL SOCIETY OF LONDON AND THE ROYAL COMMISSION ON COAL MINES

From 1879 to 1884, Clifton was doubtless as preoccupied as Lord Rayleigh with the consequences of the agricultural depression for their respective estates. Whereas Rayleigh occupied himself in Cambridge with research on the minutiae of electrical standards, Clifton became intimately involved with three institutional activities that took him away from Oxford even more often. Not only did he serve on the Council of the Royal Society, but from 1879 to 1886 he was a member, along with John Tyndall and others, of the Royal Commission on Accidents in Mines. Hardly less prestigiously, he was closely involved in the Physical Society of London, the national forum for English and latterly British physics.

Clifton's term as President of the Physical Society from 1882 to 1884 is a significant indicator of his reputation among contemporary physicists. This organization, founded by Frederick Guthrie in 1873 (Gooday 1991), was devoted to discussing new research, work in progress, and novel laboratory apparatus, as well as instruments used for teaching purposes – a primary interest of Clifton. The Society's meetings generally took place at fortnightly intervals on Saturdays in Professor Guthrie's laboratory at the 'Normal School of Science' in Exhibition Road, London, where they attracted large numbers of the country's working physicists and physics teachers. A particularly distinctive feature of the Physical Society's meetings was that experimental papers were generally accompanied by experimental demonstrations, and Guthrie acted as the Society's demonstrator until his

---

[30] Requested to referee a paper by Maxwell's former student, James E. H. Gordon, on specific inductive capacities of dielectrics, Clifton prefaced his report of 19 July 1878 with the words 'I feel very badly qualified to criticise any electrical work done under the direction of Prof[essor] James Clerk Maxwell and I consequently undertake with fear and trembling to report'; Royal Society, Referees' reports, vol. 8: 99, Royal Society Archives. For discussion of Clifton's recommendation that Gordon recalculate his results using a corrected formula see Harman (2002: 705–7).

death in 1886. In his term of office from February 1882 to February 1884 Clifton was regularly, if not invariably, in attendance to chair the meetings, despite commitments to laboratory, estate, and the Mines Commission. Accordingly he sat through most of the half dozen separate demonstrations of the stream of novel electrical apparatus, e.g. ammeters and voltmeters in 1883–4, emanating from Finsbury Technical College at which his former *bêtes noires* William Ayrton and John Perry were by then busy professors (Gooday 2004: 153–9).

During the first year of his presidency, Clifton both brought Oxford physics and physicists to London and vice versa. It was almost certainly Clifton who proposed the membership of Lazarus Fletcher in March 1882, and Bartholomew Price and John Viriamu Jones in June 1882, thereby bringing his small circle of Oxford physics into the metropolitan limelight.[31] It was also Clifton's sad duty, however, at the annual general meeting on 10 February 1883 to report the death of his friend and Oxford ally, the Savilian professor of geometry, Henry Smith. The loss of such a close supporter of Clifton and the Clarendon was to have major ramifications for Clifton, and it was almost certainly he who wrote, in the well-informed official obituary for the Physical Society:

Early in his career Professor Smith gave some attention to Experimental Science, and, although other studies drew him away from active work in the domain of Physics, he always took the greatest interest in the progress of that subject, and he used his powerful influence in promoting the study of it in Oxford. To his ever ready cooperation and his wise counsel whatever success has attended the development of Physics in the University may be largely ascribed.[32]

Before then, however, Clifton had already revealed the splendours of the Clarendon Laboratory to the members of the society at a special extra meeting of the society held on 17 June 1882 in between the scheduled fortnightly meetings in London of 10 June and 24 June. Far from being an indolent non-researcher, Clifton took the opportunity to present four of the six papers he had prepared for that occasion. The titles of these clearly indicate that these concerned his own special adaptations of lecture apparatus: 'On the action of an electrometer key', 'On a reflecting galvanometer for lecture-purposes, which can be used either as a sine or tangent galvanometer', 'On reversing and contact keys for galvanic currents', and 'On a simple form of optical bench for measuring the focal length and curvature of lenses and mirrors'.[33] Notably, it was at the meeting after this that Bartholomew Price and John Viriamu Jones joined the society. On later occasions too, Clifton gave further papers mostly on optical subjects: 'On a new spectrometer' (23 April 1883),

---

[31] See 'Proceedings of the Physical Society of London, Session 1882–3', in first appendix to *Proceedings of the Physical Society*, 5 (1882–4): 3 and 5.   [32] 'Henry John Stephen Smith', ibid.: 8–11.
[33] Ibid.: 5.

'On the complete determination of a double convex lens by measurements on the optical bench' (12 May 1883), 'On a modified form of electric insulator' (26 May 1883), and 'On the measurements of lenses, and the refractive indices of liquids by means of Newton's rings' (24 November 1883).[34] It must have been an interesting experience for the president to read out at the end of his tenure of office a report from the Society's council that said:

> The Session of the Society which has just been completed, though not marked by the communication of any great discovery, has been fully of steady work ... Several interesting communications on lenses have been made by the President, of which Council hope before long to be furnished with the MSS., so that these communications may be published in the 'Proceedings.'[35]

However, whether for reasons pertaining to his preoccupation with the Commission on Accidents in Mines or to his loss of nerve in publication after the embarrassing incident with the *Proceedings of the Royal Society* in 1877, Clifton must have implicitly or explicitly declined this request. Thus, as so often was the case with the Physical Society, only some of the papers presented at its meetings ever reached published form in its *Proceedings*, although in Clifton's case it was not for the most common reason of being deemed unworthy. Worse still for the historian, it appears that no manuscripts of these optical researches have survived to indicate how it was that Clifton enthralled the members of the Physical Society with details of the Clarendon's optical teaching apparatus in the spring of 1882 and 1883.

During and immediately after his presidency of the Physical Society, Clifton devoted great attention to researching and writing reports for the Royal Commission on Accidents in Mines. As both an Oxford professor specializing in optics, and a gentleman with a Tory sense of responsibility towards land-related matters, Clifton presumably felt a certain duty to accede to the Commission's request for his participation in 1879. This Commission sought to identify the possible ways of either preventing or at least limiting the disastrous consequences of pit explosions, poor ventilation, collapsing mineshafts, spontaneous combustion of coal and firedamp – well-known phenomena that brought death to dozens of coal-miners every year (*Final Report of Her Majesty's Commissioners appointed to inquire into Accidents in Mines* 1886). The Commissioners, mostly experts in relevant areas of science and technology, had not initially expected to spend as much as seven years in their investigations. But the wide range of tests on existing mine conditions and many modes of experimenting on innovative safety measures evidently took a great deal of time to execute and document. The issue that most concerned Clifton was the design and construction of a miner's safety lamp, and the practical

---

[34] 'Proceedings of the Physical Society of London, Session 1883–4', in second appendix to *Proceedings of the Physical Society*, 5 (1882–4): 3–5.   [35] Ibid.: 3–5 and 6–7.

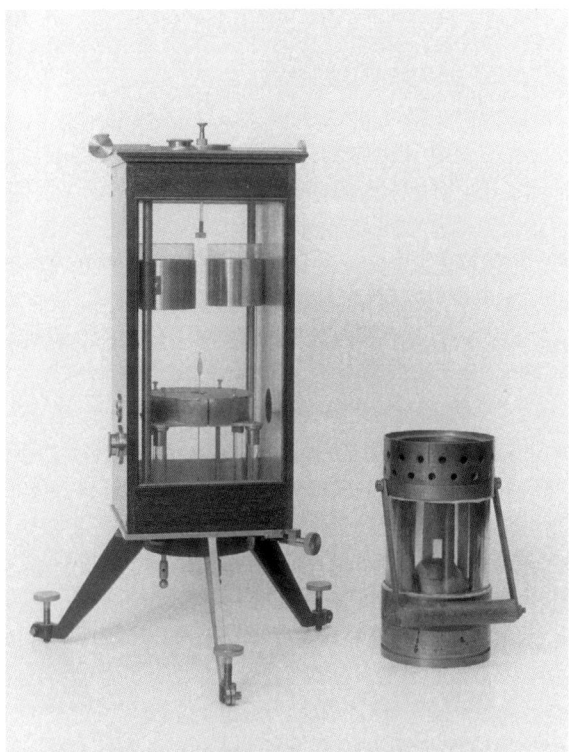

Fig. 3.8. Quadrant electrometer and miner's lamp, built to Clifton's design. On these and other instruments now in the Clarendon Laboratory Archive, see Sanders (1997).

problems that arose in its daily usage, such as the extinguishing of lamps when tilted, and the difficulties of changing or relighting *in situ* down a mineshaft.

This concern is apparent both in his questioning of witnesses from the mining industry who were summoned to give evidence to the Commissioners in November 1882, and in his development of the 'Clifton lamp'. This was an experimental modification of the standard 'Gray' lamp designed to reduce the risk that the lamp glass would be damaged by gas burning inside it and to reduce the deposition of soot on the gauze frequently used at the lamp's outlet. This adaptation was not successful however: although as safe as the original Gray lamp, it was not only rather too heavy but inconveniently sensitive to tilting and also easily extinguished by a down draft of air. Accordingly, the 'Clifton lamp' was not recommended by the Commission for further development; its subsequent fate has been to stand in a display cabinet at the entrance to the present Clarendon Laboratory far from any mineshaft.[36]

Although Clifton's influence on the work of the Commission was limited, its reciprocal influence upon him was quite marked. We can in fact trace Clifton's role in the Commission to his developing interest in technological applications of

---

[36] My thanks to Robert Fox and John Sanders for information on the extant 'Clifton lamp' in the Clarendon Laboratory.

Fig. 3.9. Group photograph of the Board of Visitors, Royal Observatory, Greenwich, 1894 or possibly later in the 1890s. This was another aspect of the public life of science in which Clifton was active. The tall figure of Clifton is at the extreme left of the group with Rücker next to him. The handwriting on the photograph is R. T. Gunther's.

electricity. When the Commissioners received an enquiry in July 1880 about the potential dangers to miners' safety posed by electrical storms and telegraphic signalling along shafts, both Clifton and Frederick Abel acted as 'electricians' to reply. They considered the dangers to be minimal or eliminable, yet deserving further investigation. During its deliberations the Commission also considered and declined to recommend the adoption of electric lighting in mineshafts, Clifton himself in November 1882 being keen to learn about miners' views of the feasibility of electric lighting (*Final Report of Her Majesty's Commissioners appointed to inquire into Accidents in Mines* 1886: 85, 91–6, 149–50, 264).[37] It is thus easier to understand why, within a year of the Commission producing its final report in 1886, Clifton asked the University for funds to extend his laboratory to electrical investigations.

---

[37] In the Minutes of Evidence (pp. 124–36) Clifton's questions are 521–31, 552–6, 568, and 586–7. Glazebrook (1921: viii) identifies Appendix XXIV of the report, in which the Clifton lamp is discussed, as the main focus of Clifton's activities for the Commission.

THE BID FOR AN ELECTRICAL LABORATORY AND THE 'POLITICS OF NUMBERS'

By 1886 it was not just his work for the Mines Commission that had redirected Clifton's attention to the important rise of electrical lighting and power technology to create teaching and research opportunities in physics. A long stream of papers on this topic was presented to the Physical Society by the physicistengineers Ayrton and Perry; moreover, the activities of Frederick Jervis-Smith at the new Millard Laboratory at Trinity College (see Simcock, Chapter 5) could hardly have escaped his attention. Clifton thus tried to persuade the University that a special new electrical laboratory was needed as an extension of the 1870 Clarendon. However, as A. J. Engel has documented, the agricultural depression that descended upon Britain in the years immediately after the Oxford Commission of 1877 to all intents and purposes nullified the projected increase in the financial resources of the University that had been earmarked for the expansion of the sciences (Engel 1983: 217–56).[38] Moreover, without the popular Henry Smith to support his case, and without large numbers of students in the Clarendon, Clifton's strategic position in the University was not strong.

Thus when Clifton formally applied in 1887 for £4800 to build a new electrical laboratory to meet the demands of students for up-to-date training in the burgeoning subjects of electromagnetism and electrical engineering, Convocation voted against it, agreeing only to pay for a new porter's lodge for the Clarendon.[39] Janet Howarth has suggested that this was almost certainly the result of what she calls 'the politics of numbers': a week before the vote in Convocation an unsigned article in the *Oxford Magazine* asked 'What do the professor and two demonstrators do?', pointing out that the Clarendon had only produced three graduates in the previous year (Anon. 1887). From Clifton's comments to the 1877 Commission about the frequency with which his students were 'bought off' by public schools, before even taking the examinations in the Natural Science School, this figure of three graduates may not have been an entirely fair reflection of his student intake; Howarth cites the figure of 20–25 students attending Clifton's class prior to graduation in 1886. Amongst the 18 courses in the Oxford Museum, this ranked midway between the extremes of 78 students for the professor of geology's lectures and 5–6 attending the lectures of Bartholomew Price, the Sedleian professor of natural philosophy (Howarth 1987: 370). The (mis)representation of his efficiency and popularity as a teacher by the anonymous writer for the *Oxford Magazine* must have been all the more galling for Clifton when five years later Convocation granted

---

[38] The long-term consequences of the agricultural depression are discussed in Brock and Curthoys (2000); see also Dunbabin (1997) and Jones (1997). [39] *OUG*, 17 (1886–7): 492 (7 June 1887).

£7250 for an anatomy laboratory after the newly appointed lecturer Arthur Thomson complained of the overcrowding of students in his makeshift anatomical shed (Howarth 1987: 343–4). Importantly though, what Clifton did not publicly acknowledge was that by 1887 there already was an electrical laboratory available to all Oxford students: the Millard Laboratory managed by Frederick Jervis-Smith at Trinity College. As Simcock suggests, the existence of the Millard Laboratory was seen by a substantial lobby of college dons to preclude the need for any additional university facility in practical applications of electricity.

The 'politics of numbers', as Howarth puts it, was clearly a matter that Clifton felt keenly, as was the close and unhelpfully elitist correlation seen commonly seen by Oxford contemporaries between students' success in the Mathematics School and their subsequent admission to the Clarendon to study experimental physics for Honours in the Natural Science School. In the year after his defeated move for an electrical laboratory, Clifton issued a plaintive statement in the *Oxford University Gazette* in which he declared:

> It is much to be regretted that the opportunities provided by the University for the study of physics are not used by a larger number of undergraduates. As I have reason to think that many persons are deterred from entering upon this study by an exaggerated estimate, which has been formed, of the extent of the mathematical knowledge required, I wish to state that a very moderate acquaintance with the most elementary branches of mathematics will enable a student to work in the laboratory with success.[40]

Clifton's complaint was soon answered by the revised regulations for the Preliminary Honour examination in the Natural Science School, which from 1891 required Preliminary students, normally in their second or third years, to take practical classes in experimental physics (Howarth 1987: 341). From that year, the numbers of students attending Clifton's laboratory classes expanded greatly, as Table 3.2 shows.

Once the Preliminary students arrived in 1891, Clifton's response to the swelling numbers indicated above soon turned to complaint. His undiplomatic handling of the subject can be seen from the way he presented his grievances very publicly in his annual reports in the *Oxford University Gazette*. At Easter 1892, Clifton protested that he could accommodate no more students in the laboratory, and that all of his demonstrators' time was so taken up with teaching that no research was possible. In 1893 he announced in the *Gazette* that he could no longer comply with students' requests to study electricity. Three years later he even announced that students seeking such instruction would have to go to the Cavendish Laboratory in

---

[40] 'Report of the professor of experimental philosophy 1888', *OUG*, 19 (1888–9): 404 (2 May 1889). This was the first of the reports that Clifton, like his fellow professors, wrote for the *Gazette*; the reports became annual features thereafter.

Table 3.2. Numbers of students registered for study in the Clarendon Laboratory, 1888–1900

Calendar year[1]	University members (male students)	Non-University members (female students)	Total number of students in laboratory	'Prelim'/'Hons' (combined m & f)
1888	12	–	12	0/12
1889	10	–	10	0/10
1890	16	1	17	0/17
1891[2]	21	2	23	5/18[3]
1892	26	3	29	13/16
1893	31	5	36	15/21
1894	40	6	46	19/27
1895	41 (inc. 1 M.A.)	6	47	19/28
1896	51 (+ 1 non-University)	6	58	21/37
1897	50 (inc. 1 M.A.)	7	57	40/17
1898	48	11	59	48/11
1899	41	8	49	35/14
1900	52	7	59	48/11

Source: Annual reports of the professor of experimental philosophy, Oxford University Gazette

[1] The numbers record the Hilary and Easter terms of one academic year and Michaelmas of the next.
[2] The numbers of Preliminary students increased markedly in Michaelmas 1891, when formal practical examinations in physics became compulsory for them.
[3] Clifton does not supply breakdown of these and subsequent numbers in this column; they are merely a rough guess.

Cambridge or London's polytechnic institutions; tellingly he did not mention the practical electricity courses offered by Jervis-Smith at the Millard Laboratory. Following that instance of institutional grand-standing, Clifton changed tactics in 1897, pointedly announcing that he would allow Honours students to take classes in electrical measurement but could only do so at the cost of 'abandoning' his own research space in the Clarendon.[41] As we see in the section after next, such tactics did not win him the electrical laboratory that he demanded. Whilst the University did introduce such a laboratory to relieve Clifton of the burden or teaching electrical physics, this was given over instead to the newly arrived Townsend.

In the meantime, however, and out of sight of almost all Clifton's students, there had been some important non-electrical research going in the basement of the Clarendon.

[41] See the successive reports by the professor of experimental philosophy for the years 1891–7, in OUG, volumes 22 to 28.

RESEARCH IN THE BASEMENT: THE BOYS DETERMINATION OF G, 1890–4

As an environment free from the disturbance of building work, and indeed free of much research, the Clarendon in the early 1890s was a strategically good place for an experimental physicist to undertake delicate high-precision measurements. Such was the view of Charles Vernon Boys when he sought a venue to extend his experiments on the Newtonian constant of gravity, G, carried out in 1889–90 as assistant professor of physics at the Normal School/Royal College of Science in London (Forgan and Gooday 1994: 153–92). Even working at night-time in the basement of his South Kensington building, Boys had found that the myriad sources of disturbance in the capital caused too much uncertainty in his 1890 value of 'G' for him to be completely satisfied with his results.[42] He, nevertheless, at least published some interim results in the *Proceedings of the Royal Society* for that year (Boys 1890: 265–8). Although no documentary record appears to have survived of correspondence between them, it is likely that Clifton and Boys knew each other well through the Physical Society of London, and certainly they shared a compulsive interest in the design and construction of new instruments. Moreover, as a fellow devotee of the ultraprecise measurement, and having charge of the most tranquil physics laboratory in the country, Clifton sympathized with Boys's aims enough to give him untrammelled access to the cellars of the Clarendon from 1890 for a new redetermination of G.

So it was that for many a Saturday and Sunday over the ensuing four-year period Boys travelled to Oxford to prepare his extraordinarily elaborate experiments. He used ultrafine apparatus designed and built by himself supplemented with some (most unusual) loans from Clifton's own stocks of optical equipment. Although Clifton gave him regular assistance with arrangements for access to the extensive cellars and the loan of apparatus, the extraordinarily delicate nature of the experiments meant that Boys had to work on his own. Not only did Boys have to avoid the effects of body heat, body mass, and bodily motion interfering with the experiment by using a telescope and remote control devices to observe the gravitational attraction of the apparatus. His main observations could only be undertaken late on Sunday night and early Monday morning (typically between midnight and 5.30 a.m.) since this was the only time when shunting activities entirely ceased at Oxford's railway station just over a mile away and when St Giles was virtually free of traffic. Even after such allowances had been made, and even after rehearsals had been undertaken throughout all of Sunday, Boys's midnight determinations were

---

[42] For more analysis on this point see Gooday (1997: 410–20).

Fig. 3.10. In 1895 Charles Vernon Boys conducted his unprecedentedly precise determination of the gravitational constant 'G' in the cellars of the Clarendon Laboratory. During the experiments he used a screen to protect the delicate apparatus from the disturbing effects of his body heat. Here the screen is withdrawn to show a wooden octagon house surrounding the apparatus that held the suspended lead balls.

still sometimes abandoned due to the disturbing effects of wind, bending trees outside, even on one occasion by tremors from an earthquake in central Europe.[43]

Boys finally submitted his long and detailed manuscript on the 'Newtonian constant of gravitation' to the Royal Society on 31 May 1894, and this was read at the Society on 4 June. Lord Rayleigh, then the Society's Physical Secretary, thereupon wrote to Clifton asking him to act as a referee for Boys's paper, and for publication in its *Philosophical Transactions*. Clifton wrote back to Lord Rayleigh on 18 July 1894 to say that it appeared to him to be a paper of 'very great interest and importance' and hence that it 'certainly merits publication in the Phil. Trans'. Clifton's report said:

Prof. Boys has designed apparatus of great perfection for this work, and by his skill in the use of it, he has obtained results which are extraordinarily consistent considering the extremely delicate nature of the research.

---

[43] For a related discussion on the difficulties of managing bodily presence in experimentation see Sibum (1995).

Fig. 3.11. This view of the Boys apparatus is from the corner behind the screen at the end of the vault. The telescope is in position to view the minute deflections of the suspended lead balls under gravitational attraction. A fixed remote scale enabled Boys to make direct telescopic measurements of the deflection.

Though the measures recorded involve in some cases figures which are open to doubt, it seems to me that accuracy of a very high order has been attained, and that the value of this important constant has been determined more exactly by Professor Boys than by any previous investigator.

He has made it very probable that the constant of gravitation lies between 6.655×10^{-8} and 6.660×10^{-8} in C.G.S. units.[44]

Clifton did offer some criticisms and suggestions for changes, however, suggesting enlargement of diagrams to afford readers more detail of the apparatus. He also wished that Boys had found even more 'leisure' for a longer series of experiments. And yet Clifton's own personal acquaintance with the experiments made him realize how great the 'strain' had been for Boys anyway. Clifton added in a 'Private note' that he would like Boys to exclude the sentence that began 'I feel that Prof. Clifton's kindness . . .' since the preceding sentence offered ample acknowledgement of the 'very small amount' of assistance that Clifton had supplied. Rather tellingly for someone who has so often been portrayed as hostile to research, Clifton added: 'I think Prof. Boys had a distinct claim upon any one who

[44] Clifton to Lord Rayleigh, 18 July 1894, Royal Society Archives, R.R. 12.64.

Fig. 3.12. Boys used a cathetometer to measure precisely the vertical distances between the tops of the large lead balls and their holders. At this stage, the lid and the large balls are all independently supported by balance weights.

was in a position to offer facilities for carrying out the important research which he had undertaken'. In fact, this request of Clifton's was not met, for in the final revised version of the paper submitted for publication on 18 October 1894, the paragraph that opened the details of the laboratory fittings was as follows:

The apparatus is set up in the vaults under the Clarendon Laboratory at Oxford, to fit which, in fact, it was specially designed. I cannot sufficiently express my obligation to Professor CLIFTON for giving up to me entirely for four years this very perfect observing room, for not only was I able to make my observations under specially favourable conditions, but I have had the advantage of having at hand the resources of his splendidly equipped laboratories, and of being allowed to make any use of them that I desired. I feel that Professor CLIFTON's kindness in the matter is the greater as I have no claim upon him whatever, and I can only hope that in so far as my work carried out in his room may represent progress in practical physics, he may feel justified in having sacrificed to this end his best observing quarters (Boys 1895: 12).

Clifton's association with the Royal Society continued to be a close one, for he acted as its Vice-President from 1896 to 1898. Even after then, Clifton entertained the presence of researchers in his laboratory, albeit not on the same scale as in

Fig. 3.13. A chronograph drum was used to keep electrical records of the times at which key stages of the experiment were conducted. This was important as a means of checking against the times of any external disturbances to the measurements of gravitational deflection.

1890–94, nor indeed on the same scale as was prosecuted in the laboratory of his newly arrived young rival.

ECLIPSE AND DECLINE: CLIFTON'S BATTLE WITH TOWNSEND

Much is said in Lelong's chapter about the circumstances that led to the arrival of John Sealey Edward Townsend in 1900 to take up the Wykeham chair of physics twenty-three years after the post had first been proposed. This section will examine the ungracious response that the 64-year-old Clifton offered to his youthful rival. At first sight it might have suited Clifton that Townsend was appointed with the specific duty of providing lectures and laboratory instruction in electricity and magnetism since Clifton claimed to have had so much difficulty in providing teaching in these areas. On the other hand, Clifton was doubtless irked that the

University moved quite quickly to offer, in association with the Drapers' Company, exactly the kinds of resources for teaching electricity and magnetism that he had been denied by Convocation in 1887. Thus, when Townsend arrived to take up the Wykeham chair in 1900, Clifton did not welcome him with open arms or offer him a room in the Clarendon. In this situation Townsend had to work in laboratory space borrowed from his colleagues the professors of astronomy and physiology in the Museum until a new electrical laboratory was built for him and opened in 1910 (von Engel 1957: 259–60).

The reports of the professor of experimental philosophy that appeared in the *University Gazette* from 1901 to 1915 show how Clifton saw his position as being gradually eroded and eclipsed by Townsend's vigorous work, liberally aided by Clifton's own inept efforts to coerce Convocation to extend its funding of the Clarendon. Recurrent themes in these reports for the years 1901–5 are complaints about the overcrowding of the Clarendon since it had to accommodate both Honours and Prelim students, accompanied by demands that the teaching of Preliminary students be shared with or given over to some other department, implicitly Townsend's. Clifton's rationale for this, as described in his report for 1903, was that:

. . . . there appears to be only one method of preventing the dead-lock with which the Department is immediately threatened, and that is by assigning a separate building, with the requisite staff of teachers, to the Preliminary classes, and by restoring the Clarendon Laboratory to the purposes for which it is alone suited, viz. the instruction of advanced students, and the provision of facilities for research by those who are qualified to undertake such work (Clifton 1902–3).

Another persistent element of Clifton's reports is the extent to which he documented in lavish detail the great care that was taken to spend Laboratory grants efficiently in maintaining and protecting the paintwork, fittings, and superstructure of the Clarendon. This, Clifton evidently did in a manner befitting a gentleman-landowner's preoccupation with the stewardship of property and its responsible upkeep.[45]

In 1905, all Preliminary classes in physics were transferred to Townsend's charge, and research was once again underway in the Clarendon. Clifton reported that one unidentified M.A. of the University was 'engaged' during a part of the year on experimental work in the laboratory, and a graduate of Edinburgh University 'commenced a research with the intention of qualifying himself for admission to the degree of Bachelor of Science in this University'. This latter person, almost certainly George Carse, worked until the summer of 1907 on the 'Rotation of the

---

[45] See Clifton's annual reports in *OUG* for the years 1900–15.

plane of polarization in certain media' when he returned to Edinburgh University to take up a lectureship there (Clifton 1905–6).[46]

The number of Honours students in physics who studied in the Clarendon was growing steadily, however, and by 1911 had reached thirty-seven, including one woman. At this point, still complaining about the lack of space in the Clarendon, and even without any Preliminary students to take up room there, Clifton's report in the *Gazette* concluded with a remarkable announcement:

All the available space in the Laboratory is now devoted to the instruction of students preparing for the examination in the School of Natural Science, and it will in future be impossible to offer facilities to advanced students wishing to engage in Research (Clifton 1910–11: 959).

In these circumstances, various speculations might be entertained about the disingenuousness or otherwise of Clifton's representation of the viability of research at the Clarendon. Whatever one infers, however, this quotation does at least offer an interesting alternative to the traditional explanation of why research ceased at the Clarendon c.1911. In his letter of 1968 to A.J. Croft concerning his demonstratorship at the Clarendon (and also in the Electrical Laboratory with Townsend) between 1910 and 1913, O.F. Brown gives this account of Clifton's management of the laboratory:

Once every summer term he demonstrated the solar spectrum with a diffraction grating spectrometer set up in the attics of the Clarendon. No one but the professor was allowed to touch the instrument. Clifton was always [sic] very perturbed at the growing 'restlessness which the desire to research was introducing into physics.' He used to tell a cautionary tale to his demonstrators about how once he had allowed a demonstrator to use the grating instrument for a research and how he had put it out of adjustment with the result that it had taken him (Clifton) some days to put it back again. Never again would he allow, he said, any research to be done in the Clarendon or must a demonstrator depart from the routine he had laid down.[47]

The extent to which this half-century retrospective offers more than a caricature of the elderly Clifton is hard to judge. But we can safely say that the last four years of Clifton's occupancy of the Clarendon were entirely free from research, and probably for more than one reason. If one of these reasons was that the Clifton was

---

[46] George A. Carse, M.A., B.Sc. was appointed as a lecturer in the Department of Natural Philosophy at the University of Edinburgh in 1907; see *Edinburgh University Calendar 1907–1908* (Edinburgh, 1907): 5. Carse later also lectured in statistics and mathematical economics. I am grateful to Irene Ferguson of Edinburgh University archives for supplying me with this information. Like Clifton, Carse does not appear to have published the result of his researches in any major national or international scientific periodical. [47] Letter from O. F. Brown to A. J. Croft (see note 1).

hostile towards research, this can only have been a disposition that took full form after the year 1907, so that it should probably be seen as characteristic of rather less than the last decade of his professorship. In support of Clifton's own account, there is evidence that the Clarendon was at least seriously taken up with undergraduate teaching: in his last full year of teaching before the First World War, there were around 17 students regularly in attendance at the Clarendon in preparation for Natural Science finals. Numbers dropped off rapidly as hostilities commenced, and as Clifton retired from the workplace of the past half-century, the number of students fell to about five per term. *Pace* Lindemann, this paltry figure was not due to Clifton's nescience, but rather more due to the flux of nascent physicists to the warfront.

CONCLUSION

From 1880 until the year when Lindemann succeeded Clifton in the chair in 1919, the Clarendon's facilities were not significantly augmented by grants from the University Chest. Thus, we see why Lindemann found the furnishing of the Clarendon so very meagre in comparison to the Cavendish. In contrast to a burgeoning research laboratory with an extensive personnel hierarchy, a dozen undergraduates were being taught by a single University demonstrator, with a minimal water supply, gas lighting, and no mains electricity. Moreover, its annual University grant of £2000 was the merest fraction of the income available to the Cavendish and to Townsend (Birkenhead 1961: 89–90; Lelong, Chapter 6, 213). To understand the 'depressing inheritance' that Lindemann acquired from Clifton we should not seek explanations that refer exclusively to Clifton's long hostility to research. This is a misconceived and evidentially unsupported notion as the above discussion has shown. Rather, we should see him as an ineffective and tactless institutional politician who was latterly without allies in Oxford and who was outmanoeuvred by younger rivals who were better able to adapt to the new research-oriented imperatives of the early twentieth century.

Admittedly it may seem hard even then to explain why Clifton chose to linger so long in an uncongenial situation, retiring from his chair at the venerable age of 79.[48] Perhaps this reflects the bitterness of a man who wanted to cause as much trouble as he could to an institution that had given away his wished-for electrical laboratory

---

[48] Clifton benefited from the absence of any enforced age of retirement, as did a number of other prominent figures in this volume, including Walker, Baynes, Townsend, and Lindemann.

planned in 1887 to his young upstart challenger from the Cavendish Laboratory in Cambridge. Thus, no matter how much the historian of Oxford science seeks to avoid gratuitous and unfair comparisons with the distinctly different institutional arrangements at the University of Cambridge, one telling point remains. Both Townsend and Clifton were Cambridge graduates. Yet the diaspora of the Cavendish Laboratory settled in The Electrical Laboratory with rather more fruitful results than Robert Bellamy Clifton's conservative harvest at the Clarendon Laboratory.

4

Laboratories and Physics in Oxford Colleges, 1848–1947

Tony Simcock

Accounts and perceptions of particular subjects at Oxford University tend to concentrate on professors and central institutions. Not surprisingly, historians and contemporary commentators alike have formed judgements of 'Oxford physics' based on what went on in the Clarendon and Electrical Laboratories. However, Oxford's collegiate nature, the diversity of its scholarly community, and local peculiarities in the definition of disciplines, all require that a true description of Oxford physics embrace activity across a much larger and more complex territory. The purpose of this chapter is to reconnoitre this uncharted hinterland, and send up flares whenever physics and physicists are encountered. The operation cannot hope to be comprehensive: some activists (like G. J. Burch) may be sighted from a distance but will escape interrogation; while some promising holes into which the grass roots have probably retreated (like the Junior Scientific Club and other university societies) will have to be flushed out on a future occasion. The exercise will illustrate, nonetheless, how in a traditionally decentralized academic landscape a discipline will have a widespread presence and even significant epicentres beyond the main facility, and sometimes utterly independent of it.

In attempting to remap Oxford's scientific geography in this way, I shall also trespass on some well-trodden ground. In particular, it seems best to concentrate on the college laboratories as obvious focal points of devolved activity, the more so as their teaching and facilities were usually open to all students, not exclusive to the one college. Yet the ample historiography of the college laboratories has all been

---

This chapter draws on basic biograpical and local information from standard and background sources that it would be tedious to cite repeatedly. They include the *Oxford University Calendar* (*OU Calendar*), *Gazette* (*OUG*), *Handbook*, and *Historical Register*, with supplements; Foster's *Alumni Oxonienses*, with supplements; the *Encyclopaedia of Oxford* (Hibbert 1988); the eight-volume *History of the University of Oxford*; and universal reference sources such as *Burke's Peerage* (*BP*), *Dictionary of National Biography* (*DNB*), *Dictionary of Scientific Biography* (*DSB*), *Royal Society Catalogue of Scientific Papers*, *Who Was Who* (*WWW*). The abbreviation MHS refers to the Museum of the History of Science, Oxford.

written by chemists concerned to document their own disciplinary ancestry, thus marginalizing the physics that once existed alongside chemistry in these places.[1] The *raison d'être* of college laboratories is often thought to have been the one implied in a famous speech by A. G. Vernon Harcourt in 1908, when he spoke of them as being 'prepared to supply any deficiency that may occur'.[2] He was alluding to the moribund state of the professoriate – in chemistry as much as in physics – at that time. But the situation was quite different fifty years before, when the first college laboratories were established at Magdalen, Balliol, and Christ Church, simultaneously with the erection of a magnificent central building for the University's scientific collections and laboratories (the University Museum). Both developments followed from the creation of the Honour School of Natural Science in 1850. The colleges that set up laboratories were at the forefront of major reforms in the seriousness of college teaching; to which end they were keen to bring the sciences, and other burgeoning subjects, into the midst of college life, making them as accessible to their members as were the traditional disciplines. They saw college and central laboratories as complementary, as with libraries. Even the founders of the University Museum – to all appearances an emphatic assertion of centralization – did not intend to segregate science from college life: when Sir Henry Acland talked of the 'study of the Kosmos' unified in one building, his thoughts were on a higher plain than institutional politics – the aim was to give students 'an abiding sense of the Unity of Nature' (Acland 1890: 23 and 22).[3]

Laboratories aside, Oxford's natural science degree required a more direct response from colleges, as the basic environment within which the new syllabus would be studied. Magdalen College took the lead, establishing the post of 'Lecturer in Natural Philosophy' in 1850 and appointing one of its fellows, and the holder of three university professorships, Charles Daubeny. Balliol likewise turned to an existing fellow, delegating the duties to its energetic young mathematical lecturer Henry Smith. Exeter College looked outside its fellowship and in 1855 appointed the chemist and mineralogist Nevil Story-Maskelyne as 'Lecturer in Chemistry and Physics' – the first instance of the use of the term 'physics' in a post at Oxford (*OU Calendar*, 1856; Morton 1987: esp. 120). Other colleges followed suit, until by one means or another virtually all of the twenty-one colleges that existed in Oxford by the end of the nineteenth century, plus various other college-like bodies (halls, private halls, delegacies, women's colleges), had made arrangements for 'armchair' teaching of the sciences, including physics.[4] During the period up to the centralization of non-tutorial physics teaching in 1905, and in a few cases later, a number of them provided fuller physics instruction through

---

[1] Brewer (1961); Hartley (1971b); Bowen (1969); Humphries (1970); Laidler (1988).
[2] Vernon Harcourt (1910); cf. Hartley (1971b: 231–2) and Laidler (1988: 239).
[3] The library analogy is implied in Daubeny (1855).
[4] I do not dispute Howarth's figure (Howarth 1987: 357–8), though I am interpreting the situation on the ground differently.

experimentally illustrated lectures, practical laboratory classes, and even participation in research. We must also be alert for activities recognizable as physics conducted by individuals who might not call themselves physicists: this being Oxford, they might easily be disguised as chemists, or as mathematicians, or even as amateur mechanics.

MAGDALEN COLLEGE LABORATORY

The laboratory at Magdalen College is the best-documented of all Oxford's scientific institutions. But while Gunther's exhaustive record of its history and personnel, *The Daubeny Laboratory Register*, illustrates the *esprit de corps* characteristic of all the college laboratories, the absence of such detailed evidence from the others makes it difficult to compare them (Gunther 1924).[5] In subject-matter Magdalen's seems to have been the most diverse, and it was the one that leaned from chemistry towards biological and geological topics, rather than physical. It was situated outside Magdalen College, at the Botanic Garden, where Charles Giles Bridle Daubeny had built himself (on land belonging to the college) a residence and in 1848 a laboratory, his aim being to concentrate his pluralistic activities as professor of chemistry, botany, and rural economy (agriculture) on this site, pursuing these interrelated subjects, along with others such as geology, meteorology, and electricity, as components of an all-embracing chemistry. With his appointment as college lecturer for the new science syllabus, the laboratory's resources became available to Magdalen students. He resigned his chemistry chair in 1854, but continued to give so-called 'catechetical' lectures in elementary chemistry, which were open to all Oxford students.[6] What in 1848 was essentially a private laboratory was thus drawn into Magdalen's ambit. Daubeny continued to develop and diversify its resources: in 1855 he donated an astronomical telescope around which a small observatory was built.[7] On his death in 1867 the whole establishment became the college's property, along with its collections of books, apparatus, and geological specimens.

There was a year's delay before a successor was appointed, suggesting the college was uncertain what to do with this bequest. The laboratory was looked after by T. H. T. Hopkins, a college fellow who had previously deputized for Daubeny, and whose main interests were in practical mechanics (Gunther 1924: 19, 305, 307).[8] Edward Chapman, a biologist, who was eventually recruited as tutor in

---

[5] For the laboratory's history in brief see Hartley (1971b: 223–5) or Laidler (1988: 222–4).
[6] Gunther (1924: esp. 4–19, 53–64, 153–5, 189, 298–303, 457–60); on Daubeny see also Phillips (1868); Willsher (1961); *DSB*; Oldroyd and Hutchings (1979); for Daubeny's college post, information from Dr Robin Darwall-Smith, Magdalen College archives.
[7] MHS, MS Museum 109, f.12; Gunther (1924: 11, 15); Hutchins (1990).
[8] On Hopkins see Simcock, Chapter 5, 173.

Fig. 4.1. Thomas Henry Toovey Hopkins. A fellow of Magdalen College with a special interest in practical mechanics, Hopkins played a leading role in the Daubeny Laboratory after Daubeny's death.

natural science in 1868, was the central figure in a dynamic group of science teachers based at Merton College, including William Esson and A. W. Reinold; he himself taught at Jesus and Wadham Colleges as well as at Merton.[9] The 'combined college lectures' in natural science originated within this group, where lecturers in the three basic sciences were provided by different colleges, giving a comprehensive elementary course open to all students. Chapman's importance has been disguised by the fact that his being married precluded him from a fellowship (until the rules changed in 1882), though at Magdalen the discrepancy was ameliorated by his happy working relationship with Hopkins. That the Daubeny Laboratory in its heyday was able to embrace, and even promote, an unusual variety of scientific subjects, to offer teaching open to the whole university as well as beyond it, and to

---

[9] *OU Calendar*, 1868; Foster (1893: column 315); Gunther (1924: esp. 414–25); obituary, *The Times*, 2 August 1906; Fox (1997: 674–5).

sustain an environment conducive to research, owed much to the hospitality and broad sympathies of Hopkins and Chapman, and from 1894 of Chapman's successor Robert William Theodore Gunther, better known as a historian of science (Gunther 1967; Simcock 1985).

Two important, if short-lived, innovations introduced by Chapman in 1874 were lectures in elementary physics and classes in practical physiology, both taught as part of the co-operative arrangement between Magdalen, Merton, and Trinity Colleges. The physiology classes conducted by Charles John Francis Yule helped establish the modern practical teaching of that subject in Oxford, and imported instrumentation from the world of physics, in particular a reflecting galvanometer for delicate electrical work and a chronograph for timing rapid events.[10] The University's new physiology department took over directly from Yule's laboratory in 1884. The physics courses were given by the Millard lecturers in physics, the first of whom was Archibald Simon Lang Mac Donald, who had been a mature student at Merton College (first, 1871). Chapman was having difficulty keeping physics on the menu of the combined college lectures, until Trinity College's Millard fund came to the rescue, linking its new lecturership to a fellowship by examination in physics at Merton, which Mac Donald won in 1873. Mac Donald lectured in different terms on mechanics, dynamics, hydrostatics, heat, and general elementary physics.[11] After he left in 1876 there were several brief successors and deputies, including J. W. Russell and Lazarus Fletcher, until 1879, when Trinity entered into a new arrangement with Balliol College and the physics lectures moved there. As with physiology, the adaptability and enthusiasm inherent in college arrangements, and the willingness of colleges to co-operate, had helped a difficult subject to achieve stability. The Magdalen College Laboratory was an appropriate venue for physics teaching because of the electrical, pneumatic, optical, and other instruments included in Daubeny's bequest (Anon. 1861); and in the 1870s these were supplemented by several new purchases, notably an Atwood machine for demonstrating Newtonian mechanics in 1876. The physical instruments increased again from 1891, when practical work became compulsory in the Preliminary physics examination.[12]

Tuition in elementary physics for Magdalen's own students was taken on by Chapman, who, although a biologist, had been a pupil of the physicist George Griffith.[13] From 1891 hands-on instruction in physics for the practical examination

---

[10] Gunther (1924: 21–3, 24–5, 51, 408–14, 431); OUG, 5 (1874–5): 14 (23 January 1874); OUG, 8 (1877–8): 218 (29 January 1878). On the chronograph see Simcock, Chapter 5, 171.

[11] For Mac Donald's fellowship, OUG, 4 (1873): 296 (4 November 1873); for his lectures, OUG, 5 (1874–5): 14, 132, 319, and 541 (23 January, 17 April, and 3 November 1874, and 9 April 1875); Gunther (1924: 111, 309–10); and see Simcock, Chapter 5, 182.

[12] Gunther (1924: esp. 15, 51–2); Magdalen College archives, MS 685; MHS, MS Museum 109, esp. f.29 (Atwood).

[13] Gunther (1924: 311, 415); Gunther (1937: 251); on Griffith see Fox (1997: 677) and Fox, Chapter 2, 41–4.

Fig. 4.2. Edward Chapman in 1888. As tutor in natural science at Magdalen College from 1869, Chapman taught elementary physics, as well as chemistry and biology, while as organizer of the inter-collegiate lectures in these core subjects he made special efforts to ensure that physics and, later, mechanics were properly represented.

was given by John Job Manley, appointed 'Daubeny Curator' in 1888 (effectively the laboratory's technician).[14] An ingenious precision mechanic and experimentalist, Manley developed a uniquely well-equipped workshop where he designed and constructed apparatus for the laboratory and lecture room. One example was a 'Lath carrying 30 pendulums to illustrate transverse vibrations of ether'.[15] A more useful invention was his differential densimeter for the analysis of seawaters, which became the focus of his own chemical research, following a long collaboration with the physical chemist Victor Herbert Veley on nitric acid.[16] His main physico-mechanical research involved the improvement of laboratory measuring techniques, especially precision weighing – Frederick Soddy considered him 'probably the greatest living authority on the chemical balance'. By 1923, through successive refinements and auxiliary devices, his standard balance was 'easily capable of

---

[14] Gunther (1924: esp. 25–6, 130, 181, 305, 313, 322); on Manley see newspaper cutting (1903) in MHS, MS Gunther Archive, Album of Printed Matter vol. III; Anon. [probably by Harold Hartley] (1946: 332–3); Gunther (1967: 38–9); Hartley (1971b: 224–5).   [15] MHS, MS Museum 109, f.60.
[16] Manley (1906–7); Campbell (1912–13); Gunther (1924: 48–50, 173–80); on Veley see obituary by Gardner (1934).

Fig. 4.3. John Job Manley, c.1910, sitting on the roof of the Daubeny Laboratory, Magdalen College. A skilful experimenter and mechanic, Manley began in the laboratory in the late 1880s as a technician. Thereafter he assumed steadily more important roles in teaching and research and earned the admiration of leading advocates of experimental precision, such as Harold Hartley and Frederick Soddy.

detecting' a 128-millionth part of a gram.[17] In 1899 Magdalen gave Manley the additional title of demonstrator, and the University awarded him an honorary M.A. four years later; in 1917 he became a research fellow of Magdalen, and his research publications earned him the Oxford D.Sc. in 1931.

The college laboratory also took pupils from Magdalen College School, where physical science was encouraged by the headmaster appointed in 1888, William Edward Sherwood, one of Daubeny's last pupils and a Christ Church mathematics graduate. Sherwood appointed Manley as part-time science master in 1889, and he taught physics to schoolboys at the Daubeny Laboratory until 1919, when with Manley's help the school established its own physics laboratory.[18] The physics that Manley taught is embodied in *An Introduction to the Study of Physics* (1901), co-written with Allan Frederick Walden, a former Magdalen student (first in chemistry, 1895).[19] The Daubeny Laboratory's outreach extended further than

---

[17] Howorth (1953: 21); Gunther (1924: 48–50, 173–6, 179–82, 449–52, 454–6 [1923 accuracy p. 451]); Hartley (1971b: 224).

[18] Gunther (1924: 25–6, 111, 141, 293–4, 321–2). The school had previously sent its boys to Clifton's lectures; see the 1867 list in Fox, Chapter 2, Table 2.2.

[19] Walden and Manley (1901); on Walden see obituary by Ing (1957); Gunther (1924: 126, 234); also see below, p. 150.

Magdalen College School, however, for Daubeny and Chapman were great champions of educational extension: on Saturdays from about 1876 Chapman and colleagues put on extramural classes for 'artisans' in biology, botany, and 'chemical physics'; and for many years Manley gave practical chemistry instruction in the evenings, aimed mainly at the university's Non-Collegiate students.[20]

From about 1880 the college relied upon a part-time chemistry tutor, the first being John Watts, who was based at Merton College where Frederick Soddy was among his pupils (Gunther 1924: 238, 311; Soddy 1933). Watts was followed from 1899 to 1903 by a Magdalen graduate (first, 1898), (Sir) Duncan Randolph Wilson, who had just spent a year working in Ostwald's laboratory at Leipzig, the continental hothouse of physical chemistry. Gunther harnessed Wilson's recent experience and Manley's technical skill in a new practical course of 'Experimental Physical Chemistry', introduced in Michaelmas term 1900. The syllabus was modelled on Ostwald's, and a pamphlet was circulated to attract students from other colleges.[21] In 1904 new arrangements for chemistry teaching moved physical chemistry to the Balliol–Trinity Laboratories and gave Manley responsibility for practical instruction in quantitative analysis at the Daubeny Laboratory. Magdalen's subsequent chemistry lecturers were the versatile Nevil Vincent Sidgwick, a fellow of Lincoln College, and his research collaborator Tom Sidney Moore, another Merton pupil of Watts. Sidgwick moved in 1907 to the new Jesus College Laboratory. Moore gave lectures on electrochemistry, but left in 1914.[22] A product of the Sidgwick–Moore period, and an excellent example of Oxford chemistry nurturing brilliant physicists, was Sir Henry Thomas Tizard, who took first-class honours in mathematics moderations before concentrating on chemistry (first, 1908), but then became a demonstrator in the Electrical Laboratory. He is best remembered as the archetypal 'boffin', working mainly in aeronautics research (Farren 1961; Clark 1965; Gunther 1924: 253–4, 509).

By the time normal teaching resumed after the First World War, Magdalen College had once more taken the lead in a new kind of relationship between college and university teaching: it had thrown its support behind the new Dyson Perrins Laboratory, and thus behind the opinion that all practical work should take place in central laboratories. In 1919 Magdalen's existing mathematics don Arthur Lionel Pedder assumed responsibility for physics and engineering science, and Edward Hope, chief demonstrator at the Dyson Perrins, was appointed chemistry

---

[20] Gunther (1924: 21–2, 130, 317); for Daubeny's advocacy of extension see Willsher (1961: 92–4).
[21] Gunther (1924: 26, 52, 127, 238–9, 499); *WWW*; Bowen (1970: 230); MHS, untitled leaflet including syllabus (1900); for Wilson, information from Dr Roger Hutchins.
[22] Gunther (1924: 29, 52, 140–1, 173–7, 238); Hartley (1971b: 225); Laidler (1988: 223–4); for Moore's lectures, *OUG*, 37 (1906–7): 537 (29 April 1907); on Sidgwick see Tizard (1954); Morrell (1997: 361–4); for the 1904 arrangements see below.

Fig. 4.4. Henry Thomas Tizard, c.1910. Tizard took first-class honours in chemistry in 1908 and went on to become a demonstrator in the Electrical Laboratory, a fellow of Oriel College (in 1911), and an influential governmental adviser, notably on the early development of radar. He was appointed reader in thermodynamics in 1920 but from the mid-1920s moved increasingly into public life. He was Rector of Imperial College from 1929 until 1942, when he became (for four years) President of Magdalen College.

tutor. Neither wished to use the college laboratory – Hope for obvious reasons, while Pedder, as Gunther noted sardonically, 'was averse from giving practical instruction' (Gunther 1924: 324). Manley temporarily revived his quantitative analysis course in 1918, but gave up teaching physics; and the Daubeny Laboratory closed in 1923. Manley's research there continued until 1929, however. The story goes that, after seeing Gunther and his research materials turned out, Manley 'filled his two rooms . . . with glass apparatus of astonishing complexity which could on no account be moved' (Gunther 1967: 125).

BALLIOL COLLEGE LABORATORY AND THE BALLIOL–TRINITY LABORATORIES

Magdalen College's laboratory having been, as it were, thrust upon it by Daubeny, the first college to build itself a laboratory was Balliol. In 1851 plans for a small laboratory and lecture room were incorporated into a new building project (the Salvin Building), and the college's mathematical lecturer, Henry John Stephen Smith, recently elected a fellow, accepted the task of teaching in them. In preparation he attended Acland's anatomy course, received instruction in chemistry from Mervyn Herbert Nevil Story-Maskelyne, and during 1853 also went through a brief course with A. W. Hofmann at the Royal College of Chemistry in London.[23] The significance of his concentration on chemistry was simply that chemistry, unlike physics, was expected to involve practical laboratory work by the students, demanding teachers with learned practical skills; whereas he presumably considered himself competent in the physics that would be required, and as a mathematics lecturer was already teaching it. From the beginning of 1854 Smith was therefore able to offer Balliol's students tuition in all three branches of the new natural science degree, together with lectures in physics and chemistry and, especially in the latter, practical laboratory instruction. While historical accounts have enjoyed portraying Smith as the mathematician who taught chemistry, in fact his pupils were more typically mathematician-physicists, several of whom went on to teach both subjects themselves – including Charles James Coverly Price, who took a double first in mathematics and natural science in 1860 and became fellow and mathematical lecturer at Exeter College; and John Wellesley Russell, who took a first in mathematics in 1872, studied physics without taking the degree, became a fellow of Merton, and succeeded Smith at Balliol.[24]

The laboratory proved useful in an unexpected way. After Daubeny's move to the Botanic Garden in 1848 the old University Chemical Laboratory in the Ashmolean Museum had been assigned to Nevil Story-Maskelyne's use. When Story-Maskelyne failed to win election to Daubeny's chemistry chair in 1855, Balliol College was in a position to preempt any embarrassment over premises by offering its hospitality to the new professor, (Sir) Benjamin Collins Brodie junior, who happened to be a Balliol graduate (in mathematics). For nearly three years Balliol's laboratory and lecture-room officiated as the University's chemistry department.[25] The construction of the University Museum had begun in the same year (1855), and Brodie moved into his new laboratory there (the so-called Glastonbury Kitchen) in 1858, two years before the official completion of the building. Brodie's appointment, incidentally, broke with the tradition of chemistry professors being medical men; and although Oxford chemistry was to remain a broad church, its special rapport with physics and mathematics had its beginnings in these years, when it fell into the hands of two mathematicians, and of their joint pupil A. G. Vernon Harcourt (soon to be placed in charge of the next college laboratory to be established, at Christ Church).

---

[23] Anon. (1894: esp. 11, 18, 44). Humphries (1970: 75–6, 92); Laidler (1988: 224–7); on Smith see also Hannabuss (2000c).

[24] On Price see Foster (1893: column 123); *The Balliol College Register . . . 1833–1933* (Oxford, 1934): 20; on Russell see below; for comments on Smith's interest in physics see Gooday, Chapter 3, 103.

[25] Smith (1982: 190–4); MHS, MS Museum 71 (syllabuses); on Brodie see Brock (1967); *DSB*.

Fig. 4.5. Benjamin Collins Brodie, junior, in a photograph taken by Nevil Story-Maskelyne in about 1847. Brodie was appointed professor of chemistry in 1855, the first non-medical man since the seventeenth century to hold the appointment. This is one of the earliest known photographs of a scientist.

The Balliol College Laboratory is usually thought to have fallen out of use after Brodie left, and Smith's involvement in science teaching at Balliol to have declined, especially once his brilliance as a mathematician earned him the Savilian chair of geometry in 1860 (for instance Hartley 1971b: 226; Bowen 1970: 228; Humphries 1970: 76; Smith 1982: 194; Laidler 1988: 226; Hannabuss 2000c: 205). But although students were not numerous, Smith continued to teach physics and chemistry at Balliol until at least 1873, when he moved to a fellowship at Corpus Christi College. He was assisted by John Andrews Dale, a Balliol mathematician who held no formal post or fellowship, though the extent of his involvement in pioneering the teaching of the new physics syllabus can be gauged from his record as an examiner: between 1857 and 1875 he was a natural science examiner on no fewer than sixteen occasions.[26] In 1873 Smith arranged for his pupil J. W. Russell to succeed him as Balliol's mathematical lecturer, also teaching physics; and shortly afterwards for Walter William Fisher to be appointed lecturer in natural science (1874–9) to teach chemistry. At the same time the college began planning a new laboratory, which is

---

[26] *Historical Register*; Anon. (1934): 1; for a glimpse of his physics interests see for instance Anon. (1855)..

evidence in itself that the existing laboratory had continued to prove useful. Smith was closely involved in these arrangements, and since he was now also Keeper of the University Museum (that is, the officer in charge of Oxford's central laboratories), his participation in the development of college facilities indicates that no contradiction was perceived between the two.

This second venture in laboratory building originated, as before, in a grander building project, the college hall, which opened in 1877. In 1874, at an early stage in its planning, the college decided to include new accommodation for science. In line with the Devonshire Commission's call for colleges to promote science and share facilities, Balliol set up committees to look into the provision of science teaching, including 'the comparative advantages of University and College teaching'; and as the rooms neared completion, discussions took place with Trinity College about the 'united teaching of Chemistry and Elementary Physical Science' (meaning physics) (Smith 1979: 33–4). The new accommodation consisted of a laboratory for hands-on practical teaching, mainly of chemistry, and the 'Physical Lecture Room', for lecture-demonstrations, mainly in physics. The cellar beneath the new hall and the 1854 laboratory across the quadrangle were used as research annexes. One of several small rooms ancillary to the main laboratory later acted as 'physics room', where Preliminary practical physics was taught from 1891 to 1904. In 1897, an extension to the main laboratory was built on Trinity's side of the 'scientific frontier' (as the boundary between the two colleges came to be named). At some later point the Physical Lecture Room was also converted into laboratory space. From 1921, several outbuildings in Trinity's Dolphin Yard (a back-yard squeezed between Balliol and St John's) were brought into use, notably the former Millard Laboratory, which from 1929 until 1941 accommodated C. N. Hinshelwood's research work.[27]

The particulars of how the two colleges staffed and financed the Balliol–Trinity Laboratories (as this archipelago came to be called) need not unduly concern us: suffice it to say that for over sixty years their collaboration sustained one of the busiest and most productive scientific institutions in Oxford. Many accounts have been written of this famous college laboratory, extolling its contributions to chemistry and conveying something of its spirited atmosphere and loveable haphazardness.[28] None of them, however, give the least indication that physics was originally an equal part of its purpose. The fact that the head of the laboratory's job-title was originally lecturer in physics seems to have aroused little curiosity. Until

---

[27] Smith (1982: esp. 217–19, including plan) and see Fig. 4.6. The Physical Lecture Room is cited in *OUG*, 10 (1879–80): 20 (10 October 1879) and *OUG*, 11 (1880–1): 22 (15 October 1880). The various Dolphin Yard annexes are covered in particular detail, with plans and photographs, by Laidler (1988: esp. 233–6, 282, and plans 225, 234); see also Hibbert (1988: 466); on the Millard Laboratory see Simcock, Chapter 5.

[28] Bowen (1970); Humphries (1970: 75–89); Laidler (1988: esp. 224–36); Smith (1982) based on Smith (1979); for the laboratory's history in brief see Hartley (1971b: 225–8).

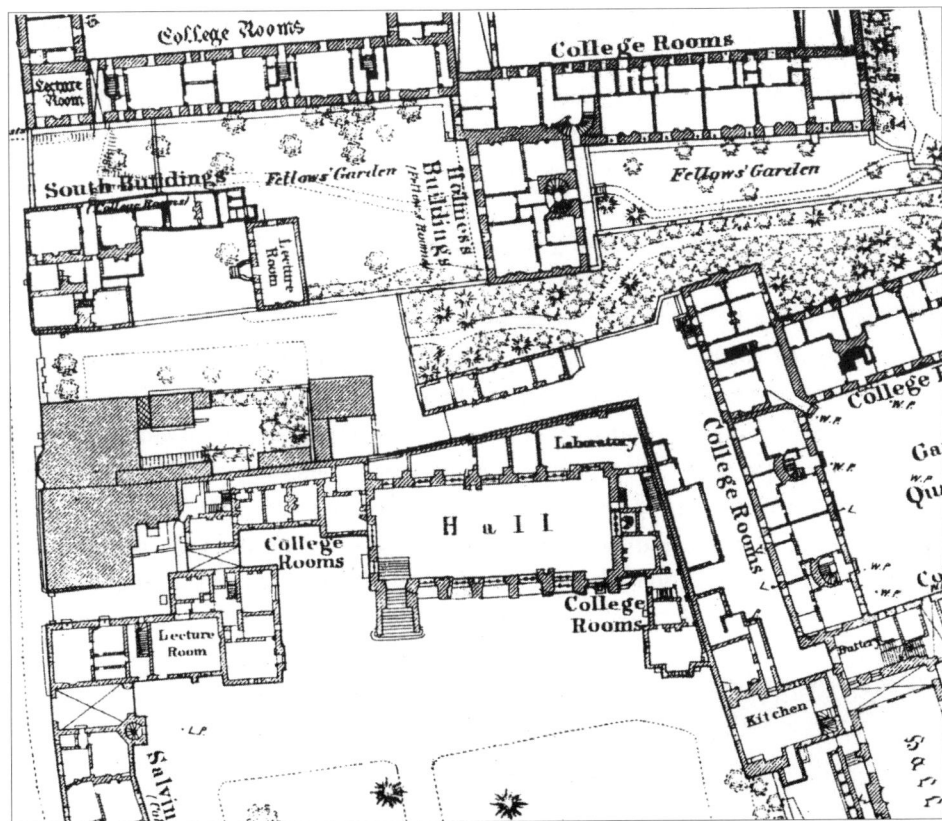

Fig. 4.6. Plan of the adjacent parts of Balliol, Trinity, and St John's Colleges, showing the vicinity that was, somewhat improbably, one of Oxford's chief centres of physical science from about 1880 until 1940. At the time of this plan (drawn in 1878) Balliol's hall was newly finished and its laboratory was about to open (1879). The main laboratory occupied at basement level an irregularly shaped space between the hall and the angle of the boundary-wall with Trinity, as shown; several adjoining basement and first-floor rooms were also part of the Balliol–Trinity Laboratories. The strip of land between this northern edge of Balliol and the southern boundary of St John's constitutes the rear entrance and service yards of Trinity, known as Dolphin Yard. The free-standing lecture room shown on the St John's side of this passage was the building that Bosanquet converted in 1880 into a physics and acoustics laboratory. In 1885–6 the Millard Laboratory (not shown) was built immediately inside the western entrance to Dolphin Yard, along the southern side. In 1897 Trinity built an extension to the Balliol–Trinity Laboratories on its own side of the wall immediately east of the Balliol laboratory. In the 1920s and 1930s the Balliol–Trinity physical chemists colonized various buildings in the Trinity isthmus as research annexes; these included the old lavatory block (shown on the plan, directly north of Balliol hall) and the former Millard Laboratory. From the 1876 Ordnance Survey 1/500 series, XXXIII.15.17.

physics teaching was centralized in 1905, chemistry and physics were very closely integrated here, the same men teaching both subjects. The 1879 scheme provided for Trinity's Millard lecturer in physics to be in charge of Balliol's laboratory and lecture-room, lecturer and premises thus being shared by the two colleges. Trinity, Balliol, and at first Exeter (until 1885 providing biology teaching in a small, exclusively biological laboratory) at the same time took over the combined college lectures previously sponsored by Trinity, Merton, and Magdalen. The lectures were open to all Oxford students for the usual £1 per term fee. From 1882 the Millard lecturership was held conjointly with (and from 1900 superseded by) Balliol's Bedford lecturership in physics, created by an endowment from the Duke of Bedford (Smith 1982: 197–8).

Harold Baily Dixon, who had graduated with a first in chemistry in 1875, was appointed Millard lecturer in physics in 1879, and threw himself with enthusiasm into the management of the new laboratory, modelled doubtless on the way of doing things at Christ Church, where he had been a pupil of Harcourt. He established a spirit and style which stayed with the Balliol–Trinity Laboratories throughout their existence. Dixon's principal duty was to lecture on 'Elementary Physics', giving three experimentally illustrated lectures per week during term, alternating termly between the topics of heat and light, and electricity and magnetism. The concession to Dixon's education as a chemist was that he did not continue the mechanics, dynamics, and hydrostatics taught by previous Millard lecturers, which

Fig. 4.7. Harold Baily Dixon in 1894, seated, in white. Detail from a group photograph of chemists gathered in Oxford for the meeting of the British Association for the Advancement of Science.

belonged to the more mathematical part of physics. His only formal attempt to give students hands-on instruction in physics was an additional practical class on electricity and magnetism in 1881. The daily provision of laboratory instruction in chemistry, however, was his other main duty, occasionally supplemented with specialized chemistry lectures.[29] Dixon vigorously promoted student participation in research, and continued his own work on gaseous explosions commenced at Christ Church, making full use of the space now available at Balliol and also working for a while at the Daubeny Laboratory (where a particularly fine chronograph was available).[30] Seven research papers appeared in his Balliol years alone, one of them in collaboration with his pupil H. F. Lowe, with whom he investigated the effect of an electric spark on carbon dioxide (Dixon and Lowe 1885). Dixon left Oxford in 1887 to become Sir Henry Roscoe's successor as professor of chemistry at Owens College, Manchester.

The physical topics dropped by Dixon as Millard lecturer were taught at first by his colleague J. W. Russell, the mathematical lecturer at Balliol, Trinity, and St John's Colleges (and also a fellow of Merton), who had deputized for previous Millard lecturers at the Daubeny Laboratory. In 1880 his mechanics course, with experiments, was accommodated in the Physical Lecture Room; subsequently he used Balliol's 1854 lecture room, mainly for pure mathematics, for such a long period that it is still known as the Russell Room.[31] Overlooked in the existing histories, he was a significant figure in the Balliol–Trinity–St John's epicentre of physical science, the picture of which is not complete without him. Russell's outstanding early pupils were Sir Lazarus Fletcher and John Viriamu Jones.[32] Joint students of Russell and Dixon included William Stroud (firsts in mathematics, 1884, and physics, 1885), later professor of physics at Leeds and co-inventor of a range-finder for warships, around which the instrument-making firm of Barr & Stroud was established in 1893; and Hubert Foster Lowe, who took mathematics moderations before taking finals in chemistry (first, 1882), and stayed on as Dixon's demonstrator and research assistant for two years, when he joined the Patent Office as an examiner.[33] Another of Dixon's pupil-demonstrators was the physical chemist Herbert Brereton Baker (first, 1883), who made the research he began at Balliol his life's work, and whom we shall encounter again upon his return to Oxford in 1903.[34]

---

[29] *OUG*, 10 (1879–80): 22 (10 October 1879), etc.; for chemistry lectures, *OUG*, 11 (1880–1): 217 (21 January 1881); and *OUG*, 15 (1884–5): 413 (20 April 1885). On Dixon see Foster (1893: column 68); Plarr (1899: 303); Hilliard (1914: 89); obituary by Baker and Bone (1931–2).

[30] MHS, MS Museum 190; Bowen (1970: 231–2); Smith (1982: 198–201); on Dixon's research see Laidler (1988: 246–50); on the chronograph see Simcock, Chapter 5, 171.

[31] *OUG*, 11 (1880–1): 22 (15 October 1880); Anon. (1934): 63; and Foster (1893: column 98); and see Simcock, Chapter 5, 182.

[32] On whom see Fox, Gooday, and Simcock, Chapter 1, 17–18.

[33] Hilliard (1914: 206–7, 330); Smith (1979: 37–8); on Barr and Stroud see Moss and Russell (1988).

[34] Thorpe (1932–5); Laidler (1988: 232, 255–6); and see below.

Fig. 4.8. William Stroud. After taking firsts in mathematics (1884) and physics (1885), Stroud went on to be professor of physics at Leeds and, in 1893, co-founder of the instrument-making firm of Barr & Stroud.

Dixon's appointment coincided with the opening of the first women's colleges (1879); and his immediate offer to admit women to his physics lectures, albeit vetoed by the two colleges, illustrates the close alliance there often was in Oxford between science and liberalization. When Harcourt joined him in taking a stand against the segregated teaching of women, the colleges backed down. The first notable beneficiary was Margaret Seward, of Somerville College, who took mathematics moderations as soon as the examinations were opened to women (in 1884), and then first class honours in chemistry in 1885. She became Somerville's first natural science tutor (1885–7), teaching chemistry and Preliminary physics, and doing research at Christ Church.[35] Starting as a mathematician, and having done Preliminary-level physics with Dixon, she was presumably – like many male students – deflected from finals physics by the readier availability of tuition in chemistry, as well as by the hospitality of the college laboratories.

Dixon's successor was an earlier protégé of Harcourt's, Sir John Conroy, who had returned to Oxford to teach chemistry and physics at Keble College. But while Dixon was a chemist capable of teaching physics, Conroy was a true physicist. His

---

[35] Adams (1996: 33, 35, 38–40, 53); Brittain (1960: 68, 82); *Somerville College. Register 1879–1971* (Oxford, [1972]): 372; Bowen (1970: 228–9).

Fig. 4.9. Sir John Conroy, c.1890, photographed against the distinctive stonework of Balliol College hall, the construction of which (in the 1870s) included at the rear a purpose-built chemical laboratory and physics lecture-room.

field of research was the optical properties of materials: a series of papers entitled 'Some experiments on metallic reflection' (1879–84), for instance, records meticulously contrived observations and measurements using (among other techniques) a photometer he invented for the purpose (Conroy 1878–9; 1880–1; 1883a; 1883b; 1883–4; 1884).[36] He also measured the reflective and transmissive properties of glass for Harcourt's work on lighthouse illumination (Conroy 1889). When Harcourt later paid tribute to 'the extreme delicacy and neatness of his experimental manipulation' and his 'endeavour to attain the highest accuracy in numerical results', he was presenting Conroy as the epitome of the qualities that the Oxford experimentalists most cherished.[37] Conroy's routine physics lectures continued Dixon's pattern. In Michaelmas term 1887 he gave an additional course entitled 'Chemical Physics', following which V. H. Veley's regular physical chemistry lectures commenced in the University Chemistry Department.[38] Dissatisfied with the complacent routine of

[36] Conroy's papers are listed with comments in Vernon Harcourt (1905).
[37] Obituary by Vernon Harcourt (1905: quoting 252); on Conroy see also Foster (1893: column 64); WWW; Bowen (1970: 229); Smith (1982: 201–6); for Conroy at Keble see below.
[38] For lectures on physics topics, OUG, 17 (1886–7): 215 and 369 (17 January and 25 April 1887); for lectures on chemical physics, OUG, 18 (1887–8): 47 (18 October 1887).

Fig. 4.10. David Henry Nagel, 1910, in the main teaching laboratory of the Balliol–Trinity Laboratories, of which he was in charge. Nagel, who specialized in chemistry as an undergraduate, also taught physics and did much to promote the Balliol–Trinity strength in physical chemistry.

college teaching (Smith 1979: 45–7), Conroy later developed an innovative lecture course (with experiments and a separate practical class) on the optical physics of materials, based on his research. 'The Study of Light in its Relation to Chemistry' was listed under physics in the lecture list for Lent term 1896; but the following term and thereafter it appeared in the chemistry list as 'Physical Chemistry, Optics' – a revealingly explicit instance of the Oxford tendency to redefine physics as chemistry.[39]

In Conroy's time only four Balliol men specialized in physics at final honours level, chemistry and physiology remaining as always the favourite subjects, even students keen on physics often preferring to graduate in chemistry. One such was Edgar Ford Morris (first in chemistry, 1895), who as an undergraduate gave one of the world's earliest demonstrations of Hertz waves (to the Junior Scientific Club on 7 February 1894), but then turned to physical chemistry.[40] As well as encouraging the best students to try research, Conroy also extended his hospitality to outsiders.

---

[39] *OUG*, 26 (1895–6): 229, 409 (27 January and 27 April 1896); *OUG*, 27 (1896–7): 39 (19 October 1896).
[40] *Journal of the Oxford University Junior Scientific Club*, 1, no.13 (1894): 190–2.

Lord Berkeley was allowed to use the 1854 laboratory for his own research (1897–8), as too was the young Frederick Soddy for a while after graduating (1898–9).[41] The vigorous impetus imparted by Dixon was thus continued. Yet personally Conroy was completely unlike Dixon – conservative, religious, shy, and in delicate health. Accordingly, he placed much of the laboratory administration in the hands of Dixon's last graduate assistant, David Henry Nagel (first in chemistry, 1886); the two colleges soon recognized Nagel's position and created the post of Millard demonstrator for him (1888), and in 1890 he became a fellow of Trinity.[42] His earliest lecture course in 1888 added a new topic to the repertoire of physics lectures, under the title 'Elementary Physics – Sound'.[43] He also provided the demonstrations and practical instruction for Conroy's courses. When V. H. Veley retired in 1903, Nagel took over responsibility for giving the University's main lecture series on physical chemistry, and the associated practical classes, both of which he gave at Balliol.[44] It was thus Nagel, and not as is usually supposed Harold Hartley, who determined the Balliol–Trinity specialization in physical chemistry that became formalized in 1904.

While Nagel was not a researcher, he was a highly effective teacher of both chemistry and physics, and a trusted adviser in the research of others. He was principal tutor for several able physicists, the most distinguished being Henry Gwyn Jeffreys Moseley (second in physics, 1910), whose experimental verification of the atomic number of the elements has been fundamental to the modern progress of physics and chemistry alike.[45] Robert Tabor Lattey graduated in 1903 with a first in chemistry, but after two years demonstrating for Baker at Christ Church he moved to Townsend's Electrical Laboratory to assist in the new centralized preliminary physics course, and spent the rest of his career there. Richard Littlehailes was tutored for physics finals by Nagel and by C. E. Haselfoot after Conroy's death, taking a second in 1902. After serving as demonstrator in the Clarendon Laboratory (1902–3) he joined the Indian Education Service, becoming *inter alia* deputy director of the Madras Observatory, professor of mathematics and physics at Presidency College, Madras, and finally Vice-Chancellor of Madras University (1934–7).[46] Nagel was also part-time lecturer in natural science for Jesus College (1890–6), which produced a high number of chemistry and physics graduates.

---

[41] Hartley (1971b: 227); Smith (1982: 204–6, 221); Hartley (1942–4: 169–70); Howorth (1953: 27); on Berkeley see below.

[42] *OU Calendar* for 1888, 1889; obituary by Dixon (1920a); Gunther (1937: 74); Howorth (1953: 22–3); Bowen (1970: 229); Smith (1979: 42–4, 46, 50).

[43] *OUG*, 19 (1888–9): 29 (15 October 1888), etc. The lectures might have used apparatus from Bosanquet; see below.

[44] *OUG*, 33 (1902–3): 450 (27 April 1903) and *OUG*, 34 (1903–4): 40 (19 October 1903); see Hartley (1971b: 228).

[45] Hartley (1971b: 227–8); Heilbron (1974: esp. (for Nagel) 36–7); on Moseley see also obituary by Rutherford (1915–16); Sarton (1927); Turner (1965); and see Lelong, Chapter 6, and Hughes, Chapter 8, 221–2 and 277–9.   [46] *WWW*; Anon. (1934: 245).

Fig. 4.11. Robert Tabor Lattey, c.1910, cycling away from the doorway at the south-west corner of the University Museum, the main entrance to the central chemical laboratories. Lattey, a chemistry graduate, was demonstrator in the Electrical Laboratory, which until 1910 was also situated within the University Museum. He was still lecturing in the Electrical Laboratory during the interregnum between Townsend's retirement in 1941 and the appointment of Maurice Pryce as Wykeham professor in 1946. The lady is Florence Isaac, a mineralogist.

After Conroy's death at the end of 1900, Balliol and Trinity again (as in 1877–9) seriously considered their position within Oxford science, deciding to continue their existing provision for the physical sciences. Harold Brewer Hartley, appointed in 1901, had been Conroy's last pupil, graduating with a first in chemistry and mineralogy in 1900. In biographical accounts the young Hartley has been eclipsed by Brigadier-General Sir Harold Hartley, the First World War boffin, physical chemistry discipline builder, and (after leaving Oxford in 1930) captain of industry.[47] But his original leaning was towards crystallography; while the first lectures he gave at Balliol in 1901 took a historical approach, including a general 'History of Chemistry' course. Other early lecture topics ranged from organic chemistry to stoichiometry.[48] His interest in electrolysis began at the same period, and slowly became the focus of his mature research in physical chemistry, his dedication to which evolved in response to the new arrangements for chemistry practical teaching of 1904. Along with the simultaneous centralization of physics teaching (to which we shall return), this development allowed Oxford physical chemistry to consolidate

---

[47] Smith (1979: 55); obituaries by Ogston (1973) and Bowen (1973); Morrell (1997: 330–1).

[48] For historical approach, OUG, 31 (1900–1): 267 (21 January 1901); for organic chemistry, OUG, 32 (1901–2): 40–1 (14 October 1901); for history of chemistry, OUG, 34 (1903–4): 40–1 and 284–5 (19 October 1903 and 25 January 1904); for stoichiometry, OUG, 39 (1908–9): 51 (10 October 1908). Some of the historical lectures emerged seventy years later in Hartley (1971a: see esp. preface vii–viii).

Fig. 4.12. Harold Brewer Hartley in the Balliol–Trinity Laboratories, 1910. At this time, Hartley had held a tutorial fellowship at Balliol for almost a decade and was assuming the role of champion of physical chemistry in Oxford, which he led to its great achievements of the 1920s and 1930s.

a unique college power-base, which suited Hartley's discipline-building inclinations. In the same year his post as Bedford lecturer in physics was renamed Bedford lecturer in physical chemistry (Smith 1982: 207).

Nevertheless, as college tutor he nurtured students in physics throughout his career. In 1901 his very first pupil was Idwal Owain Griffith (double first, 1901 and 1903), who went on to be a demonstrator at the Clarendon Laboratory, research fellow of St John's College, lecturer at Wadham, Exeter, and St John's, and after the First World War Lindemann's senior demonstrator at the Clarendon.[49] Men like Griffith and R. T. Lattey illustrate how far the new centralized laboratory physics of the twentieth century relied on an earlier culture of college physics training. The subsequent history of the Balliol–Trinity Laboratories and their important contributions to physical chemistry under Hartley and his successor Sir Cyril Norman Hinshelwood, whose work earned him a Nobel Prize in 1956, have been well documented elsewhere.[50] During this period the Balliol–Trinity Laboratories

[49] Obituary Anon. (1941b); see Morrell, Chapter 7, 238–9.
[50] Most recently and thoroughly by Laidler (1988), one of Hinshelwood's pupils; see also Bowen (1970: 233–6); on Hinshelwood see obituary by Thompson (1973).

amounted to a university department of physical chemistry in all but name, a situation recognized when Hinshelwood was appointed professor in 1937. A new Physical Chemistry Laboratory was built by the University (and paid for by Lord Nuffield) almost simultaneously with the new Clarendon Laboratory, and the cluster of laboratories and research annexes around the 'scientific frontier' between Balliol and Trinity was abandoned in 1941 (Barrow and Danby 1991).

CHRIST CHURCH LABORATORY

For Christ Church more directly than for Magdalen and Balliol, the development of the University Museum had the effect of stimulating independent college facilities, rather than obviating the need for them. Christ Church possessed an Anatomy School, built in 1767, together with extensive anatomical collections and an endowed lecturer in the subject, known as 'Dr Lee's Reader' (after the benefactor). The college had been persuaded by Henry Acland (the Dr Lee's reader from 1845 to 1857) to allow its collections and the reader's biological teaching to be absorbed into the new University Museum. Following their transfer in 1860, the Anatomy School building was converted into a laboratory and lecture room for teaching the other two disciplines of the new natural science syllabus. The Dr Lee fund was sufficient to finance a readership in chemistry in 1859, and a readership in physics ten years later. Although these were college appointments, their lectures were open to the entire university, and the University's new statutes of 1882 effectively promoted them to the status of university readerships.[51]

While most historical accounts imply that the Christ Church Laboratory was exclusively chemical, this is no more the case than at the Balliol–Trinity Laboratories. Both readers were based in the laboratory building, and used the same public rooms; but the fame of its distinguished school of chemists has long eclipsed the presence of the less popular subject of physics. The first Dr Lee's reader in chemistry was Augustus George Vernon Harcourt, whose research work was to place him in the vanguard of the emerging field of physical chemistry, and who was a dominant figure in Oxford chemistry up to his retirement in 1902.[52] Arriving at Balliol in 1854 expecting to study classics, he became instead one of the two first chemistry students taken on by Henry Smith. He was then one of Brodie's

---

[51] For the advertisement of open lectures see *OUG*, 1 (1870): 13 (1 February 1870), etc.; *New Statutes 1877*: 44–7.

[52] Obituary by Dixon (1920b); Shorter (1980); King (1984); Laidler (1988: esp. 202–3, 217–22, 239–46); Kent (2001: esp. 31–4); and see Vernon Harcourt (1910); for the laboratory's history in brief see Hartley (1971b: 228–30) or Laidler (1988: 220–2).

first pupils, and (even before graduating) his assistant in the newly built chemistry department at the University Museum. Appointed to the Christ Church post in 1859, he continued working with Brodie until the Anatomy School conversion was completed, about 1866.[53]

Responsibility for a brilliant physics student, John Conroy, who graduated in 1868, probably convinced Harcourt that provision for physics was a desideratum. In the same year the physicist Robert Holford Macdowall Bosanquet was appointed lecturer; some 'physical apparatus of an elementary kind' was acquired;[54] and in 1869 the readership in physics was created, Arnold William Reinold being appointed. One of Clifton's earliest Oxford pupils, he had graduated from Brasenose College with firsts in mathematics (1866) and natural science (1867), and was already lecturing on elementary physics as a Fellow of Merton. As Fox has pointed out, by hiring Harcourt and Reinold Christ Church had harnessed the talents of 'two of the ablest early products of the school of natural science'.[55] In 1873, however, it lost Reinold to the chair of physics at the Royal Naval College, Greenwich. The Dr Lee's readership in physics was advertised at an annual salary of £300, rising to £500 in the eighth year, with residential rent-free rooms. The duties were to give instruction in 'the subjects recognised in the Physical department of the Natural Science School' and also in 'the Mechanics, Hydromechanics, and Optics required for the Final Honour School of Mathematics'. The advertisement went on to state that apparatus was provided 'to be employed either in lecturing or for purposes of original research – to which the Reader will be expected to devote a portion of his time'.[56]

In December 1873 Robert Edward Baynes was elected, and the following month he began his first lecture course, on heat.[57] He had recently obtained his double first (1871 and 1872) at Wadham, having been one of the first students to benefit from the new Clarendon Laboratory. At Christ Church he later held high administrative office; and as a Freemason he rose to the position of Deputy Provincial Grand Master of Oxfordshire in 1915 (Baker 1921–2; Foster 1893: column 408; *WWW*; Gunther 1937: 225, 271; Kent 2001: 34). In other respects, however, Baynes was surprisingly unambitious. Notwithstanding the terms of his appointment, he did not pursue academic recognition through research: he was reluctant to publish papers in the scientific journals, so what little original work he did went largely unnoticed.[58] Yet he was highly dedicated to physics education, not only as a teacher

---

[53] No record can be found of the date of the laboratory's completion or opening (information from Mrs Judith Curthoys, Christ Church archives); secondary accounts vary from 1863 to 1866; Vernon Harcourt (1905) implies the later date.

[54] *OU Calendar* for 1869; Gunther (1924: 12 footnote); for Bosanquet see below.

[55] Fox (1997: 674); for Reinold see Fox, Chapter 2, and Gooday, Chapter 3, 65–6 and 91–2.

[56] *OUG*, 4 (1873): 297 (4 November 1873).

[57] *OUG*, 4 (1873): 343 (23 December 1873) and *OUG*, 5 (1874–5): 24 (27 January 1874).

[58] *Royal Society Catalogue of Scientific Papers* lists only five short notes published by Baynes; some original work is implied in his revisions to Stewart's book on heat (Stewart 1895).

but as a frequent examiner and as the author of several textbooks. Early in his career he wrote two for publishers who were catering for the growing demand from school and college science teachers: *Lessons on Thermodynamics* (1878) for Oxford's Clarendon Press Series, and *The Book of Heat* (1878) for Stewart's Local Examination Series, the latter containing essentially the course he gave at Christ Church. Later his revisions and additions to the sixth edition (1895) of Balfour Stewart's *An Elementary Treatise on Heat* prolonged its life as the classic textbook in the same field.[59]

Baynes's devotion to lecturing was extraordinary. In contrast to Harcourt, who seldom took on more than his minimum lecturing commitments, Baynes at the height of his powers routinely gave two concurrent courses per term, lecturing on Monday, Wednesday, and Friday, with practical classes in the laboratory on Tuesday and Thursday (when Harcourt was in the lecture-room).[60] His practical classes were always in electricity and magnetism, and concentrated on techniques of measurement; in Michaelmas term 1882 they were linked to lectures on 'Electrical Testing'.[61] His staple lecture topics were those of the preliminary physics syllabus: elementary treatments, illustrated with experiments, of heat and light, electricity and magnetism, mechanics, and dynamics and hydrostatics. These lectures continued throughout his career, and were the only elementary physics lectures not absorbed or suppressed by Townsend when he centralized preliminary physics teaching in 1905. In addition, Baynes developed the subjects of heat and electricity, his main personal interests, in more advanced supplementary courses, especially in the middle part of his career (1880s and 1890s), a period characterized by some bold innovations of theme. Boldest of all, perhaps, was a one-off course in Lent term 1879 entitled 'Chemical Physics', the first formal lectures on this subject ever given in the university, and a precursor of the courses that Conroy and Veley started almost a decade later. Subsequent years saw regular courses on thermodynamics and on (kinetic) theory of gases, as well as other interesting titles, including (from 1885) 'Fourier's Theorem and its application to Conduction of Heat . . .' and (from 1893) 'Mathematical Physics'.[62] From the mid-1890s most of these non-experimental lectures, containing mathematical and theoretical treatments, were given in the Clarendon Laboratory, a privilege that Clifton hardly ever allowed to others.

Baynes's first pupil at final honours level, William Nelson Stocker (double first, 1873 and 1874), became a fellow of Brasenose in 1877, and professor of physics at

---

[59] *The Book of Heat* is cited as the textbook for his heat course in *OUG*, 9 (1878–9): 305 (18 April 1879); and see Stewart (1895: esp. Baynes's preface ix). He also translated Oskar E. Meyer's *Die kinetische Theorie der Gase* (Meyer 1877; 1899); see Fox, Chapter 2, 65.

[60] Baynes's lectures are advertised in the body of the *OUG* as well as listed in the lecture lists (*OUG* supplements), virtually every term from 1874 to 1921. [61] *OUG*, 13 (1882–3): 26 (13 October 1882).

[62] For chemical physics, *OUG*, 9 (1878–9): 184 (24 January 1879); for thermodynamics, *OUG*, 10 (1979–80): 221 (20 January 1880), etc.; for lectures on gases, *OUG*, 11 (1880–1): 358 (22 April 1881), etc.; for Fourier's theorem, *OUG*, 16 (1885–6): 30 (16 October 1885), etc.; for mathematical physics, *OUG*, 23 (1892–3): 397 (24 April 1893), etc.

Fig. 4.13. Robert Edward Baynes. As Dr Lee's reader in physics at Christ Church from 1873 until his death in 1921, Baynes played a prominent role in the teaching of the subject in Oxford. His textbook, *Lessons on Thermodynamics* (1878), in particular, reflected his understanding of the latest developments in that field. But he made little contribution to research.

the Royal Indian Engineering College, Cooper's Hill, in 1883. He had been a demonstrator at the Clarendon Laboratory from 1874 to 1883, and returned to this post for the period 1904–9 after the closure of Cooper's Hill. A much later pupil was Ernest Walter Brudenell Gill (double first, 1905 and 1907), the electrical physicist and radio enthusiast, who became a fellow of Merton and demonstrator in the Electrical Laboratory.[63] Such students, taking both mathematics and physics, were shared with the college's pure mathematics don, who at first was Charles Lutwidge Dodgson. Indeed, one of the most important Christ Church physicists of this period was a pupil of Dodgson and Baynes in mathematics only: James Walker (first, 1880) presumably studied further with Baynes and with Clifton without taking the physics degree. In 1883 he followed Stocker as a demonstrator (and

---

[63] On Stocker see Foster (1893: column 349); Fox, Chapter 2, 65–6; Gooday, Chapter 3, 92; Fox and Gooday, Epilogue, 310. On Gill see *WWW*; Gill (1934).

from 1884 the senior demonstrator) in the Clarendon Laboratory, and remained in this post until 1919. Walker also succeeded Dodgson as mathematical lecturer of Christ Church (1881–5) but evidently preferred his duties in the Clarendon.[64]

If Baynes (in common with all Oxford's physics dons) produced only a small number of honours graduates in physics, the greater part of his college work was intentionally devoted to the physical component of the mathematics degree and to the elementary physics that was compulsory for all natural science students. In this way many students came briefly under his tutelage. His influence is most interesting in respect of the chemists, for it was in the Christ Church Laboratory that the characteristically Oxonian fusion of chemistry and physics, as well as the research field of chemical kinetics that became its trademark, had their beginnings. But while 'Harcourt's laboratory' has become legendary, an important ingredient in its success has been forgotten. Harcourt's collaboration with the mathematician William Esson is well known.[65] But the presence of a full-time physics teacher working alongside Harcourt for most of his career surely helps to account for the origins of his famous dynasty of physical chemists; and Baynes's 1879 lecture course on 'Chemical Physics' suggests that this interplay was deliberate.

The exigencies of wartime brought an unanticipated new role for the unassuming Baynes. After Clifton retired in 1915, Baynes moved his teaching to the Clarendon Laboratory, working alongside his former pupil James Walker. The lecture lists, scrupulously set out in order of status, place Baynes above Walker[66] – Baynes was effectively the acting professor in the interregnum between Clifton and Lindemann. Ironically a Statute of 1915, transferring responsibility for funding Clifton's chair in experimental philosophy to Christ Church's Dr Lee fund – in effect merging the chair and readership – required the existing Dr Lee's readership in physics to be vacated before the arrangement could come into force (*Historical Register First Supplement*: 56–9, esp. 57). But having been appointed for life, with no pension, Baynes had no intention of retiring. In 1919, with the war crisis over and parallel arrangements for the anatomy and chemistry chairs going ahead, the University felt compelled to advertise the physics chair under the old arrangements, appointing Lindemann. Baynes meanwhile, at the age of 70, resumed his teaching in the Christ Church Laboratory, giving three open lectures per week during term, until his death in 1921.[67] Since then Baynes's long career of conscientious teaching has barely registered in the historical record – except where Howarth dismisses him as 'somewhat ineffectual'. His colleague H. B. Baker (Harcourt's successor) was more sympathetic, testifying in an obituary to his diligence as a teacher,

---

[64] *OU Calendar* for 1884, 1885; Foster (1893: column 426); for a glimpse of Baynes and Dodgson's acquaintance see Cohen (1995: 426).

[65] Hartley (1971b: 229); Shorter (1980: 411, 413); King (1984: 21–2); Laidler (1988: 243–5); on Esson see Fox, Gooday, and Simcock, Chapter 1, 10.   [66] For example *OUG*, 48 (1917–18): 229 (17 January 1918).

[67] *OUG*, 51 (1920–1): 48, 322, and 540 (8 October 1920, 14 January and 21 April 1921).

and to the reputation he had, in his prime, as the best physics tutor in Oxford (Howarth 1987: 341; Baker 1921–2: 48; cf. Fox 1997: 686).

PHYSICS LABORATORIES OF THE 1880s

Undoubtedly influenced by the early glamour and promise of the Clarendon Laboratory, a good deal of interest in physics is evident in Oxford's academic community in the 1870s and 1880s. College initiatives, bolstered by the endorsement given to college involvement in science by the Devonshire Commission, included Trinity College's funding from 1873 onwards of several Millard lecturerships and demonstratorships in physics, whose instruction was open to all students. Equally, of course, activities in colleges could be isolated and quirky, and the historical record of them limited, contingent as they often were on the presence of enthusiastic or eccentric individuals – from R. H. M. Bosanquet's course on musical physics at St John's College (1873–4) to E. G. Spencer-Churchill's demonstrations of X-ray apparatus in his college room at Magdalen (1897–8) (Gunther 1924: 499; 1939). The Balliol–Trinity project of 1879 was the most visible of these activities. Three further laboratories came into existence in the early 1880s, and were in fact the only laboratories in collegiate Oxford ever to be devoted exclusively to physics. A remarkable laboratory for experimental acoustics at St John's College was set up for his private research by Bosanquet in 1880. It gained a more public role after it was taken over as an annexe of the Millard Laboratory in neighbouring Trinity College. The latter, established in 1885, was a formal college laboratory for experimental physics and mechanics, on the one hand extending the provision for physics while on the other it was meant as a stepping-stone to the new subject of engineering. Consideration of the Millard Laboratory and its role will be deferred to the next chapter. Finally, a small laboratory was built at Keble College about 1882 for Sir John Conroy's research, and later used for teaching practical preliminary physics.

R. H. M Bosanquet studied at Balliol College under Henry Smith, obtaining firsts in the new natural science school (1862) and then in mathematics (1863). His precocious talent was already in evidence, for very few students were clever enough to take the two degrees in that order. After qualifying as a barrister, he returned to Oxford in 1868 to become a lecturer at Christ Church and, in 1870, fellow and lecturer at St John's College, a position he retained for the rest of his life.[68] Bosanquet represents some of the scientific vitality of his tutor, or perhaps of that particular moment in Oxford science, for coming up in 1859 he was part of the first cohort of students to

---

[68] Hilliard (1914): 33–4; *St John's College Biographical Register 1875–1919* (Oxford, 1981): 6; Foster (1885: 45); Foster (1893: column 477); Brown and Stratton (1897).

use the facilities of the University Museum and associated science departments. He was influenced in an unexpected way there by the astronomy professor W. F. Donkin, who planned a *magnum opus* on theoretical and practical acoustics but only completed the first part (of three planned), which was in the press at the time of his death in 1869 (Donkin 1870). It was from Donkin that Bosanquet acquired his aspirations for a perfected science of musical acoustics. But while this was the most ambitious theme to which he applied his considerable intellect, his inclination seems to have been to pursue briefly (and publish quickly) a diversity of research quests and practical challenges in the branches of physical science that interested him – mainly acoustics, electricity and magnetism, and astronomy – sustaining each only until captivated by the next.

Between 1872 and 1895 Bosanquet was a prolific producer of journal articles and of papers read before scientific and musical societies, many of them published in the *Philosophical Magazine*. The Royal Society bibliography lists forty-six titles, several of these being in multiple parts – his 'Notes on the theory of sound' (Bosanquet 1877b, c), for example, was a series of six, and the title 'On electromagnets' (Bosanquet 1884; 1885a, b, c; 1886a, b; 1887) covers seven notes of research in progress on that subject. A series of three historical papers on Babylonian astronomy was co-written with A. H. Sayce, later Oxford's professor of Assyriology (Bosanquet and Sayce 1878–9; 1879–80a, b). For all his mathematical brilliance, Bosanquet's expertise was not merely theoretical: he was an enthusiastic experimentalist and a skilled mechanician, and his projects frequently revolved around an instrument or component which he had not only invented but had made with his own hands. At the South Kensington Loan Exhibition of Scientific Instruments in 1876 he exhibited both his astronomical polariscope, and an 'enharmonic harmonium', described in his book *An Elementary Treatise on Musical Intervals and Temperament . . .* and in the paper 'On instruments of just intonation' which he gave at the conferences held in conjunction with the Loan Exhibition (Anon. 1877: 191, 223; Bosanquet 1876a; 1876b; 1877a; see also Bosanquet 1873). His purpose was to explore musical intonation with true or 'just' temperament, the harmonically accurate keyboard that resulted having many more keys than any practical instrument – the version exhibited in 1876 had fifty-three keys per octave; a later version had eighty-four. As well as a research tool, he intended it to be of didactic use, contending that music 'should be taught experimentally' and not from 'mere book knowledge' (Bosanquet 1877a: 57). He also felt that composers should try writing for such 'just systems'. These were not Bosanquet's only radical suggestions for the study or performance of music: he also proposed that scores be printed in different colours for each instrument, representing the psychological association between colour and timbre.[69]

---

[69] Bosanquet's ideas are mentioned several times in Scholes (1938 and subsequent editions, e.g. 1945: 110, 182 [colours], 495, 924).

That music at this period had serious standing as a subject of study, and close links to mathematics, is indicated by how frequently those taking the B.Mus. degree were mathematics graduates, and by the high praise expressed by Bartholomew Price in reviewing Donkin's posthumous work on acoustics (1870).[70] The usual crop of music teachers in Oxford was joined in the 1870s by several who offered theory rather than performance. At the same time, outside Oxford, acoustics was attaining a respectable place as a branch of physics through the work of Helmholtz, Rayleigh, Tyndall, and the instrument manufacturer Koenig.[71] Although Bosanquet taught physics and mathematics within St John's College, the only open lectures he gave were a series in 1873–4 on 'the Elements of the Theory of Harmonious Music, as founded on the Physical Theory of Sound . . . treated with the view of making it intelligible to those who are not acquainted with Mathematics'.[72] They were delivered on Thursday evenings at St John's, for an expensive fee of three guineas for the course of twelve. Partly aimed at B.Mus. students, scheduling the lectures in the evening implies that a more general audience of cultured academics may also have been anticipated.

Growing from all this, Bosanquet's grand project was to revolutionize the subject of experimental acoustics, establishing an ideal laboratory in which to perform an intensive programme of research. In some measure the design and fitting out of the laboratory was to be an experiment in itself. This idea was published in the *Philosophical Magazine* in October 1879 in a paper entitled 'On the present state of experimental acoustics, with suggestions for the arrangement of an acoustic laboratory, and a sketch of research' (Bosanquet 1879). It was immediately put before the governing body of St John's College, which assigned a room and £50 for the proposed laboratory; this was supplemented by a Royal Society grant of £152. That an Oxford college at this period should host and sponsor such a venture, exclusively concerned with advanced research, runs contrary to historiographical expectations. Stranger still, St John's was the conservative college *par excellence*, and its President, James Bellamy, a famous opponent of university reform as well as the actual leader of the Conservative political party in Oxford (*DNB*; Hibbert 1988: 406). So our assumption that the patronage of science was an expression of liberal and reforming tendencies is called into question.

The room assigned was a disused free-standing lecture room at the southern edge of the college adjoining Trinity College.[73] During 1879–80 enormous trouble

---

[70] See advertisement for Donkin's book in *OUG*, 1 (1870): 16 (3 May 1870).

[71] Scholes (1947: vol. 2, 764–5) comments on this activity and mentions Bosanquet and Donkin alongside Rayleigh *et al.*; for a charming introduction to the acoustic experiments and apparatus of the time see Wormell (1883).

[72] *OUG*, 4 (1873): 258 and 281 (10 and 21 October 1873); *OUG*, 5 (1874–5): 26 and 35 (27 January and 3 February 1874).

[73] It is not the room initially stipulated in college records; St John's College archives, College Register, vol.10: 187 (11 October 1879); Millard Laboratory brochures (see Simcock, Chapter 5, 187n, 189); Anon. (1911*a*); information from Dr Malcolm Vale, St John's College archives, and from Professor Keith Laidler.

was taken to convert and equip this room, laying gas and water supplies and even a telegraphic line to a master clock in Bosanquet's college room, installing a steam engine, and fitting overhead drive-shafts across the laboratory to take power to various tools and experiments. The main tool was a lathe, with additional features of his own design; with it he intended to make the unique instruments needed for his programme of experiments. The key experimental apparatus was a large pneumatic bellows, adapted from organ bellows but more powerful, and blown directly by the steam engine; it was capable of creating 'a wind of the most perfect evenness'. The progress of this installation and specifications of the equipment were recorded in extraordinary detail in two *Philosophical Magazine* articles (Bosanquet 1880, 1881b), which give a clear impression of Bosanquet's predilection for the technological infrastructure and workshop aspects of the enterprise. They also reveal how the project's complexity and Bosanquet's perfectionism precipitated endless practical difficulties, from the rapid deterioration of the bellows because of heat to the discovery that the engine and drive-shafts created sufficient noise to disturb the core acoustic experiments. In 1881 an additional room or outbuilding was built to overcome this problem by isolating the bellows from the steam engine; later the steam engine seems to have been replaced with a gas engine.[74]

Some acoustic experiments were commenced while these arrangements were still in train. But other things were done too. Bosanquet used the lathe to make the circles of a precessional globe, which he exhibited to the Royal Astronomical Society in 1880, as well as to make the electromagnets for his slave clock dial (Bosanquet 1880: 222, 225). It is clear from the ensuing publications that his customary pattern of research was not disciplined by all this effort, and that the ambitious programme of experiments by which he hoped to reform acoustical physics was not performed. A series of experiments on the beats of consonances was published (1881–2), after which, apart from a brief note on electromagnetic tuning forks in 1883, the subject of acoustics makes no further appearance in his publications (Bosanquet 1881a; 1883; and see *Royal Society Catalogue of Scientific Papers*). His enthusiasm had been hijacked by the construction of electromagnets, and magnetism in general, and the laboratory was now the scene of experiments on these topics, with other digressions. His last two papers, in 1895, returned to his earlier interest in astronomy. His appointment as professor of acoustics at the Royal Academy of Music, in London, in 1881, owed something to his laboratory project, but had little effect on its outcome. In 1888 he transferred the laboratory to Jervis-Smith to form an extension of the Millard Laboratory, in Trinity College, giving Jervis-Smith all his expensive plant and equipment.[75] It seems probable that some of the instruments were made

---

[74] St John's College archives, College Register, vol.10: 203 (24 June 1881); Bosanquet (1881: 178–9); Woods (1887) (gas engine).

[75] For Jervis-Smith and the Millard Laboratory see Simcock, Chapter 5.

available to Trinity's physics don D. H. Nagel, who began his acoustics course later the same year. Bosanquet was elected an FRS in 1890, but from about this time he lived most of the year in retirement on Tenerife.

Little is known of the other physics laboratory that came into existence in the early 1880s, at Keble College. In 1880 the Warden of Keble, Edward Talbot, invited his Christ Church contemporary Sir John Conroy to become a lecturer at Keble, following the departure of the chemist W. F. Donkin (junior) and the appointment of the young biologist E. B. Poulton to succeed him. The wealthy Conroy had been carrying out research in his private laboratory (in London), and also at Christ Church. Soon after taking up the Keble post in 1881, he built and equipped a small laboratory at his own expense so that he could continue his research there.[76] In 1885 he was promoted to tutor (equivalent to fellow), and in 1887 he became lecturer in physics at Balliol and Trinity Colleges, moving his research activities there, though he did not relinquish his Keble post until 1891.[77] The fact that practical work became compulsory in the Preliminary examination in that year is not coincidental. Conroy donated the laboratory to Keble, and for the following thirteen years it was used for practical classes in Preliminary physics. Keble itself had few science students (and none, at this period, who took finals in physics); but in the event the laboratory served a much wider constituency, for the man who taught in it was E. S. Craig, a figure who has escaped the attention of historians of Oxford science.

A mathematics graduate from University College, Edwin Stewart Craig returned to Oxford in 1890 after a short period as a schoolteacher and began advertising his services as a private tutor.[78] A highly entrepreneurial character, he quickly built up a practice as Oxford's most active and successful freelance tutor in mathematical, military, and related subjects. Craig held no post at Keble College but was affiliated by some informal arrangement in connection with his post (from 1891 to 1908) as lecturer in mathematics and natural science at Marcon's Hall, one of the private teaching establishments that the University licensed at this period, which was situated in a house next-door to Keble (the present 10 Parks Road). In Michaelmas term 1892 the *Oxford University Gazette* carried an addition to his usual advertisement, announcing that 'Mr. E. S. Craig . . . will be glad to receive pupils for the Preliminary Examination in Physics. Classes for practical work (according to the new regulations) will be conducted in the Laboratory at Keble College'. Similar notices appeared for several years thereafter.[79] The laboratory was closed in 1904 in consequence of the centralization of physics teaching

---

[76] Vernon Harcourt (1905: 246); Hartley (1971b: 227); Smith (1979: 42); for Conroy see above and Vernon Harcourt (1905).
[77] *OU Calendar* for 1891; *The Keble College Centenary Register 1870–1970* (Oxford, 1970) says that he retained it until 1900. [78] *OUG*, 21 (1890–1): 51 (14 October 1890).
[79] *OUG*, 23 (1892–3): 30 and 231 (14 October 1892 and 20 January 1893), etc.

Preliminary, and its apparatus was sold to the Daubeny Laboratory. Craig transferred his physics teaching to Townsend's laboratory, taking charge from the beginning of 1905 of the lectures and practical instruction for the Preliminary examination.[80] He continued in this role until 1913, even after becoming assistant registrar of the University in 1907 and lecturer in physics and engineering science at University College in 1909. Later Craig was a fellow of Magdalen, and finally registrar of the University from 1924 to 1930.[81]

LATER COLLEGE LABORATORIES

Two further college laboratories were established at the beginning of the twentieth century, at The Queen's College in 1900 and at Jesus College in 1907. The former was the only such laboratory at which physics was not taught alongside chemistry. Remembered, indeed, for its organic chemistry research, it acquired this specialism under Frederick Daniel Chattaway, another of Harcourt's pupils, who ran it from 1919 until it closed in 1934.[82] But the Queen's College Laboratory originated with two physical chemists, George Bernard Cronshaw, chaplain of Queen's from 1898, and his tutor V. H. Veley, whom he succeeded as college lecturer in chemistry in October 1900, when the new laboratory was opened. It served as a general teaching laboratory for students from Queen's and certain other colleges, and A. F. Walden, lecturer in chemistry at New College, worked there as demonstrator, bringing with him the largest contingent of students. Under the arrangement of 1904 the Queen's laboratory took responsibility for general Preliminary and 'post Preliminary' practical teaching for all Oxford chemistry students.[83] Both Cronshaw and Walden were devoted to teaching rather than research, and had wide interests. Walden had earlier taught physics with J. J. Manley; and Cronshaw's talents also extended to physics: for example, the talk he gave to the Ashmolean Natural History Society in 1905 was on 'surface waves' (Bellamy 1908: 71, 196; for Walden see note 19). In 1918 it was, significantly, Cronshaw and D. H. Nagel, rather than any mainstream physicists, who appealed to the University to unfreeze the vacant physics chair as a matter of urgency (Morrell 1997: 383).

---

[80] Gunther (1924: 181); for the 1904 arrangements see below.
[81] Newspaper cutting [1918] in MHS, MS Gunther Archive, Album of Printed Matter vol.II; Foster (1893: columns 35–6); *WWW*; see Lelong, Chapter 6, 215.
[82] Hartley (1971b: 230–1); Humphries (1970: 100–2); Laidler (1988: 236–8); Morrell (1997: 319); most accounts assume incorrectly that Chattaway ran the laboratory from an earlier date.
[83] MHS, MS Museum 49; on Cronshaw see *St. Edmund Hall Magazine*, December 1928: esp. 3, 35–7; obituary in *The Times*, 21 December 1928.

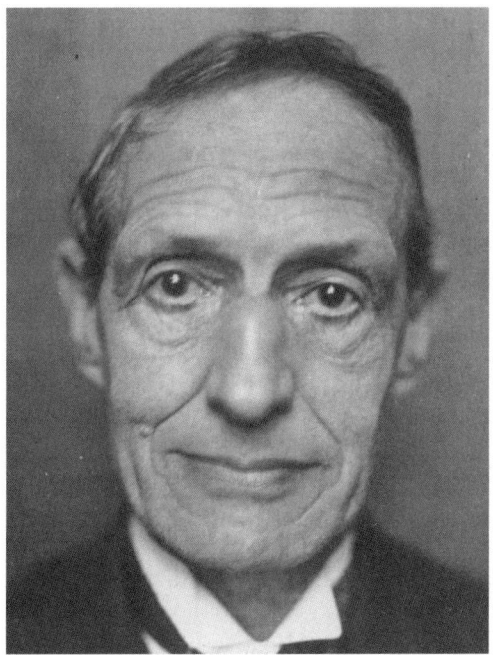

Fig. 4.14. David Leonard Chapman, c.1927, one of a series of photographs taken of staff in the Chemistry School. Chapman was appointed as a fellow to take charge of the new Jesus College Laboratory in 1907. He worked there until 1944, three years before the laboratory finally closed.

In 1907 Jesus College made the last and the most extravagant gesture in favour of college laboratories, erecting a substantial three-storey building, equipping it generously, and electing David Leonard Chapman to a fellowship to take charge of it (Long 1989; 1995–6; Hartley 1971b: 231–2; Laidler 1988: 238–9). The feeling at Jesus was still that colleges should provide virtually comprehensive tuition and facilities for their science students. An unbroken fifty-year succession of capable science lecturers had ensured its place among the most successful scientific colleges. In the 1890s the lectureship was held by D. H. Nagel, through whom Jesus students had access to the Balliol–Trinity Laboratories, and then by Arthur Harry Church, a botanist with great sympathy for the physical and mathematical sciences. Church, in fact, played a direct role in bringing the new laboratory about and even had a hand in designing the building, being rewarded with a research fellowship when Chapman succeeded him.[84]

The Jesus laboratory was intended 'for the teaching of chemistry and physics', and was not expected to serve any chemical specialism. N. V. Sidgwick, who moved from Magdalen, lectured on 'The Organic Chemistry of Nitrogen' and 'Tautomerism'; and Chapman himself taught a course of chemistry for Oxford's Diploma in Public Health.[85] But it was physical chemistry for which it

---

[84] Gunther (1937: 252–3); Church obituary by Tansley (1936–8); for Church's highly mathematical treatment of botany see Church and Hayes (1901–4).

[85] Anon. (1907: 453); OUG, 38 (1907–8): 303 (20 January 1908); OUG, 39 (1908–9): 565 (23 April 1909), etc.

became noted, a bias inherent in the appointment of Chapman, one of the most mathematically talented physical chemists of his time (Bowen 1958; Morrell 1997: 61–2, 327). Chapman was a pupil of Baynes and Harcourt at Christ Church, and was unusual in having taken finals in both chemistry (first, 1893) and physics (second, 1894). He then worked under H. B. Dixon at Manchester, by whom (like others) he was imbued with a lifelong passion for studying the kinetics of explosions. His first research paper (in 1899) was a theoretical and mathematical consideration of Dixon's findings on explosion of gases, in which he devised a theory of detonation, still known as the Chapman–Jouguet theory. His 1913 paper on a more purely physical matter, electrocapillarity, was also of recognized theoretical importance (Chapman 1899; 1913).[86] His most extensive work subsequently concerned the interaction of hydrogen and chlorine, especially their photochemical combination – classic Harcourtite physical chemistry, combining minute experimentation with mathematical analysis.

The laboratory contained a room devoted to physics, Chapman's job description requiring him to teach physics as well as chemistry (Anon. 1907; Humphries 1970: 103) – although the fact that he did so, and with conspicuous success, has gone unremarked. Amongst its routinely high number of science graduates, Jesus College now achieved a statistically astonishing record of honours finalists in physics – thirteen in the eight years 1907–14 alone.[87] This is compelling evidence that Chapman and the laboratory boosted student interest and achievement in physics at least as much as they did in chemistry. So successful was Jesus's venture that from 1914 an additional lecturer was engaged, the first being Malcolm Percival Applebey, a Trinity graduate and fellow of St John's. Another collaborator was Chapman's wife Muriel Catherine Canning Holmes, whom he married while she was doing graduate research under his supervision; she gained her B.Sc. in 1923. A characteristic of the Jesus Laboratory was that undergraduates participated in research projects, and frequently shared publication credits with Chapman.[88]

From the point of view of centralizing forces, the Jesus College Laboratory, like that at Balliol–Trinity, was superfluous once the University's new Physical Chemistry Laboratory was completed in 1941. But it was resilient, and became the last college laboratory to close. At first it was sustained by war research for the secret Tube Alloys project, the code-name for the British atomic bomb: Chapman and a small research team worked there on the technique of isotope separation by gaseous diffusion. Chapman retired in 1944, but his assistant Leonard Ary Woodward continued to use the Jesus laboratory until 1947, being compensated for

---

[86] For further discussion see Laidler (1988: 238–9, 250).
[87] Statistic counted from honours lists in *Historical Register First Supplement*.
[88] See bibliography in Bowen (1958: 43–4); on Applebey see obituary by Steel (1957); for Mrs Chapman, information from Professor Keith Laidler.

its closure by appointment to the University posts of demonstrator and lecturer in chemistry (based at the Inorganic Chemistry Department). He remained fellow and chemistry tutor at Jesus until 1970.[89] Woodward, like Chapman, was a very physical physical chemist. As a student he had taken mathematics moderations before turning to chemistry (under Sidgwick; first, 1926), and then went on to work at the intersection of physics and chemistry in his doctoral studies with Petrus Debye at Leipzig. When he gave a talk on the Raman effect to the University's Physical & Radio Society in 1941, Woodward was introduced light-heartedly as 'a physicist who had unfortunately degenerated into a chemist'. He responded that 'It was not easy nowadays . . . to distinguish clearly between physics and chemistry'.[90]

THE BAKER AND TOWNSEND SCHEMES OF 1904–5

Between the opening of the new laboratories at Queen's and Jesus Colleges, there were moves to curtail the autonomy of college teaching in physics and chemistry. Harcourt's retirement in 1902 opened the way for a more assertive role for the Dr Lee's reader in chemistry, because the university statutes of 1882 had made it a university readership as well as a college post, giving it (in effect) professorial status. Harcourt, choosing not to avail himself of this status, had instead been influential by peopling Oxford chemistry with his own carefully nurtured dynasty of experimentalists. By contrast, his chemical grandson H. B. Baker (Dixon's pupil) returned to Oxford in 1903 after half a career as a schoolmaster, determined to bring discipline to his disorderly siblings.[91] He immediately used his authority to co-ordinate the practical chemistry instruction provided by the various laboratories, giving each its speciality which it would teach to all Oxford chemistry students, Preliminary and advanced. One of the ironies of Baker's scheme was that it treated the University's central laboratory on equal terms with each college laboratory.

Although the arrangement was only concerned with practical work, the effect was that lecture topics became localized too, as did the growing culture of research. It was said that one of Baker's aims was to encourage the development of research schools through the concentration of practical expertise. The scheme came into force at the beginning of 1904 (Hartley 1971b: 229–30).[92] Inorganic

[89] MHS, MS Bowen B.28; Morrell (1997: 327, 335).
[90] MHS, Physical & Radio Society Minute Book 1940–1, Trinity Term 1941.
[91] For Baker see Thorpe (1932–5); he was in fact the only Dr Lee's reader (in any subject) to be appointed under the 1882 terms.
[92] For the first appearance of the Baker scheme in lecture lists see OUG, 34 (1903–4): 500 (25 April 1904), and MHS, MS Museum 49 and MS Museum 83 for confirmation that it started the previous term.

chemistry was assigned to Baker's own laboratory at Christ Church, where he taught it himself along with successive graduate demonstrators, the first of whom was R. T. Lattey. Physical chemistry was allocated to the Balliol–Trinity Laboratories, where it was provided by D. H. Nagel and Harold Hartley, joined over the years by various assistants. Organic chemistry was left to the central University Chemistry Department, where the existing team of demonstrators were all organic chemists (the physical chemist V. H. Veley having left in 1903). General Preliminary practical work became the responsibility of G. B. Cronshaw and A. F. Walden at the Queen's College Laboratory. And quantitative analysis was to be the contribution of the Daubeny Laboratory at Magdalen College, where J. J. Manley taught it single-handedly.

At about this time J. S. E. Townsend became acquainted with Baker – among other things, their areas of research had much in common – and in similar spirit devised a scheme to rationalize the teaching of his subject. In 1903 he offered to take responsibility not just for his statutory area (electricity and magnetism) but for all physics teaching at Preliminary level, of which Clifton so often complained, and for which additional provision was scattered amongst the colleges. The college laboratories at Christ Church, Balliol–Trinity, and Keble all provided Preliminary physics teaching that was open to students of other colleges, as did the Millard Laboratory at Trinity. Apart from Baynes's lectures at Christ Church, they ceased to cater for the subject at the end of 1904: the Balliol–Trinity Laboratories were formally requested to do so by the Vice-Chancellor (Humphries 1970: 83). A clever aspect of Townsend's scheme was his recruitment of the king of the freelance physics coaches, E. S. Craig, from the nearby Keble College Laboratory – a professional crammer rather than an advocate of experimental finesse. Craig was to provide a comprehensive course (both lectures and practical) for Preliminary physics, assisted by R. T. Lattey and H. E. Hurst, Lattey being transferred from Baker's chemistry laboratory for this purpose. The course is embodied in *A Text-Book of Physics* (1910), which was 'largely written' by Craig but published under Hurst and Lattey's names.[93] The new centralized provision came into force at the beginning of 1905, in Townsend's original Electrical Laboratory in the University Museum.[94]

Its effect was to eliminate most non-tutorial physics teaching from Oxford colleges. Through Craig it even embraced private tuition. The map of Oxford physics thus changes dramatically in 1904–5, and much of the devolved world of physics activities that has been explored above disappears. The only surviving college-based physics course ended in 1921, with the death of Baynes. The teaching associated with college posts and fellowships was now developing closer links with

---

[93] Hurst and Lattey (1910 and 1912, esp. preface); for Hurst see below and Lelong, Chapter 6, 217–19.

[94] For the first appearance of the Townsend scheme in the lecture lists, *OUG*, 35 (1904–5): 281 (23 January 1905). For this scheme in Townsend's annual report, *OUG*, 35 (1904–5): 573 (16 May 1905); *OU Calendar* for 1906.

the central departments; one can see this happening in the careers of men like I. O. Griffith and C. H. Collie. Following immediately upon Baker's orchestration of practical chemistry, Townsend's centralization of Preliminary physics had a profound effect on Oxford chemistry too. Removing physics from the college laboratories and physics lecturing duties from their staff had particular impact at the Balliol–Trinity Laboratories, allowing the long cohabitation of physics and chemistry to resolve itself into the healthy new discipline of physical chemistry.

THE HERTFORD COLLEGE PHYSICISTS

Around the turn of the twentieth century, a statistical curiosity emerges from examination of Oxford's graduation lists. Over the period 1890 to 1909, when two or three would have been a respectable figure, twelve honours students in physics graduated from Hertford College, one of Oxford's smaller colleges, and one not especially noted for science. Only two colleges produced greater numbers in those years: New College, one of the two largest, and Jesus College, which had deliberately courted scientific success over several decades. Even the colleges with laboratories were less productive. During the same years Hertford College was also the single largest source of demonstrators in the Clarendon and Electrical Laboratories, providing six in comparison with five from Christ Church and three from Balliol.[95] The particular advantages that allowed physics to flourish to this extent seem to come down to the strong mathematical grounding available at Hertford, and the services as physics tutor of Charles Edward Haselfoot. A mathematics scholar at New College, where he obtained a first in 1886, he went on to take a first in physics in 1888, becoming lecturer in mathematics and physics at Wadham College in the following term and a fellow of Hertford at the end of the year. In addition to tutorial duties at the two colleges, he gave regular lectures which were open to all students as part of the combined college lectures in mathematics. From 1897 Haselfoot was tutor in honours physics to the Non-Collegiate students. He also took private pupils, and helped other colleges with supplementary physics teaching on a more occasional basis, for instance at Balliol after Conroy died.[96] In his heyday, Haselfoot was more than just a busy and effective teacher, he was evidently a vigorous promoter of physics and the physico-mathematical sciences.

---

[95] Student statistics counted from honours lists in *Historical Register* and *Historical Register First Supplement*; demonstrator statistics counted from *OU Calendar*, 1891–1910.

[96] Haselfoot obituary, *Oxford Magazine*, 55 (1936–7): 183; Foster (1893: column 598); for Haselfoot's private tuition, *OUG*, 19 (1888–9): 23 (12 October 1888); for his lectures, *OUG*, 19 (1888–9): 242 (22 January 1889); *OUG*, 23 (1892–3): 396 (24 April 1893); *OUG*, 31 (1900–01): 42 (15 October 1900); *OUG* 43 (1912–13): 55 (10 October 1912); *OU Calendar* for 1898 (Non-Collegiate); Hilliard (1914: 419).

Crucial to Haselfoot's success was the thoroughness of Hertford College's provision for its mathematical students. John Edward Campbell had become a fellow of Hertford the year before Haselfoot, and their careers seem to have been destined to run in parallel. As undergraduates, both had won the Herschel Astronomical Prize, given to runners-up in the University's prestigious mathematical scholarships. Both held tutorial fellowships with lecturing duties for thirty-seven years, remaining active in college life until the 1920s.[97] Furthermore, both taught mathematics; but the division of labour between them was that Campbell was the pure mathematician while Haselfoot was the applied mathematician, teaching physics both for the physical parts of the mathematics syllabus and for the natural science degree. The commitment of Hertford College to mathematics was underlined further by its engagement of the well-known pure mathematics teacher William Esson as part-time lecturer. The college's intention was presumably to strengthen its mathematical reputation; but the combined effect of the arrangement was to enable Haselfoot to promote physics within the mathematics course, encouraging mathematics students to proceed to the physics degree too.

Several of the students taught by Haselfoot and colleagues went on to distinguished and successful careers. John Leigh Smeathman Hatton (first in mathematics, 1889; second in physics, 1890) was briefly a demonstrator in the Clarendon Laboratory in 1891, although his main interest was geometry. He became Principal of East London College in 1896, and spent the rest of his career there, holding various high-ranking positions in London University, including Dean of the Faculty of Science (1922–6) (*WWW*). More explicitly devoted to physics was Paul Jerome Kirkby (double first, 1891 and 1893), who became a fellow of New College in 1895 and later was the first resident scientist to join Townsend during his early years at Oxford. He served as Townsend's first demonstrator (1901–10), and collaborated in his pioneering research into electrical conductivity in ionized gases. Kirkby also taught at Exeter College, deputized for Jervis-Smith at the Millard Laboratory, and did research with the physical chemist H. B. Baker.[98] The astronomer Henry Crozier Keating Plummer (first in mathematics, 1897; second in physics, 1898) was second assistant at the University Observatory, Oxford (1901–12), and from 1912 to 1921 the last Royal Astronomer for Ireland; after Irish independence he became professor of mathematics at Woolwich (*DNB*).

For two decades from 1891, Haselfoot was the most prolific supplier of junior demonstrators in the Clarendon Laboratory. Hatton was followed in 1892 by Haselfoot's Wadham pupil Stuart Arthur Frank White (double first, 1890 and 1892), who went on to become demonstrator of physics and later professor of

---

[97] Foster (1893: column 598); obituaries of Campbell in *Oxford Magazine*, 43 (1924–5): 8, and *Proceedings of the Royal Society of London*, series A, 107 (1925): ix–xii; Morrell (1997: 56, 308).

[98] *OU Calendar* for 1900 (Exeter); *OUG*, 34 (1903–4): 284 (25 January 1904); and see Lelong, Chapter 6, 213–14, 216, and 219–22.

mathematics at King's College, London (*WWW*). Plummer was at the Clarendon Laboratory for part of 1901, before moving to the Observatory. In the 1900s there were two further Clarendon demonstrators drawn from Hertford College, while Richard Littlehailes (from Balliol) was also partly taught by Haselfoot. After Kirkby's early association with it, a special bond formed between Haselfoot's circle and Townsend's new department. Harold Edwin Hurst, who obtained a first in physics at Hertford in 1903, immediately joined Kirkby as a demonstrator in the Electrical Laboratory and at the same time became Townsend's first Oxford research student. In 1905 he was part of the team of demonstrators who provided the new centralized Preliminary physics course.[99] Naturally enough, Haselfoot himself became acquainted with Townsend and interested in his research work: his collection of authors' offprints presented to him by Townsend and collaborators was proudly kept and is still extant. Haselfoot even briefly collaborated with Kirkby in a related piece of research, published under their joint names in 1904.[100]

With this single exception, Haselfoot did not make original contributions to research. Of the two intersecting Oxford traditions, he belonged to the Baynes school of devoted and undramatic teaching, rather than the Conroy model of fastidious experimentalism. Indeed, he has been held up as an example by Morrell of those dons who retreated into the 'protected epicurean enclaves' of college society and preferred to be 'good college men' than to contribute to the wider university or to the world of research (Morrell 1997: 55–6). I have tried to show Haselfoot in a different light. He does of course illustrate the fact that, even in what was becoming a laboratory-intensive science, Oxford remained a college-based system of education, and for better or worse (be he dull or inspiring) a student's principal teacher and advisor was his college tutor. It seems reasonable to conclude that Haselfoot belonged to the inspiring category; and *pace* Morrell, his services were available, and his influence felt, widely beyond his own college. His most notable achievement as a physics teacher was not simply to obtain respectable results at Hertford College, but to turn it for a while into the leading Oxford college for producing the next generation of physicists.

CHRIST CHURCH AND THE NEW PHYSICS

In Baynes's later years, as in Haselfoot's, a new physics emerged which was characterized by a mathematical and conceptual sophistication that made the 'classical' physics taught in universities seem very elementary. Oxford is often assumed to

---

[99] See above and Hurst and Lattey (1910); Hurst (1969–70); on Hurst see also *WWW*; for a further link between Hertford College and the Electrical Laboratory, through the Drapers' Company, see Lelong, Chapter 6, 223–5.

[100] MHS, binder of offprints by Townsend and others; Haselfoot and Kirkby (1904); Townsend's annual report for 1904 in *OUG*, 35 (1904–5): 573 (16 May 1905); and see Lelong, Chapter 6, 219.

have been out of touch with the new ideas of quantum theory and relativity; but in truth, the same challenge confronted most of academia as it emerged from the hiatus of the First World War. With Moseley dead, who could shepherd Oxford into these new pastures? The simultaneous appointments of Soddy and Lindemann in 1919 seemed to provide an answer; and Lindemann's career as professor was to consist of a protracted effort, in a scientific culture so resolutely experimental (and now so dominated by chemistry), not to be left too far behind during this fundamental revolution in physical science.[101] By the 1930s, Schrödinger thought the Oxford physicists were 'quite mad' when they did not 'drop their trivial researches & enter at once in to the new & promising fields he had pioneered'.[102] Yet paradoxically his very presence in Oxford was a conscious homage to the new learning, as too were the honorary degrees that Oxford bestowed upon physicists of the new order, notably Einstein and (as early as 1926) Bohr and Eddington, the leading English exponent of relativity (*Historical Register Supplement 1901–1930*; *Historical Register Supplement 1931–1950*).

Because of the new funding arrangements for their chairs, both Soddy and Lindemann were attached to Christ Church. Soddy's Glasgow pupil Alexander Smith Russell, a radio-chemist, quickly followed him to Oxford when the Christ Church chemistry readership was revived (1919), an appointment (in preference to the local candidate, Sidgwick) which gave this college a strong association with modern chemical physics.[103] Curiously enough, Christ Church already housed one of the most influential early champions of the new physics, although he is now largely forgotten. Henry Herman Leopold Adolph Bröse, generally known as Henry L. Brose, came to Christ Church as a graduate Rhodes Scholar from Australia in 1913, spent the entire war as a civilian prisoner in Germany, returning to Oxford in 1919, and then became one of Lindemann's early researchers in the Clarendon Laboratory. He was also one of the first to enrol for the new D.Phil., though under Townsend's supervision. It was a traditional experimental doctorate, entitled 'On the Motion of Electrons in Oxygen', which he completed in 1925.[104]

In December 1919 Blackwells published Brose's little book *The Theory of Relativity. An Introductory Sketch based on Einstein's Original Writings*, which in the space of only thirty-two pages provided one of the first clear and comprehensive expositions of relativity (the special, general, and mechanical theories) for an

---

[101] See Morrell, Chapter 7; Hughes, Chapter 8; and cf. Birkenhead (1961: 34–6) (Collie's account of the revolution in physics).

[102] Collie's words recollecting his conversations with Schrödinger: personal communication (Collie to the author, 1989).

[103] Obituary, *The Times*, 17 March 1972; Humphries (1970: 98–9); Morrell (1997: esp. 321, 327, 353); MHS, Collie MSS.

[104] Subsequently published as Brose (1925); see also Bannon and Brose (1928); the thesis is not extant; information from Mr Simon Bailey, Oxford University Archives.

English-speaking audience. It went through several reprints and revisions in the space of a few months.[105] Its inspiration may have been Lindemann's none-too-successful attempts earlier in the year to defend relativity on Oxford's debating circuit (Morrell 1997: 401). There followed during the 1920s and 1930s Brose's translations from the German of monographs and textbooks on all three areas of the new physics: relativity, quantum theory, and atomic structure, notably the early books on Einstein's theory by Erwin Freundlich and by Moritz Schlick (both 1920, with introductions by H. H. Turner, Oxford's Savilian professor of astronomy, and by Lindemann, respectively); Fritz Reiche's *The Quantum Theory* (1922), the classic textbook for the pre-Heisenberg/Schrödinger theory; and works by Born, Planck, Sommerfeld, and Weyl.[106] The translation of these important texts was life-blood to the new physics in the English-speaking world. That this crucial contribution to the modernization of physics emanated from the supposedly inhospitable cloisters of an Oxford college seems to have gone unnoticed, and in common with other translators Brose's name is not celebrated. He left Oxford in 1927 as assistant professor and then professor of physics at Nottingham, and in 1936 returned permanently to Australia, where he worked in biophysics (Home 1990: 34–7).[107]

Baynes's successor at Christ Church was Claude Henry Bosanquet, who graduated with a first in physics in 1921, a pupil of Nagel and Hartley at Balliol. He was appointed Dr Lee's reader in physics in 1922 and held the post until 1929, when he left to work in industry. He gave tuition within Christ Church, and conducted his experimental work in the Clarendon Laboratory, looking at the extinction of polarized X-rays.[108] His successor, Carl Howard Collie, had considerably greater impact, and Collie's career is of special interest not just as another example of a physicist emerging from the Oxford chemistry school, but because he also became Oxford's first home-grown nuclear physicist, and had informal charge of that subject for thirty years before it gained its own chair and department. He read chemistry at New College (first, 1925), his tutor there being A. F. Walden, though he described himself as 'trained under Soddy', from whom he derived his lifelong interest in radio-chemistry.[109] On graduation he stayed with Soddy as a demonstrator, and simultaneously did research on radioactivity under A. S. Russell in the Christ Church Laboratory, obtaining his B.Sc. in 1927. Coming in this way to

---

[105] Reprinted the same month, December 1919; 2nd edition (revised and with a biographical note), February 1920; Einstein's theory was reported and discussed in English-language publications from about 1914, but the first English translation of Einstein's own 'popular exposition' was not published until later in 1920.

[106] Freundlich (1920); Schlick (1920); Reiche (1922); he also translated Einstein's new article on 'Space-time' for the supplementary volumes of the *Encyclopaedia Britannica*, 13th edition (London, 1926).

[107] Information from Mr Peter Budek; Brose is mentioned in a footnote but not indexed in Morrell (1997: 387) and not mentioned at all by Kent (2001).

[108] Anon. (1934: 344); *The Balliol College Register . . . 1916–1967* (Oxford, 1969): 53; MHS, Collie MSS.

[109] MHS, Collie MSS; obituary, *The Times*, 28 August 1991; Morrell (1997: 422–5); Kent (2001: 41–2).

Fig. 4.15. Carl Howard Collie in 1927. Collie read chemistry at New College and then did research in radioactivity. As Dr Lee's reader in physics at Christ Church from 1930, he inaugurated work in nuclear physics, eventually using the Clarendon Laboratory's 1MV Cockcroft–Walton accelerator.

Lindemann's attention, he was recruited in 1926 in a first attempt to develop nuclear physics in the Clarendon Laboratory. His first task was to prepare pure uranium; after which he applied himself to the problem of measuring radioactivity, trying Lindemann's idea of 'a super-sensitive Lindemann Keeley electrometer' before going on to explore other methods of detecting nuclear particles that were more successful (at the same time as Geiger and Müller in Berlin were perfecting their particle 'counter').[110] These were the modest beginnings of atomic physics in Oxford, and in their wake Christ Church appointed Collie lecturer in physics in 1929 and Dr Lee's reader in physics in 1930, a post he held until retiring in 1971. He was also made a Research Student (that is, fellow), indicating the college's commitment to his research work.

By now the real power in Christ Church science was Lindemann, who from 1922 actually lived in the college. It was he who promoted Collie's career, and who persuaded Christ Church to retain its laboratory when the college authorities thought of closing it in 1930. Expectations that a school of radio-chemistry might form around Soddy and Russell were disappointed; but Lindemann's presence ensured that Christ Church became a focus for physicists. He used both his famous

---

[110] MHS, Collie MSS; Morrell (1997) omits this 1920s phase; on counters see Trenn (1986).

hospitality and a research fellowship there (1931–6) to tempt Einstein, though in fact he only paid brief visits.[111] A more lasting impression was made by Erwin Schrödinger, whose sojourn in Oxford at this same period (1933–6) has become something of a *cause célèbre*, and throws a number of issues relating to Oxford science into exaggerated focus (Hoch and Yoxen 1987; Morrell 1997: 409–10). A clash between two worlds of social etiquette – the Bohemian Schrödinger with his wife and mistress on the one hand, the polite conventions of Magdalen College (to which he was attached) on the other – was as much to blame for its failure as the absence in Oxford of a bridge between the two worlds of experimental and theoretical physics. Collie, one of those who *was* interested in what Schrödinger had to say, regretted the impression historians have formed 'that Schrödinger had no part in Oxford Physics', having himself been much influenced by him, and having treasured some notes the great man wrote in answer to a query Collie raised after one of his lectures.[112]

In 1935 Collie was placed in charge of a second, more stage-managed attempt to introduce nuclear physics to the Clarendon Laboratory. This time, Lindemann installed the powerful equipment necessary, and imported the Hungarian atomic expert Leo Szilard. Also on the team was James Howard Eagle Griffiths, who owed his fellowship at Magdalen to Schrödinger's recommendation, and ended his career as President of the college, the first physics graduate to become head of an Oxford college (Morrell 1997: 422–3).[113] Establishing a research school in nuclear physics from scratch was not going to be easy, and would never match the glories of the Cambridge school which had inspired it. But some respectable progress was made, and Collie had published several papers on atomic physics and on work with Geiger–Müller counters by the time the war intervened. Equally important, in its way, there was now great curiosity about the subject among science students, which he and his protégés were able to satisfy, for instance, by giving talks to university societies.[114] During the war Collie worked mainly on radar and microwave research with others in the Clarendon Laboratory, and on the dielectric properties of water with a Christ Church chemistry undergraduate David Ritson, the latter work leading to a series of papers.[115] After the war Collie and his students continued to promote the subject of nuclear physics, forged links with Harwell (the British government's atomic research facility), and became involved in the development of cyclotrons. A department of nuclear physics was created at Oxford in 1959.

---

[111] Morrell (1997: esp. 401, 407, 408–9); Birkenhead (1961: esp. 159–62); on Lindemann see Morrell, Chapter 7.
[112] Personal communication (Collie to the author, 1989); MHS, MS Museum 161 (Schrödinger's notes).
[113] MHS, Collie MSS; see Morrell, Chapter 7, 257–60; on Griffiths see obituary by ter Haar (1982); Bagguley and Bleaney (1990).
[114] Collie and Griffiths (1936); Collie and Roaf (1940); MHS, Physical Society Minute Book 1937–46, talks in 1937 (Duckworth) and 1939 (Collie).   [115] MHS, Collie MSS; Collie *et al.* (1948); Collie (1946–7).

Throughout his long career Collie was a busy teacher and supervisor, whether of Christ Church undergraduates or of doctoral researchers at the Clarendon Laboratory. His broad range of interests – in radioactivity, atomic physics, nuclear instrumentation, and electronics – allowed him to encourage or inspire a variety of distinguished careers. He was immediately confronted at Christ Church with his first doctoral student, Orvald Arthur Gratias, who obtained his D.Phil. (on radioactivity) in 1932, and with whom he published several collaborative papers. Another early doctoral student was Douglas Roaf, a Brasenose graduate who joined Collie at Christ Church in 1936 and thereafter shared his teaching burden.[116] One of Collie's early undergraduates, whose interest in radio and electronics he encouraged, was (Sir) Martin Ryle, who gained a first in 1939 and went on to become professor of radio astronomy at Cambridge, Astronomer Royal, and in 1974 joint winner of the Nobel Prize for Physics for the discovery of pulsars (Smith 1986; *DNB*; Kent, 2001: 41–2). The Christ Church Laboratory closed in 1941, and Russell gave up research but remained a busy chemistry teacher until retiring in 1955. In 1945 Collie became a university demonstrator and lecturer at the Clarendon Laboratory, as well as remaining reader in physics at Christ Church, giving him the now more usual affiliations to both college and central laboratory. Christ Church's flirtation with Einstein may have amounted to little more than public relations, and its attraction for Lindemann have had more to do with good living; but through the quieter work of Brose and Collie, Christ Church gave staunch support during much of the twentieth century to the new physics, which one should beware of assuming was anathema to traditional, collegiate Oxford.

PHYSICS BEYOND THE COLLEGES

There remain a number of physicists in our period who operated beyond even the broad landscape of collegiate Oxford surveyed above. Out on the margins there were always a few self-sufficient individualists, exiled (in one sense or another) in their private laboratories. One of the most celebrated was the pioneering atmospheric physicist Gordon Miller Bourne Dobson, university lecturer in meteorology from 1920. He was nominally attached to the Clarendon Laboratory, but his fundamental advances in the study of the upper atmosphere, including the discovery of the ozone layer, were made at meteorological research stations he set up at his own expense at two successive country homes (at Boars Hill and then at Shotover Hill)

---

[116] For these, and some other Collie (and Roaf) students, who are too numerous to mention, see Morrell (1997: 422, 424–5) and Kent (2001: 41–2).

Fig. 4.16. Sprengel-type mercury pump by Alvergniat frères of Paris, made in the 1880s. This was a standard piece of physics laboratory equipment at this period. It was used, for instance, to evacuate Crookes tubes and subsequently X-ray tubes. This example is the one belonging to the Duke of Marlborough, whose apparatus was presented to the University of Oxford after his death in 1892 and mostly used for the teaching of physical chemistry. The pump is now in the Museum of the History of Science (inventory no. 87524).

just outside Oxford.[117] Our reconnaisance of the devolved world of 'Oxford physics' reaches the outer fringes with two remarkable aristocrats whose scientific inclinations responded, in different ways, to the pull of the nearby university: the eighth Duke of Marlborough and the eighth Earl of Berkeley.

The Duke's case is flimsy, and would be forgotten but for its curious legacy. After his death in 1892 his widow presented the University with a magnificent collection of expensive, and in several cases ostentatiously large, physical instruments, apparently purchased during the 1880s. For the Clarendon Laboratory Clifton accepted 'some electrical apparatus, and . . . two large and costly spectroscopes specially constructed for the late Duke'. But most of the collection went to the Chemistry Department: as described in its annual report, it 'comprised a very complete set of laboratory fittings and general apparatus; a stock of chemicals, including compounds of rare metals; a large direct vision spectroscope; two balances;

---

[117] Dobson (1968: with interesting illustrations); Houghton and Walshaw (1977); Morrell (1997: esp. 381, 389, 393–5); see also Little (1995) and Morrell, Chapter 7, 239.

a mercury-pump; air-pump; Gramme machine; two Ruhmkorff induction coils; sets of standard thermometers; and a cabinet with various electrical and other apparatus; altogether a most valuable addition to the equipment of the department'.[118] The last phrase was not a platitude: the gift immediately enabled the Chemistry Department to introduce regular practical instruction in physical chemistry, conducted by V. H. Veley as an extension of his existing lecture course (Gardner 1934: 570). In 1904 some of the Marlborough instruments were borrowed by the Balliol–Trinity Laboratories when they took over the physical chemistry teaching. One of the spectroscopes was also used in studies of the spectra of argon and helium by G. J. Burch, demonstrated to the Junior Scientific Club in 1895 and 1896.[119] Some of the instruments are preserved in the Museum of the History of Science.

George Charles Spencer-Churchill succeeded to the dukedom in 1883, and lived partly in London and partly at Blenheim Palace, eight miles north of Oxford. His father had been a prominent political figure, as was his brother Lord Randolph Churchill; but the eighth Duke led a less public life (BP; Gunther 1937: 309). He had not received a university education, but was presumably taught by private tutors. He may have been inspired by his family's interest in astronomy, a tradition that went back to the eighteenth century. Too little is known about the Duke of Marlborough to form an impression of what kind of enterprise he might reasonably have attempted with his extravagant apparatus, or how far he progressed in it before his early death. Gunther believed that there was 'a well-equipped scientific laboratory at Blenheim Palace, where advanced work was being carried on', though I suspect he was extrapolating from the mere existence of the instruments.[120] Elsewhere he quotes Lord Redesdale as saying of the Duke that he came 'within measurable distance of reaching conspicuous success in science, mathematics and mechanics' (Gunther 1937: 309). Among his business interests in London he was Chairman of the city's electric light and telephone companies, which may suggest where his scientific interests lay. The list of apparatus implies a relatively sophisticated laboratory with provision for research on electricity, spectroscopy, and chemical physics – not dissimilar from that ventured upon more successfully by another wealthy nobleman, the Earl of Berkeley.

A unique figure in the orbit of Oxford science, but excluded from membership of the University, Randal Thomas Mowbray Rawdon Berkeley had been educated in France, served in the Royal Navy, and on leaving in 1887 enrolled in chemistry courses in London, keen to join the world of experimental science. His original interests were mathematics and chemistry. He succeeded to his family title in 1891

---

[118] OUG, 23 (1892–3): 626 (13 June 1893).

[119] MHS, MS Museum 191, f. 22; Journal of the Oxford University Junior Scientific Club, 1, no. 28 (1895): 369–70 (meeting of 20 May); 2, no. 36 (1896): 32 (meeting of 28 February).

[120] '250th Anniversary of England's Oldest Museum . . .', Oxford Mail, 19 May 1933: 5 (where the Marlborough Gramme machine is also illustrated).

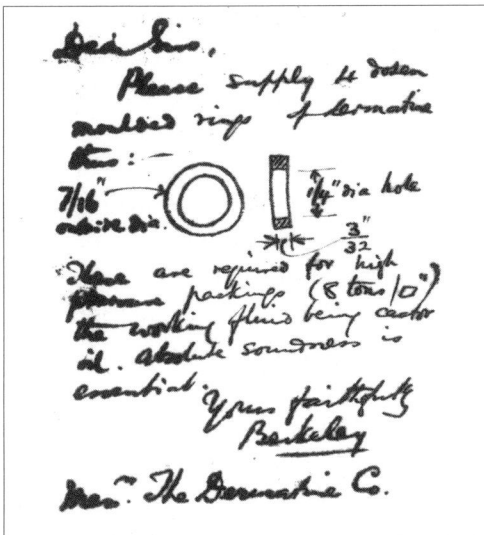

Fig. 4.17. This letter from Lord Berkeley's copy-book is a typical example of his correspondence with apparatus manufacturers. Addressed to The Dermatine Co. and dated 6 August 1912, it orders rings or washers of the company's proprietary synthetic material 'dermatine' (presumably a leather substitute) to be made to very exact specifications for his experiments on osmotic pressure: 'Absolute soundness is essential'. Museum of the History of Science, MS Museum 53, f.70v.

(Hartley 1942–4; *DNB*; *ODNB*; *BP*; Pine 1972: 25–7; New 1990; Anon. 1994; Laidler 1988: 227–8). In 1893 he came to live at Foxcombe, a country estate on Boars Hill, a few miles outside Oxford, and immediately established a relationship with the Christ Church Laboratory, working there as Harcourt's guest. Early in 1894 he was admitted as a member of the Junior Scientific Club as of 'Ch.Ch.', but in fact he was never a matriculated member of either the college or the University.[121] It seems that he wished to join but he was not willing to conform to the residence requirements. In 1895 he hoped to enrol for the newly established degree of B.Sc. – Oxford's first graduate research degree – but in the end the University attached the same rigid regulations to it. One of his mentors, the mineralogist Sir Henry Miers, believed that he 'intended to establish a research laboratory for Oxford' in physical chemistry, but once it became evident that it would not relax its rules for him he 'declined to have any more dealings with the University' (quoted in Hartley 1942–4: 169).

Berkeley's work at Christ Church consisted of measuring the densities and specific gravities of crystals, for which he devised techniques and instruments capable of far greater accuracy than before. His first published paper appeared in *The Mineralogical Magazine* in 1895, having been presented at both the Mineralogical Society in London and the Junior Scientific Club in Oxford, where Harcourt led the discussion afterwards.[122] In the Christ Church Laboratory he was assisted by one of Harcourt's students, Ernald George Justinian Hartley, who graduated in 1897 and

---

[121] *Journal of the Oxford University Junior Scientific Club*, 1, no.14 (1894) (meeting of 16 February – proposed); no.15 (1894) (meeting of 2 March – elected); information from Mrs Judith Curthoys, Christ Church archives.
[122] *Journal of the Oxford University Junior Scientific Club*, 1, no.25 (1895): 321, 331 (meeting of 22 February); Berkeley (1895–7).

then worked with Berkeley at Balliol College, where Conroy allowed them to use the old (1854) laboratory for experiments on the electrolysis of glass. By this time, Berkeley was preparing a laboratory in the grounds of his home on Boars Hill, and it was in intensive use from 1898 until 1918 (when he moved to Berkeley Castle), and did not finally close until he sold Foxcombe Hall in 1928.

In spite of having fallen out with the University, his good relationship with the physical chemists in the college laboratories continued, and a number of them (or their pupils) worked with him at the Foxcombe laboratory. E. G. J. Hartley joined him there after a brief period in industry, and from 1902 to 1916 worked as his full-time assistant. Afterwards he became a demonstrator (1919–45) in the University's Chemistry Department (Applebey 1948; New 1990; Morrell 1997: 357). In 1910 M. P. Applebey was recruited from the Balliol–Trinity group of chemists; he was already a fellow of St John's College, and subsequently worked at the Jesus College Laboratory and then as a demonstrator in the Chemistry Department. Sidgwick's pupil Bernard Howell Wilsdon joined Berkeley soon after graduating from Lincoln College in 1911, and afterwards went as professor of chemistry to Lahore.[123] Berkeley's other important collaborator, Charles Vandeleur Burton, was head-hunted in 1907 from the Cambridge Scientific Instrument Company, where he had worked on engineering research and instrument design with Horace Darwin. Burton contributed particularly to the theoretical aspects of the work; Applebey described him as 'a mathematical physicist of great and overflowing genius'.[124]

These collaborations amounted to true team research, which was not yet a familiar feature of Oxford science. Taking what he had learned in the collaborative atmosphere of Harcourt's laboratory a step further, Berkeley became a deliberate and very effective team-leader. His 'programme of research' was imaginatively planned, and he would regularly adjourn from the laboratory bench to the blackboard, where he 'loved to gather the team around him to thrash out some aspect of the work' (Applebey 1948: 901). His approach was nevertheless fundamentally experimental, and he was another of those scientists who were attracted as much by the laboratory craft as by the quest for knowledge as such. His Oxford friends admired him for possessing the very quality that they themselves most esteemed – 'an unusually fine sense of technique'.[125] He was ingenious in the design of very precise experiments, and untiringly patient in their performance – or in his collaborators' performance of them. His maxim was: 'The only thing that matters is accuracy. In our work we must behave as if we expected to live for ever'.[126] The design, acquisition,

---

[123] Hartley (1942–4); MHS, MS Museum 53; on Applebey see Steel (1957); on Wilsdon see Gunther (1924: 284).

[124] Hartley (1942–4); New (1990); the quotation is Applebey as quoted in former: 174; on Burton see also Cattermole and Wolfe (1987: 126).

[125] Hartley (1942–4: 170); cf. Applebey (1948: 899) and *DNB*.

[126] Berkeley speaking to E. Stenhouse quoted in Hartley (1942–4: 178).

and calibration of instruments were consequently of major concern, and were managed very professionally. Instruments were tested and certified at the National Physical Laboratory: his 'Thermometer Book' records the care taken, for instance, to calibrate a collection of thirty-six numbered thermometers against the Kew standard between 1895 and 1917.[127] As well as employing his own mechanic, Berkeley corresponded continuously with instrument manufacturers and engineering firms, seeking out the best sources for his specialized needs, and frequently asking them to make bespoke items to his own designs and to very exacting specifications. Not limiting himself to the usual scientific suppliers such as Elliott or Hilger, he explored more esoteric sources too – from André Citroën & Co. of Paris to 'Mr. Jackson, Organ Builder' of Oxford.[128]

The approach paid off, and between 1904 and 1919 there appeared a flood of twenty-two papers, mostly published in the Royal Society's journals and the *Philosophical Magazine*, reporting meticulous experimental work on osmotic pressures, densities of crystals, boiling points, solubility, and various instruments devised for these analyses and measurements.[129] Usually co-written with E. G. J. Hartley or other collaborators, they represent an achievement in research unmatched by any Oxford laboratory of the period. The most sustained and important research was that on osmosis and the theory of solutions, and in particular osmotic pressure of solutions, of which the most accurate measurements thus far were obtained. Berkeley and Hartley developed the indirect vapour-pressure method of measurement, and established its validity. This core work on osmotic pressure was recognized, immediately it was published, as a model of its kind, and led to Berkeley's election as FRS in 1908. His other passion was for golf, but typically he applied his analytical mind to this too, studying 'the physics of the interaction of club and ball' on his private golf course at Foxcombe, which has been described as 'a second laboratory'.[130]

CONCLUSION

When the American physical chemist T. W. Richards visited Oxford in 1911, his host, Harold Hartley, keen that he should see the showpiece of Oxford's growing renown in chemical physics, took him not to a university laboratory, such as Townsend's magnificent new building, and not to a college laboratory, such as his

---

[127] MHS, MS Museum 172 (2); see also 172 (1).   [128] MHS, MS Museum 53.
[129] They are listed in Hartley (1942–4); three final papers appeared in the 1930s.
[130] New (1990); see also Hartley (1942–4: 175). Berkeley later published a book, *Sound Golf by Applying Principles to Practice* (London, 1936).

own busy centre for teaching and research at Balliol; instead he took him several miles out into the countryside, to a small, private laboratory without academic affiliation, conducted by a man who was not even a member of the University.[131] I am not concerned to present this as an indictment of the University; it is simply an illustration of how things were in Oxford at this period. Lord Berkeley's Foxcombe laboratory was an extreme manifestation of the devolved and individualistic nature of Oxford's scientific culture, more usually expressed within the college environment, whether by the solitary efforts of a Bosanquet or the corporate initiatives of a college such as Trinity.

The worst victim of Oxford's peculiarities at this time was reputation, or perhaps what would now be called 'public understanding'. John Perry's famous attack on Oxford science in 1903 crystallized an 'understanding' that has proved very durable, and has carried into the historiography.[132] Clifton, and his equally lacklustre contemporary as professor of chemistry, William Odling, were and remain easy targets for criticism; that one should not tar the entire discipline with the same brush is logical enough, but in practice the misunderstanding endures because the alternative is too complex to present a coherent image. Perry's kindred spirits in Oxford – men and women as passionate as he was about educational extension, or about the promotion of physics and engineering as university subjects – have generally been as ill-served by historiography as they were by the contemporary debate. We have encountered some of them above; and the next chapter centres around a particularly interesting example.

---

[131] Hartley (1942–4: 173); Hartley says 1912 but Richards's visit to Oxford was 1911, when he was given an Honorary D.Sc. and a dinner by the Alembic Club: *Historical Register First Supplement* and MHS, MS Museum 141 respectively.

[132] Perry (1903–4); Howarth (1987: 335); and see Simcock, Chapter 5, 183–4.

5

Mechanical Physicists, the Millard Laboratory, and the Transition from Physics to Engineering

Tony Simcock

The Millard Laboratory at Trinity College was the principal epicentre of physics in collegiate Oxford, and the man with whom it is identified, Frederick John Jervis-Smith, university lecturer in mechanics from 1888 to 1908, was the only university-employed physics teacher outside the central laboratories. This chapter focuses upon this uniquely Oxonian manifestation of the expansion and evolution of physics in the late nineteenth century. Jervis-Smith was a transitional character in several ways. He belonged to the last generation in which it would not seem odd for a science teacher to be an ordained clergyman. Less trivially, he represented in Oxford the transformation of experimental mechanics and applied physics into a robust and coherent discipline, that of 'engineering science'. He attempted to steer his nascent subject through this transition with limited resources and only modest recognition, while at the same time contributing to mainstream physics teaching and conducting constant research.

Jervis-Smith has tended to be seen as 'a somewhat marginal enthusiast' (Fox 1997: 687), and it would certainly be easy to dismiss a kindly Victorian clergyman-inventor, tinkering away in a shed in a college back-yard, as nothing more than an irrelevant eccentric. Yet it would also be possible to argue conversely that he was, or anyway attempted to be, a key modernizing influence just at the time that the momentum of Oxford science was fading. I shall be content to present him as the central figure in Oxford physics outside the centre, so to speak. The broad collegiate and devolved context in which to situate him has been portrayed in the preceding chapter. The more particular context of scientific mechanics and the

For standard and background sources used for basic biographical and local information see preliminary note in the preceding chapter. In this chapter, I use the following abbreviations for frequently cited archive repositories: MHS (Museum of the History of Science, Oxford) and TCA (Trinity College archives, Oxford).

emerging discipline of engineering is the other theme of the present chapter; though as with chemistry, my purpose in trespassing upon the territory of engineering is simply to discover physics and physicists, both at the roots of engineering as a university subject, and at the heart of the work of some of Oxford's early engineers.

OXFORD AND THE CULTURE OF PRECISION

We have seen ample evidence in the previous chapters of the characteristic, not to say overwhelming, preoccupation of physical science in Oxford not just with experimentation but with experimental finesse: a concern for precise measurement, for minute experimental procedure, and for the techniques and technology used to achieve them. The highest compliment an Oxford scientist could pay to a colleague was to praise his 'fine sense of technique' (Hartley 1942–4: 170, on Berkeley) or 'the extreme delicacy and neatness of his experimental manipulation' (Harcourt 1905: 252, on Conroy). Some scientists devoted so much attention to their technique or to their laboratory technology that the results of scientific research took second place, or were stifled altogether. (Clifton and Bosanquet, in their different ways, suffered this fate.) Conversely some non-graduate technicians fitted so well into this milieu and were so admired by their superiors that they rose to the status of academic scientists (like J. J. Manley).[1] Such concerns were by no means peculiar to Oxford, and have been identified by Gooday as a defining feature of the emergence of physics laboratories, and physics education more broadly, in the heyday of Victorian Britain (Gooday 1990: esp. 46–7). Oxford not only participated in this trend in the 1870s and 1880s, but in some respects (such as the building of the Clarendon Laboratory) led it.

Another characteristic of Oxford science was how the definition of physics could be fluid in relation to the adjacent disciplines of mathematics, mechanics, and chemistry: the long-established interplay between physics and mathematics, and the symbiosis of physics and chemistry that arose in our period, have been explored in earlier chapters. The shared culture of experimental precision played a large part in this fluidity, and the individual skills of its adherents crossed disciplinary boundaries, allowing physicists and biologists to find common ground in precision mechanics. Emerging subjects like astrophysics and crystallography benefited from this feature of Oxford science: the new astronomical observatory which Oxford opened in 1875 was able to boast 'one of the most perfect and powerful Spectroscopes in the world', presented by the physicist J. P. Gassiot (Moore 1878: 283); and we have already encountered Oxford's pioneer crystallographers (Nevil Story-Maskelyne and Sir Lazarus Fletcher) in the guise of

---

[1] For Clifton see Gooday, Chapter 3; for Bosanquet and Manley see Simcock, Chapter 4, 145–9 and 124–7.

physics teachers. The principal biological science, physiology, shared the characteristic to a high degree. Preliminary physics was much more relevant than might be supposed to Oxford physiology students, and also compulsory for the medical degree. The senior Oxford physiologists all did collaborative research with physicists – notably, in the case of successive professors J. S. Burdon Sanderson and Francis Gotch, with George James Burch, a natural genius in the precision techniques of the laboratory and workshop, who studied for his undergraduate degree (in chemistry, I need hardly say) at the age of 37 while working on Sanderson's capillary electrometer. Twenty years later, at the height of his career, Burch had the unusual distinction of being a demonstrator in the Oxford physiology department and at the same time professor of physics at Oxford's satellite university at Reading.[2]

Part of the symbolic relevance of C. J. F. Yule's chronograph, a celebrated instrument in the annals of physical chemistry, is that it was part of the apparatus of the new experimental physiology (introduced in courses at the Magdalen College Laboratory in 1874) when it was spotted by the chemist H. B. Dixon and recruited in connection with his timing of gaseous explosions. Eventually he purchased the instrument for use at the Balliol–Trinity Laboratories, and took it with him to Manchester in 1887 (Gunther 1924: 22, 24–5; Laidler 1988: 249). In the meantime Jervis-Smith had joined the Balliol–Trinity community and been inspired by Dixon's work to experiment with his own form of chronograph, having physiological and physical measurements chiefly in mind. The electromagnetic styli from Jervis-Smith's chronograph were adopted in Dixon's later research, while in commercially marketed form the Smith 'tram' chronograph was used in the explosives industry and in ballistic research, and was capable of modification for various laboratory purposes (Jervis-Smith 1890b; 1903; 1910b; and see below). It is a vivid instance of how a technique or instrument can take a journey through seemingly disparate sectors of science and industry, united in their quest for precision.

Physicists such as Manley, Burch, and Jervis-Smith, whose primary research interest was the development and perfection of such techniques and instruments, constitute an interesting category in their own right. If Oxford routinely deflected potential physics students into chemistry (which we have clearly seen happening), or moulded a distinctive breed of chemical physicist, there were surely going to be others who were drawn towards the mechanical and technological side of physics, or who might (if it existed, or as it evolved) be deflected into the discipline of engineering. These mechanical physicists (as I presume to call them) were more peripheral than the chemical physicists, and often assumed ancillary roles, at least partly because there was no academic discipline or career to embrace them until late in the nineteenth century.

[2] MHS, Burch MSS; Bristol University Library, C. R. Burch MSS, esp. A.31, A.34–A.52; *Oxford Chronicle* (supplement), 11 August 1905; obituaries by Vernon (1914), Anon. (1913–14), and Anon. (1914).

Against this disadvantage could be set the fact that the common concern for precision and technical finesse made the existing scientific disciplines very welcoming to them.

All this should have made Oxford fertile ground for the cultivation of applied mechanics as a subject of study. Engineering chairs under one name or another were a new thing in most universities, but were widely accepted. Cambridge's chair of 'mechanism' was founded in 1875; and Oxford's agreement in principle to such a move dates from the same decade, and (along with a second physics professorship) was formalized in the new Statutes of 1882. A professor of applied mechanics, or of mechanics and civil engineering, was to teach 'the principles of Civil and Mechanical Engineering'.[3] The timing of the chair's establishment, however, was left to the two colleges nominated (as alternatives) to fund it, and this was the fatal flaw. For during the 1880s even these wealthy colleges, St John's and Magdalen, reconsidered their priorities in the face of economic depression in agriculture (rent from agricultural land being the basis of their endowed income). While the physics chair had to wait until 1900 (Lelong, Chapter 6), the 1882 provisions for an engineering chair never took effect. Different arrangements were made in order to bring it about in 1908 – but the delay made a crucial difference. The timely adoption of the new subject by a university whose provision for science was maintaining a brisk pace, as it might have seemed in 1882, could scarcely appear better, twenty-six years on, than the grudging acceptance of engineering by a university which had 'stumbled at the threshold of the modern world'.[4]

AMATEUR MECHANICS AND PROFESSIONAL ENGINEERS

By the standards of the late nineteenth century, workshop activities were either genteel hobbies or professional trades, so we shall expect to find some of Oxford's mechanical physicists in either context. The amateur mechanic was a characteristic Victorian figure, easily recognized if hard to define, for in the amateur's workshop electrical experiments and carpentry might happily coexist, and the owners of these workshops might be clergymen, wealthy ladies, biology students, or shopkeepers. In Oxford at the end of the nineteenth century the town watchmaker's son Henry Minn, the chemistry student H. E. Stapleton, and the science don and zoologist R. T. Gunther were all workshop enthusiasts, and avid readers of the weekly magazine that bound this disparate community together, *The English*

---

[3] *New Statutes 1877*: 38–9. On the term 'applied mechanics' see note 35 below.
[4] Fox (1997: 691). The concluding paragraphs of Fox (1997) contain a carefully considered summary of Oxford science's predicament at this period.

Mechanic.⁵ An excellent example of the recreational mechanic was Thomas Henry Toovey Hopkins, a clergyman, classicist, and Fellow of Magdalen College, where he assisted in running the Daubeny Laboratory and had also taught chemistry as Daubeny's deputy. But his main interest (apart from rowing) was mechanics, and his particular love the lathe, his Holtzapffel lathe being the centre-piece of an attic workshop above his college room, where he proudly demonstrated it to interested colleagues and students. He was a highly skilled ornamental turner and invented various lathe accessories (Gunther 1924: 13, 19, 96, 305, 307, 401–6; Hopkins 1878). His scrapbook, illustrating the range of his mechanical interests, is dominated by lathes and by geometric and ornamental turning, but also features other gadgets, industrial machinery, and even cranes and excavators.⁶ He had a small printing press, which interested him as a mechanism as well as printing useful things, like meteorological records. About 1869 Hopkins fitted out a formal meteorological station, where a strict routine of daily observations continued until 1923. In addition to the usual readings, the temperature and height of the adjacent River Cherwell were taken, the height being measured by a floating gauge made in 1876 to Hopkins's design.⁷

In the older world of college fellowships there is not a large gulf between a scientific amateur such as Hopkins and a more serious yet still dilettantish experimentalist such as Bosanquet. Likewise, there is only the collegiate world he inhabited to distinguish Hopkins from many a clergyman or country gentleman outside academia. Equally independent of the universities, the profession of engineering originated in the world of the artisan (and also the military engineer), and a graduate who wished to pursue such a career had to learn it there – either by apprenticeship or, in the upper echelons of the profession, through a system of pupillage which had evolved. It began to be taught as an academic subject in the middle of the nineteenth century. At Oxford, the first engineering course was announced by Jervis-Smith in 1885. It cannot be argued that Oxford-educated engineers were other than a rarity before this; but looking at several examples will at least prove that they did exist, and that they were not mavericks in terms of their place in Oxford culture⁸ – for it has come to be assumed that Oxford bred a disdain for engineering careers, not to say for any utilitarian application of knowledge. It will also be useful to see the paths that might lead a graduate into engineering, and to note the range of their activities, 'engineering' embracing a gamut of endeavours, from large-scale construction to the minute precision physics already discussed.

A conventional career pattern for a graduate engineer is represented by John George Gamble, who read mathematics at Magdalen College (second in

---

⁵ Examples are named on the basis of scrapbooks or similar evidence in MHS. *The English Mechanic* began publication in 1865; see MacLeod (1972: 156). ⁶ MHS, MS Gunther 80.

⁷ Gunther (1924: 30–5, 51, 172, 404, 444); Magdalen College archives, MSS 375 and 376.

⁸ Samuel L. Maverick (d. 1870) was an engineer (though it was to his cattle that the word was originally applied).

classics, 1863; first in mathematics, 1864), was briefly mathematical lecturer for Lincoln and Merton Colleges, and then in 1866 entered into professional training under Sir John Hawkshaw (best known for certain London railways and bridges). He was Hawkshaw's assistant on several construction and surveying projects involving docks and sewers. In the heyday of the Empire most opportunities for British engineers lay in the colonies, and in 1875 Gamble headed for South Africa as Hydraulic Engineer to the Colony of Cape of Good Hope, remaining there until his occupationally related death of typhoid. He was responsible for the water supply of Cape Town and Port Elizabeth, and the utilization of water from Table Mountain. His other interest, appropriately, was meteorology, especially rainfall. He published papers in both fields in the *Transactions* of the South African Philosophical Society, of which he was President in 1883 (Gunther 1924: 482–4). His connections with Oxford were not severed on entering the engineering profession, and in 1871 he gained the Johnson Prize (one of the university's few scientific prizes) for an essay 'On the Laws of Wind', indicating that his meteorological interests arose early – presumably at Magdalen (*Historical Register*: 181).

A previous pupil of Hawkshaw, from 1862 to 1865, was Leveson Francis Vernon-Harcourt, who became a hydraulic engineer of international standing at both professional and academic levels. In the practical phase of his career he worked on docks and harbours, and surveyed rivers in connection with water supply; he also wrote standard books on these subjects. He then moved into the academic arena, as professor of civil engineering at University College, London from 1882 to 1905 (*DNB*). Although he hyphenated his name, he was in fact the brother of the chemist A. G. Vernon Harcourt, whom he followed to Balliol College, studying under Henry Smith and taking a first in mathematics in 1861 and a first in natural science the following year. Vernon-Harcourt was a contemporary of Bosanquet: they took the natural science degree together. His subsequent career choice did not make him the black sheep of his family, however, since his brother leaned strongly towards utilitarian applications of his chemical knowledge. All of Harcourt's research outside the famous work on chemical reactions was in applied chemistry for public service. His experiments with coal-gas arose from his position as a Metropolitan Gas Referee (monitoring the purity of the London gas supply), and led in 1877 to his invention of the pentane lamp as a photometric standard; experimental and advisory work on lighthouse illumination followed from this. It was also Harcourt the chemist who invented the Harcourt chloroform inhaler, which was adopted by the British Medical Association in 1910 (Shorter 1980: 413–14; Harcourt 1880; Griffin 1910).

The engineer and naval architect William Froude read classics and mathematics at Oxford under the old system (before the creation of the Natural Science School), and then pursued a professional career under Brunel, working on railways before moving with Brunel into steam ships. His subsequent work on ship design, and his creation and conduct of the first experimental tank at Torquay, famous in its day, won him acclaim in both engineering and scientific circles, leading to his FRS in 1870 and a Royal Society Gold Medal in 1876. Among other things, he investigated resistance and rolling-angles, and designed dynamometers for marine experiments, which inspired the work in this field of his young admirer Jervis-Smith (Jervis-Smith 1915). Like the Harcourts, Froude was not an outsider to Oxford culture: both his brothers were also Oxford-educated and also achieved distinction in their chosen fields, one (J. A. Froude) a famous historian and the other (Richard Hurrell Froude) a theologian connected with the Oxford Movement.[9] The tradition was continued by their nephew, Arnulph Mallock, whose early attraction to engineering did not prevent him being sent to Oxford, where, it seems reasonable to speculate, his vocation was not dampened but refined – for Mallock was to become one of the pre-eminent seekers after precision of his time.

Henry Reginald Arnulph Mallock obtained a second in natural science from Hertford College in 1875, and gave his first research paper to the Ashmolean Society later that year, accuracy already being his watchword ('On a method of obtaining an accurate meridian shadow').[10] From Clifton and the new Clarendon Laboratory, and from other physicists whom he encountered in Oxford (such as Bosanquet, whom he met in the Ashmolean Society meetings), he absorbed a life-long passion for extremely precise measurement and the development of instruments and experimental techniques to achieve it. Leaving Oxford, he worked with Froude towards the end of his uncle's life, and was Lord Rayleigh's assistant at the Cavendish Laboratory in Cambridge shortly after (about 1880–1). But he spent much of his career as a consultant precision engineer working for business and railway concerns, or more typically for military, government, and other public bodies. He was Consulting Engineer of the Ordnance Board, and a civilian member of the Ordnance Committee and the Aeronautical Research Committee. The Admiralty and latterly the Air Force found regular need of his services, hence much of the work he did was secret. As for the rest, he published forty-eight papers in the *Proceedings of the Royal Society*, and numerous notes and letters in *Nature*. Mallock became a Fellow of the Royal Society in 1903, and nine years later gave the highly prestigious James Forrest Lecture to the Institution of Civil Engineers, speaking on the topic of 'aerial flight'. His admirers were those scientists attuned to his particular brand of applied physics: such as his Cambridge friends Lord Rayleigh and Sir Joseph Larmor; Sir Charles Vernon Boys, who wrote his obituary; and among the scientific antiquaries R. T. Gunther, who acquired a few of his instruments for preservation in Oxford. Mallock was shy and evaded public recognition, even

[9] All three are in *DNB*; on William see also Anon. (1879); Anon. (1880); Smith (1937: 356–7).
[10] *Report of the Ashmolean Society*, 1875: 3 (the paper was not printed).

Fig. 5.1. Henry Reginald Arnulph Mallock, c.1910 but no later than 1917. This is a copy made after Mallock's death from a group photograph. No other likeness is available, as Mallock avoided publicity and refused to be photographed. Museum of the History of Science, MS Museum 279.

refusing to be photographed. Had Boys and Gunther not championed him, his name would be virtually lost from the historical record.[11]

The problems that became the focus of Mallock's genius were not simply those of ultra-accurate measurement: they typically required the detection and investigation of extremely small or slow movements, changes, or transient disturbances. Tremors caused by underground railways in London were one such subject. The most exacting measurement that he was called upon to make was the rate of expansion of cracks in the structure of St Paul's Cathedral, a process so gradual that to monitor its hourly rate was estimated to require measurements to one ten-millionth of an inch. Mallock achieved an accuracy of one millionth of an inch using wavelengths of light observed from the movement of interference bands. He then applied this technique to the more difficult problem of measuring the growth rate of trees (by trunk girth) – more difficult because, although faster,

---

[11] Obituary by Boys (1932–5); Gunther (1933: 3; 1937: 260); the portrait accompanying the Royal Society obituary is a heavily retouched enlargement from a group photograph.

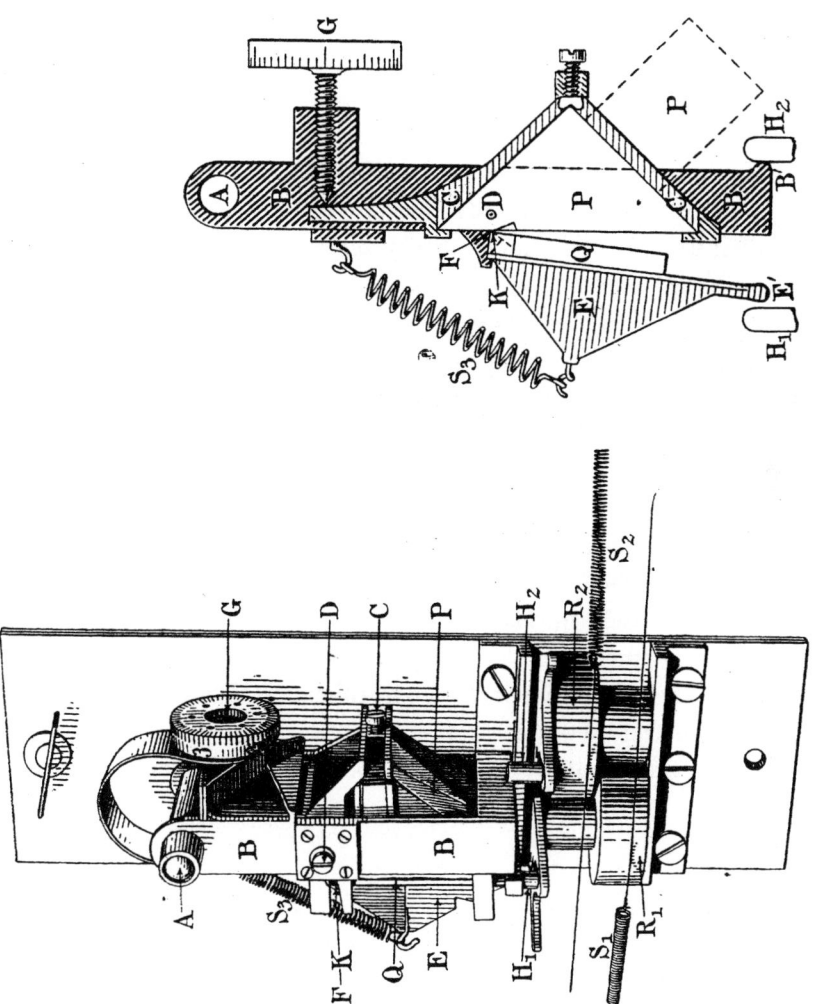

Fig. 5.2. Line illustration of Arnulph Mallock's device for measuring the growth-rate of trees by trunk girth, from his paper in *Proceedings of the Royal Society*, series B, 90 (1917–19): 189. The expansion of the trunk and in contact with the device at positions $R1$ and $R2$, increases the angle between the two optical surfaces Q (a glass plate) and P (a right-angled prism). The principle, established in measuring the expansion of cracks in buildings, is that the intervals between interference bands in reflected light are directly proportional to the wavelength of the light, and inversely proportional to the angle between the two reflecting surfaces; so that if lateral shifts in these bands can be detected to an accuracy of one tenth of the band, a movement of a millionth of an inch can be measured. Such accuracy is neither required nor possible in measuring trees, but the experiment has had to be embodied in a cleverly designed portable device.

it fluctuates with several variables and does not permit the same stability of set-up. He made an ingenious adaptation of his optical method, mounted on a delicate mechanical device, and achieved measurements in the required region of a hundred-thousandth of an inch per hour. The invention was tested at Kew Gardens in 1917.[12] In other spheres, he worked for instance on ballistics and the speed of bullets, including collaboration with the engineer W. E. Metford; investigated various optical and colour phenomena; and invented a machine for ruling very accurate diffraction gratings. He was also a skilful dissector under the microscope: according to Boys, his dissections of the muscles in a butterfly wing 'astonished the Zoological Society' (Boys 1932–5: 97).

The vocabulary of 'precision engineering' hardly conveys the kind of ingenuity in the pursuit of minute phenomena and extreme precision that was Mallock's forte. It involved the patient application of his knowledge of physics, combined with mechanical skills and an imaginative boldness in the design of experiments, to seemingly intractable problems at the very limits of the precision that was attainable at the time. Like the best mechanics and inventors, he attacked these problems from 'first principles'; and the apparatus and experiments he devised were always 'simple but mechanically perfect'.[13] Mallock's individualistic career is very difficult to categorize; he called himself (in *Who's Who*) a 'Consulting Engineer and Physicist'. But he was only an 'engineer' in the sense that he addressed his expertise to problems located outside the laboratory, or set him by others: in all other respects he stood at the very heart of Oxford physicists' preoccupation with the minutiae of precision physics (*WWW*; Gooday 1989: 50, 54; 1990: 25; and see Gooday, chapter 3, 88, 90–1).

FREDERICK J. JERVIS-SMITH

It was another man of this kind, a contemporary of Mallock, who came back to Oxford in 1885 and established there the laboratory-based teaching of mechanical physics, experimental mechanics, and the physical principles of engineering. This was Frederick John Smith, who in 1897 in honour of his kinsman Admiral John Jervis, a hero of the Napoleonic Wars, adopted the surname Jervis-Smith (though for clarity I refer to him at all periods of his life as Jervis-Smith).[14] His grandfather had derived a fortune from his coal-rich land in the Staffordshire Potteries, such that

---

[12] Mallock (1917–19); Mallock's own copy, with annotations, is in MHS, MS Museum 279.

[13] Boys (1932–5: 96); both Boys and Larmor (letter to Mallock in MHS, MS Museum 279) comment on his first principles approach.

[14] Obituaries by Boys (1913), Anon. (1911a, probably by A. G. Vernon Harcourt), Anon. (1911b), and Anon. (1911c); cuttings of other obituaries in MHS, MS Museum 62; Foster (1893: column 456); Plarr (1899: 572); *WWW*; *DNB Missing Persons*: 353–4; and see below.

even his youngest son, Frederick Jeremiah Smith, was able to be a noted philanthropist in religious and educational causes. Having entered the church and become Prebendary of Wells, he founded and paid for two new churches in the suburbs of Taunton, and promoted the cause of elementary education on Church of England principles. His only son, Frederick John, was intended for the church and given a single-mindedly classical education by private tutors. Thus it was that Jervis-Smith came to Oxford rather late, at the age of 20, read for the Pass degree at Pembroke College, graduating in 1872, and then attended Wells Theological College. Upon being ordained he became his father's curate and organist at St John's, Bishop's Hull. He had been taught to play the organ by Sir John Stainer, one of those who revitalized the serious teaching of music in Oxford at this period. Jervis-Smith's friends considered him a highly accomplished keyboard player.[15]

But his passion and genius from an early age were for experimental physics and mechanics. He was acquainted from his youth with the engineer William Ellis Metford, who lived in Taunton and had his laboratory there. Metford had worked as a railway engineer but then became distinguished in experimental precision engineering, especially in the field of firearms. He was a great influence on the young Jervis-Smith, as was the work of Metford's own mentor William Froude. These were the two names universally admired in English mechanical-engineering circles, amateur and professional, at this period. It seems likely that Jervis-Smith was present at Froude's important talk on his turbine dynamometer at Bristol in 1877, later describing it as 'a monument to mathematical acumen and its practical application'.[16] Jervis-Smith learned so many secrets of the workshop from Metford, and had such natural aptitude, that friends who knew him later in Oxford mistakenly assumed that he had been an apprenticed mechanical engineer before studying for the church. In fact, his formal education had been so entirely classical that he had to be coached in arithmetic for Responsions, and later, in the mid-1870s, sought private tuition in more advanced mathematics and physics from one of Oxford's freelance coaches, Henry Tootell.[17]

In his Taunton workshop Jervis-Smith conducted serious experiments in precision mechanics and applied electricity, his main early interests being dynamometers and telephones. As amateur mechanics typically did in this inventive era, he replicated and improved new inventions as they were announced. The telephone commanded his attention very soon after its invention, the details being published in *The English Mechanic* in 1877. Experiments with microphones naturally followed. In 1879 he invented the liquid microphone, which was of little more than academic

---

[15] Boys (1913: iv); Anon. (1911b); on Stainer see Charlton (1984); on music and physics in Oxford see Simcock, chapter 4, 145–7.

[16] Jervis-Smith (1915: 98–9); Froude (1877); on Metford see *DNB*; for Froude see above.

[17] Anon. (1911a: 318, 390); notes by Mrs Jervis-Smith in MHS, MS Museum 62.

interest at the time but proved important several decades later when a more powerful microphone was needed for wireless telephony (radio). He revived the idea in the 1890s. In the 1880s he took out two patents relating to telephones.[18] The improved dynamometers he had constructed were shown at the Paris Electrical Exhibition of 1881 and the Inventions Exhibition at South Kensington in 1885, winning medals at both. In 1881 he invented the 'cradle dynamometer', for testing small electromotors. The mechanical integrator (1882) and the integrating planimeter (1884) followed, the former making possible his integrating transmission dynamometer, exhibited in 1882, which he considered a true ergometer.[19] What he was doing was turning the dynamometer into a more accurate composite device, which would integrate the several measurements needed to produce an 'estimate' (as he preferred to call it) of the actual work done by different types of prime-movers, motors, and other machines. The integrators were essentially calculating mechanisms. For this reason he promoted the word 'ergometer', his classical education offended by 'dynamometer', a term that implies only force is measured. In 1884 the technical publisher Spon issued his pamphlet *On Some New Forms of Work-Measuring Machines . . .* , describing the ergometers of various inventors, but mainly his own (Jervis-Smith 1884).

On 6 February 1884, at the persuasive invitation of the Bishop, Jervis-Smith gave his first public lecture to the working men's club associated with Wells Cathedral, though it was given in the town hall. He demonstrated a series of experiments to illustrate 'the forces in nature' and 'the storage of energy', including at the Bishop's request 'your wonderful phonograph'.[20] Edison's invention, announced in 1877, was not perfected and commercially marketed until 1887 – so this was Jervis-Smith's own replication or improvement of Edison's design. Popular science lectures were well attended at this period, and the audience response to Jervis-Smith's was most appreciative. His wife later remembered it as a turning point in his life.[21] It seems that he discovered an unexpected aptitude, and delight, in communicating his scientific knowledge and enthusiasm. Moreover, word of it probably reached the local clergyman-schoolmaster who was now President of Trinity College, Oxford, John Percival, who had many acquaintances in the diocese. As the idea of a mechanical lecturership and laboratory at Oxford came under discussion, Percival appears to have sought out Jervis-Smith and invited him to collaborate in the enterprise: taking the post at a modest salary and removing his research activities to

---

[18] Jervis-Smith (1879; 1910a); O'Dea (1934: 84); MHS, leaflet entitled *Notes on a Liquid Rheostat* . . . (dated 1892, but actually later); patent nos. 4758, 1881 and 18715, 1888.

[19] Jervis-Smith (1885: read November 1883); Thompson (1884: 383–4); Jervis-Smith (1915: ix, 47–50, 58–61, 125–7, 168–71, 218–20); MHS, leaflets on integrator (1882) and integrating planimeter (1884); Jervis-Smith (1883); for later work see below.

[20] MHS, MS Museum 62, letters from Bishop to Jervis-Smith, 10 and 22 December 1883, with Jervis-Smith's annotations.   [21] Notes by Mrs Jervis-Smith in MHS, MS Museum 62.

the laboratory that Trinity College would build. If the redoubtable Prebendary Smith had not died later in 1884, it is questionable whether his son would have felt able to abandon the ecclesiastical career he had been groomed for. He dutifully succeeded his father as Vicar, but relinquished the post two years later, having in the meantime accepted the Oxford appointment.

TRINITY COLLEGE AND THE MILLARD BEQUEST

In 1873 Trinity College received an unexpected bequest of £8000 from one Thomas Millard of Bristol, the proceeds from its investment to be used 'to advance Mathematical and General Science in the College'. There is no record of how the bequest came about, and nothing more is known of the benefactor, who had not been a student at Oxford.[22] At the time, Trinity was a very small college with no recent reputation in science and few science students. Science in Oxford, however, was in a vigorously developing state, with the Clarendon Laboratory just opened and a new observatory about to be built. The undergraduate syllabus in natural science had recently been reformed, introducing preliminary examinations and the freedom to specialize for finals. The best way for a small college such as Trinity to keep pace with these opportunities and to cultivate scientific studies was to co-operate with other colleges; the notion of the colleges as bastions of privilege and selfishness, embattled in mutual unco-operativeness, is a myth. The rising ethos of the time, for the liberal and reforming colleges anyway, was not simply to collaborate amongst themselves but to use their financial resources in ways that would benefit the university, or student body, as a whole. This was the philosophy embodied in the third report of the Devonshire Commission in 1873, which explicitly encouraged college involvement in science (*Devonshire Commission Third Report* 1873). In this regard, the Millard bequest allowed Trinity College to take the lead in Oxford.

The immediate uses to which the college devoted its new income were to finance scholarships for its science students, and to fund a new physics teaching post, the attachment of which to Trinity was largely nominal. The Millard lecturership in physics was conceived in discussion with other colleges, or their science dons, and its main purpose was to provide open lectures in elementary physics and mechanics, subjects that all natural science students were now required to study for the preliminary examination. Trinity entered a co-operative arrangement with two

[22] *OU Calendar*, 1874; information from Mrs Clare Hopkins, Trinity College archives.

of the strongest scientific colleges, Magdalen and Merton, for the shared teaching of science, each providing one teacher: Magdalen the biologist (Edward Chapman) and Merton the chemist (initially T. H. G. Wyndham). Merton was also in a position to provide fellowships in science: in 1873 Archibald S. L. Mac Donald obtained the physics fellowship by examination, and was then appointed the first Millard lecturer in physics. He left in 1876, and was followed by several deputies and successors, each only briefly: J. W. Russell, who lectured as 'Deputy of the Millard Lecturer' in 1876; Russell's Balliol pupil Lazarus Fletcher, 1877–8; Herbert Basil Jupp, deputizing in 1878; and Alexander Macdonell, 1878–9.[23] Trinity entered into a new arrangement with Balliol College (and also with Exeter) in 1879, which involved sharing Balliol's newly built laboratory and physical lecture room under the direction of the Millard lecturer in physics, the chemist H. B. Dixon being appointed. Elementary physics lectures open to all students continued to be his main duty.

As we saw in the preceding chapter, Dixon not unreasonably limited himself to the areas of physics within the competence of a physical chemist, leaving aside the more mathematical and mechanical topics. Since his were the main college lectures in elementary physics, the want of these other subjects, especially mechanics and hydrostatics, was felt generally around the University. In 1884 – significantly – Edward Chapman, the most influential science tutor in Oxford, reported to his college that 'In the Preliminary branch a great want is felt for instruction in Mechanics, a serious stumbling block to many beginners . . .' (Gunther 1924: 311, quoting Chapman report of 1884). J. W. Russell had at first continued the mechanics lectures in the Balliol–Trinity lecture-room (1880); but his teaching burden as mathematical lecturer to the three neighbouring colleges – Balliol, Trinity, and St John's – was already heavy.[24] The idea of a mechanical lecturership and laboratory offered independence and continuity for the subject. A natural progression from this was to appoint not a mathematical physicist but an experimental mechanic, who would be interested in extending teaching in the direction of practical work and applied mechanics; this would complement the existing provision for the physical sciences in the Balliol–Trinity–St John's enclave. Here then is the immediate local inspiration for what came to be the Millard Laboratory.

A number of other strands came together in Trinity's bold decision to establish a mechanical laboratory. These included, as we have seen, the encouragement given by the Devonshire Commission to college initiatives in science; and the establishment in principle (by the statutes of 1882) of new professorships in physics and in engineering, the funding for which did not materialize. Clifton's original ambition

---

[23] See Simcock, Chapter 4, 123; for Russell, OUG, 7 (1876–7): 48 and 111 (24 October and 21 November 1876); for Fletcher, OUG, 7 (1876–7): 315 (13 April 1877); for Jupp, OUG, 8 (1877–8): 335 (26 April 1878); for Macdonell, OUG, 9 (1878–9): 25 (11 October 1878); Gunther (1924: 23, 119, 215, 309–10).

[24] OUG, 11 (1880–1): 111 (15 October 1880); on Russell see Simcock, Chapter 4, 128, 133.

(in 1874) had been for three new chairs in physics, one in 'experimental mechanics', and one in engineering.[25] In 1884 he and Chapman were lobbying Merton College to establish a mechanics chair. Much has been said of how frustrating the failure of these aspirations was for Clifton, but it was equally so for those trying to promote science in the colleges. In the circumstances, it seems not implausible that Trinity, having identified itself with the cause of physics, felt that, while its windfall from Millard was not sufficient to allow any grand gesture, it could in a modest and interim way carry forward these intentions.

The progressive philosophy of Trinity College had been both signalled and reinforced by its choice of a new President in 1878. The renowned educational reformer John Percival was 'known throughout the country, and especially in the west, for his exertions for the spread of university education among the middle classes'.[26] He advocated university colleges in the major towns (helping to establish that at Bristol), and links through technical education 'between the centres of culture and the centres of practical activity' (Goldman 1995: 25, quoting Percival speech of 1870). Being an Oxford graduate himself, and having already cultivated support from Jowett and others for his local projects, it was not ridiculous for Percival to think that by 1878 his work could be translated on to a higher stage. In Oxford he allied himself to the friends of reform, and assumed an active role in the spheres of science, women's education, and university extension (Brittain 1960: 44, 49, 59, 69, 71; Goldman 1995: 25, 26–7, 61). At the same time Trinity College experienced a large increase in student numbers and began a programme of new building. Even so, Percival's presidency (1878–87) is remembered as a disappointment, and he soon returned to his preferred arena, as headmaster of Rugby School.

For it was in school teaching that he had formed these ideals and acquired his reputation. He was the first Headmaster of Clifton College, near Bristol (1862–78), which he quickly established as one of the most reputable English public schools. Its modern outlook and liberal ethos were his; more distinctively, it was one of the few schools that made serious provision for science at this period. Percival had engaged a brilliant and charismatic young physics master (from 1870 to 1874) who established the school's physics laboratory and also set up a mechanical workshop, which remained in use for many years. This was John Perry, later the scourge of Oxford science, whose famous accusation that science was 'trifled with, feared, and hated at Oxford' is still debated and misquoted a century later.[27] His crusade on behalf of scientific and technical education made him impatient with the slow pace at which the older English universities were modernizing. But the roots of his

[25] MHS, MS Gunther 65; see Fox, Chapter 2, 62 and 76.
[26] Plarr (1899: 846); on Percival see also Foster (1893: column 448); Temple (1921); *DNB*.
[27] Perry (1903–4: 214, 270); Howarth (1987: 335); on Perry see Plarr (1899: 849–50); for the influence of Clifton College see, for instance, Vernon and Vernon (1909: 112–13 (T. H. Warren speaking in 1908)).

particular hatred of Oxford went much deeper, as Gooday has shown.[28] Distinguished for his research and inventions at the interface of physics and engineering, Perry's interests were extremely similar to those of his contemporary Jervis-Smith, whose work he seems to have either overlooked or dismissed as insignificant. Yet ironically, both began their academic careers under the patronage of the same man.

Percival brought to Trinity not just a commitment to making education relevant to the citizens of an industrialized society, but a vivid sense of how the laboratories at Clifton College had invigorated the school's scientific side and embodied this ideal. One of his first moves was to propose the collaboration with Balliol in 1879. By 1884–5 the success of this venture was apparent, and its financial position improved by the Duke of Bedford's gift to Balliol.[29] In the same years there had emerged an acknowledged need in the university for physics to be expanded, firstly by catering for the mechanical areas of the subject, in the long term by fuller provision for electricity and engineering. One option was thus for the Millard Laboratory to be a university laboratory in the full sense, but built at the expense of the Millard fund. Trinity College broached this idea with the university in 1885. Guided by the science professors who were collectively responsible for the site, the university declined the offer, stating that there was no space in the 'Museum grounds' (the Science Area) that was not already earmarked for future developments (Woods 1887). The absurdity of this is demonstrated not only by the extremely small space into which Trinity was able to squeeze its new laboratory, but by the fact that within two years the physics professor was applying for funds to build annexes to the Clarendon Laboratory for elementary physics and electricity – funds which shortly before he might have negotiated from Trinity.

Clifton's ambitious 1887 proposal illustrates the tactlessness which Gooday has commented upon. Trinity College was understandably piqued. The new President, H. G. Woods, circulated a pamphlet opposing it on the grounds that the university need not have been burdened with such a cost had Trinity's original offer been accepted, and that the needs which Clifton identified were 'already provided for, without cost to the University' by the Millard Laboratory and the Balliol–Trinity Laboratories, even down to the detail of having the electrical power plant Clifton required. The implication was that Clifton was being disingenuous about the pressures upon his department's resources, in order to reclaim for it territory lost to the colleges. Heads of colleges being the most powerful lobby in the university, and finance usually being a decisive issue, Woods had a persuasive case: that 'the trust funds at their [Trinity and Balliol's] disposal have, as they believe, been laid out for the benefit of the University at large', so

---

[28] See Gooday, chapter 3, 100.   [29] Smith (1982: 195, 197–8); see Simcock, Chapter 4, 130–4.

that a virtually duplicate facility 'at a considerable cost' could not be justified.[30] In an Oxford where centralization was not in itself a compelling argument, Woods's revelation was devastating to Clifton's case. On the other hand, of course, it unintentionally added another obstacle to the full-scale evolution of physics and engineering, which elsewhere was to proceed at a pace that would make Oxford's tardiness a matter of public shame.

THE MILLARD LABORATORY

In 1885 Frederick J. Jervis-Smith was appointed Millard lecturer in experimental mechanics and engineering at Trinity College. A long, narrow, single-storey building was erected in Dolphin Yard, essentially a passage squeezed between Balliol and St John's Colleges, forming a rear entrance to Trinity from St Giles. Initially referred to as the 'Millard Laboratory and Workshop for Experimental Mechanics and Engineering', the un-academic term 'workshop' was rapidly dropped. Its modest dimensions were Jervis-Smith's first challenge: his colleague Dixon observed that 'the space at his disposal was utilized with the greatest ingenuity'.[31] It was fitted out with the existing equipment of his workshop at Taunton. He also engaged a laboratory assistant at his own expense. The building and installation were due to be completed by the beginning of 1886, and a printed notice or prospectus was circulated as early as September 1885 (Fig 5.4a). But it was May 1886, the start of Trinity term, before the *Oxford University Gazette* announced that the laboratory was open for practical work and instruction.[32] These early notices give few details of the proposed course or syllabus; and such instruction as was given in this first year (only one name being entered in the student register) was probably of an unsystematic kind. A formal lecture course was first given in Lent term, 1887, and routinely included in university lecture lists thereafter.[33]

The Millard Laboratory's initially stated purpose was 'for instruction in Theoretical and Practical Mechanics and Engineering'.[34] It seems to have been the intention from the outset to teach the 'experimental mechanics' part of the existing physics syllabus, and use this as a bridge to the new subject of 'applied mechanics',

---

[30] Woods (1887); Clifton's defeat is usually attributed to the other argument levelled against him, that of low student numbers: see Howarth (1987: 343) and Gooday, Chapter 3, including his concluding comments on Clifton.
[31] TCA, Order Book, 16 October 1885; TCA, 1885 leaflet; MHS, 1900 testimonials (Dixon's letter); on Dolphin Yard see Laidler (1988: 233–6 and plans: 225, 234).
[32] TCA, 1885 leaflet; *OUG*, 16 (1885–6): 498, 514, and 532 (4, 11, and 18 May 1886).
[33] MHS, MS Gunther 55 (see below for analysis of this document); *OUG*, 17 (1886–7): 369 (25 April 1887) and *OUG*, 18 (1887–8): 23 (14 October 1887), etc. [34] TCA, 1885 leaflet; Woods (1887).

Fig. 5.3. Frederick John Jervis-Smith, in a photograph probably dating from 1885, the year of his appointment as Millard lecturer in experimental mechanics and engineering at Trinity College.

or engineering.[35] Entering academia from the amateur's workshop, however, Jervis-Smith required some acclimatization. In his first leaflet he misguidedly targeted (after prospective engineers, physics students, and students of electricity) 'Those who desire to learn the use of tools, general carpentry and handicraft . . .'.[36] He doubtless gave instruction of this kind informally; but he quickly learned that woodwork was not something to be included in an Oxford syllabus. Instead (he tells us) he visited 'many Science Schools and Laboratories in England and on the Continent', comparing their courses and teaching methods. Jervis-Smith finally modelled his upon that devised by (Sir) Alexander Blackie William Kennedy, the

---

[35] Experimental mechanics (e.g. in Jervis-Smith's job title) was not an *ad hoc* term: it was a well-established name for the subject that taught mechanics non-mathematically by means of apparatus and demonstration-experiments, as exemplified in the textbook by Ball (1871). Applied mechanics similarly was not an Oxford affectation but the name for mechanical engineering in the Science and Art Department syllabus used in technical schools and extramural classes, for example Cryer and Jordan (1888), which also reproduces the syllabus and recent examination papers.   [36] TCA, 1885 leaflet.

pioneer of the concept of the academic 'engineering laboratory', who established the laboratory at University College, London, in 1878 (Gooday 1990: 47–8). Although Kennedy's background was as a practising engineer, his approach was not to provide professional training but to teach the scientific techniques of precise measurement, the use of instruments, and 'the art of making experiments'.[37] The difference from a practical physics course was largely one of emphasis.

Kennedy's plan transferred perfectly to Oxford's world of experimental finesse: 'In a University course', Jervis-Smith concluded, 'the best method of teaching appears to me to be that which has to do with Accurate Measurements, Testing the Strength of Materials, and the determination of the efficiency of Prime Movers . . .'.[38] The last, of course, was his own special field of dynamometry. Oxford's limited definition, during the twentieth century, of its engineering syllabus as 'engineering science' has sometimes been considered symptomatic of a snobbish and anti-vocational attitude, but in fact it represents the predominant approach to engineering as a university subject evolved by Kennedy in London and, previously, by W. J. M. Rankine in Glasgow from the 1850s, mediated to Oxford by Jervis-Smith (Channell 1982; Buchanan 1985; Marsden 1992). So on mature reflection he came up with a carefully worded manifesto for the new subject:

The object of the course of instruction is to place before students those fundamental principles on which Mechanical and Electrical Engineering, as well as certain branches of Physics, are based, and without which the manual training of a workshop appears to be incomplete. The principal training is in the experimental methods by which the constants most frequently employed in Mechanical and Electrical Engineering are determined: in this branch of work modern graphic methods are largely used, mechanical work and fitting is taught only so far as is necessary to show how machine and other tools are used, and processes of construction conducted . . . [39]

A brief syllabus of 1888 lists the four sections of the course as: Machine Construction, Mechanical Drawing, Machine Tools and Engine and Dynamo Tests, and Graphical Statics (Figure 5.4b).[40] The last (a body of graphical and computational methods which remained a standard part of proper engineering training) was sometimes given as a separate course. The outline was flexible and evolutionary over the years, greater emphasis being introduced on electrical

---

[37] Jervis-Smith's post-1897 brochure; Kennedy (1886: 70–3), a long extract from which is quoted in Jervis-Smith's brochures; see Gooday (1990: 47–8). A leaflet promoting the Millard Laboratory and its engineering course exists in MHS and TCA in several variants, that headed *The Millard Laboratory, Oxford* under the name of Smith dating from before 1897 and that headed *The Millard Mechanical and Engineering Laboratory, Oxford* under the name Jervis-Smith from after (cited here as Jervis-Smith's pre-1897 and post-1897 brochures, respectively).

[38] MHS, 1890 testimonials; reiterated in Jervis-Smith's post-1897 brochure.

[39] Jervis-Smith's pre-1897 brochure.   [40] TCA, 1888 syllabus leaflet.

engineering for instance. It was a laboratory-based course of practical instruction, essentially given on demand, sometimes to individual students and tailored to their needs or interests. The numbers attracted, however, were not large. The difficulty in drawing a sustained attendance to an extra-curricular course was one of the inevitable drawbacks to this back-door method of introducing a new subject: 'A race without a goal is never popular, not even in a University', as he later wrote.[41]

Redeeming this situation to a great extent, the staple teaching that appeared in the lecture lists consisted of an experimentally – illustrated lecture course in elementary mechanics and physics tailored to the natural science preliminary course, and supplemented by practical instruction. This brought students to the Millard Laboratory, while fulfilling an acknowledged need in the existing curriculum. Additional advantages of making it the core of the laboratory's teaching were that the same course was suitable for Final Pass School candidates, for medical students (usually reading physiology), and (seemingly) could serve as the starting point for the engineering course too. These lectures were given on two or three days of the week each term from 1887 until 1904 (when Preliminary physics teaching was centralized by Townsend), under slightly varying headings such as 'Experimental Mechanics' (1887), 'Mechanics and Hydrostatics' (1888–90), and 'Mechanics and Physics' (1890 onwards). Practical work and instruction were available each weekday. The fee for the lectures was the usual £1 per term, and use of the laboratory (inclusive of lectures and instruction) was three guineas per term (one guinea for members of Trinity College).[42]

The fact that his teaching was open to all members of the university was recognized by Jervis-Smith's appointment in 1888 as university lecturer in mechanics (classified as part of the subject of physics), 'to give lectures in Practical Mechanics and Experimental Physics'.[43] It brought him a supplementary salary from the recently established Common University Fund, which he used towards running costs of the laboratory. Although a post of modest status, this was an important acknowledgement not just that a unique extension to the curriculum was being offered, but that the university accepted the need for it. It was also an answer to Clifton's application of the previous year, though not one calculated to placate the professor, endorsing as it did the college solution that had helped to defeat him. And it made Jervis-Smith at the time the only university-employed teacher in physics outside the Clarendon Laboratory. His integration into the mainstream of Oxford science had begun before this, 1887–8 being the first of various years in

---

[41] Jervis-Smith, 'Engineering at Oxford', cutting from newspaper (title unspecified) dated 11 December 1907 in MHS.

[42] *OUG*, 16 (1885–6): 498 (4 May 1886); *OUG*, 17 (1886–7): 369 (25 April 1887); *OUG*, 18 (1887–8): 391 (23 April 1888); *OUG*, 21 (1890–1): 33 (13 October 1890), etc. Cf. letter on fees from Jervis-Smith to Gunther, 2 July 1903, MHS, MS Museum 120, item 3. [43] *OUG*, 18 (1887–8): 554 (12 June 1888).

which he acted as examiner in physics and mechanics for the pass school. In 1889 he served on the committee preparing a new Preliminary physics syllabus, ready for the introduction of practical work into the examinations in 1891. And in his last years (1905–7) he sat on the Board of the Faculty of Natural Science during the time that engineering science gained formal recognition.[44]

Another important expression of confidence in the new venture came from St John's College, one of the colleges nominated to fund a future engineering chair. The acoustic laboratory that R. H. M. Bosanquet had set up there in 1880–1 stood close to the Millard Laboratory, being one of the buildings forming the opposite wall of Dolphin Yard. Clearly the two laboratories – and the two men – had much in common. In 1887 Bosanquet decided to abandon his laboratory, obviously prompted by seeing an alternative future for it. He gave Jervis-Smith his extensive and costly apparatus, much of which was concerned with the workshop and engineering aspects, though it also included acoustic and electromagnetic instruments; while the two rooms – Bosanquet's original laboratory and the extension he had built against it – were lent by St John's College to Trinity as an annexe to the Millard Laboratory.[45] The transfer began in the summer vacation of 1887, and the new rooms were used from the beginning of 1888.

Jervis-Smith's brochures of the 1890s describe the accommodation and equipment of the extended Millard Laboratory. There were six rooms: Lecture Room, Experimental Room, Engine Room (containing the steam and gas engines), Machine Room (containing the lathes and other engineering tools), Instrument Room 'devoted to electrical testing', and Book Room (also containing drawing, graphical, and calculating facilities).[46] Jervis-Smith had moved his lectures to one of the two St John's rooms, Bosanquet's laboratory having originally been a lecture-theatre. The Millard building itself consisted in the 1930s (when it was a research annexe of the Balliol–Trinity Laboratories) of four rooms in series plus a store room.[47] The three-horse-power steam engine with boiler and dynamos, and Bosanquet's gas engine, were objects of study themselves, and at the same time functioning power sources, operating other machinery, running electric lights, and supplying electricity by wire to the Balliol–Trinity Laboratories about fifty yards away (Woods 1887). Mains electricity was connected later in the 1890s. The usual didactic apparatus and measuring instruments were supplemented by Bosanquet's gift, and around the turn of the century by additions to cater for the

---

[44] *OU Calendar* for 1888 (examiner); printed report in MHS, MS Gunther 65 (committee); MHS, MS Museum 89: 60 (board).
[45] Jervis-Smith's pre-1897 and post-1897 brochures; Woods (1887); TCA, Order Book 1885–94 (14 December 1887) (information from Dr Joanie Kennedy).
[46] Jervis-Smith's pre-1897 and post-1897 brochures.
[47] TCA, 1888 syllabus leaflet; Laidler (1988: 234); information from Professor Keith Laidler; his photograph of the interior of the main room in 1937 is reproduced in Laidler (1988: 236) and Morrell (1997: 329).

> # TRINITY COLLEGE, OXFORD.
>
> *Millard Laboratory and Workshop for Experimental Mechanics and Engineering.*
>
> This Laboratory will be opened in January, 1886, for instruction in Theoretical and Practical Mechanics and Engineering.
>
> **THEORETICAL INSTRUCTION.** The principles on which the strength, arrangement, and proportions of machines are determined.
>
> **PRACTICAL INSTRUCTION.** The construction of machines, engines, and instruments of precision. The use of hand and machine tools, working in wood and iron, &c.
>
> ---
>
> It is hoped that the instruction provided will be of special use to the following classes of students:—
>
> 1. Those who look forward to engaging in manufacturing or engineering work as a profession, and who wish to begin their preparation for it during their University course.
>
> 2. Those students in Physical Science who may wish to work out Physical Science questions, either with instruments of their own construction, or with instruments provided in the Laboratory.
>
> 3. Those who wish to carry out electrical testing, as required in electric lighting engineering, and telegraphic and telephonic engineering.
>
> 4. Those who desire to learn the use of tools, general carpentry and handicraft, the elements of practical building or engineering, with a view to emigrating, or going into some form of practical business or industry at home or in the colonies.
>
> ---
>
> The Laboratory is furnished with Models for shewing the construction of machines: Lathes: Planing, Shaping, and Drilling machines driven by steam power; and all the usual tools employed in mechanical work in wood and metal: Dynamo-electric machines: Motors, arc lamps, and incandescent lamps: Mercurial pumps used in the manufacture of incandescent lamps: Ergometers, or work-measuring machines: Photometric apparatus: Electrical test instruments: Wheatstone's bridge, resistance coils, large standard galvanometers, voltmeters, ammeters, reflecting galvanometers, and all the usual instruments required for electrical investigations.
>
> ---
>
> **TERMS.** The Laboratory is open to all Members of the University during the University Terms, and at such other times as may be arranged.
>
> The minimum fee charged will be £3 3s. per Term. This fee admits to the Laboratory and all lectures or other instruction given there, on three forenoons or three afternoons every week during the Term of eight weeks. A larger number of weekly attendances is charged proportionally. Each student pays for the materials he uses, but there are no other extra charges. He is at liberty to take away his own work when finished. The minimum fee for Members of Trinity College will be one guinea per Term.
>
> The Laboratory is under the superintendence and direction of the Rev. Fred. J. Smith, B.A., M.S.T.E., Millard Lecturer in Experimental Mechanics and Engineering.
>
> Any gentlemen wishing to become students are requested either to write to Mr. Smith, or to call upon him at the Laboratory, Trinity College (St. Giles's entrance).
>
> *September, 1885.*

Fig. 5.4a. The first printed notice or prospectus describing Trinity College's Millard Laboratory and the courses and other instruction that Jervis-Smith intended to offer there. The notice was issued in September 1885, before the laboratory was even built. Trinity College archives.

> # MILLARD LABORATORY.
>
> ### TRINITY COLLEGE, OXFORD,
> Lent Term, 1888:
>
> #### SYLLABUS.
>
> *Subjects*—I. Machine Construction.
>   II. Mechanical Drawing.
>   III. Machine Tools and Engine and Dynamo tests.
>
>   IV. Graphical Statics.
>
> Introduction.—Scalar, Vector, and Rotor Quantities.—The Geometrical Representation of Quantities by Lengths, Vectors, and Rotors.
>
> Graphical Mensuration, or the determination of areas by graphical methods.
>
> Graphical Determination of Mass-Centres, Areas, and Volumes.—The Planimeter. Engine test diagrams.
>
> THE LECTURES ON MECHANICS, TREATED EXPERIMENTALLY, WILL BE GIVEN IN THE NEW SECTION OF THE MILLARD LABORATORY IN ROOM A, ON MONDAYS, WEDNESDAYS AND FRIDAYS, AT 11 A.M., BEGINNING ON MONDAY, 16TH JANUARY.

Fig. 5.4b. Brief printed syllabus of Jervis-Smith's engineering course, issued at the beginning of 1888. Trinity College archives.

increasing interest in electrical rather than mechanical engineering. Jervis-Smith's own inventions and adaptations, notably dynamometers and chronographs, had a prominent place in the laboratory of course. But there was little money to spend on new items, and where possible they were made on site. In 1890 the apparatus was valued at £1500.[48] It can be assumed that it was not treated with the protective attitude for which Clifton was notorious, for the Millard Laboratory was also a working environment for Jervis-Smith's own research, with which his students were encouraged to feel involved. According to Dixon, himself a pioneer of student participation, Jervis-Smith's 'mode of teaching was a direct incitement to research'.[49]

[48] MHS, 1890 testimonials (Woods's letter).
[49] MHS, 1890 testimonials (Dixon's letter); cf. Soddy's anecdote below, and Jervis-Smith (1915: 3).

THE MILLARD LABORATORY REGISTER

The Millard Laboratory student register covers the years 1886 to 1901 inclusive.[50] Only one student – Charles Luard, a Christ Church undergraduate – is recorded in 1886 (Michaelmas term). Although the laboratory was first advertised in Trinity term of that year, the formal courses, in particular in elementary physics, began in Lent Term, 1887. There were eleven students in that term, and in the following term twenty-one; the average over the next four academic years was twenty per term. In Michaelmas 1891, however, the numbers dropped suddenly, and thereafter remained lower: nine that term, and an average over the next ten years of about ten per term, between unusual extremes of twenty and three (the latter annotated '7 really' by Jervis-Smith). This dramatic difference between the attendances for 1887–91 and 1891–1901 is obviously connected with the introduction of practical work in examinations for Preliminary physics. While this would lead one to expect an increase in demand, perhaps improved provision at the Clarendon Laboratory and elsewhere, coupled with the discouraging effect often found when an option becomes compulsory, accounts for the drop in numbers at the Millard Laboratory.

The total attendance over the sixteen years counted from termly figures (meaning that individual students attending for two or more terms are counted two or more times) comes to 595, which explains Jervis-Smith's own claim that '512 men' (some of them were actually not men) had studied in the Millard Laboratory in the twelve years 1888–1900.[51] The actual figure reduced to real individuals is 303. In status these divide as follows: 149 (49 per cent) were undergraduates who subsequently took honours degrees; 115 (38 per cent) were undergraduates who did not take honours; 24 (about 8 per cent) were graduates; and 15 (5 per cent) were recorded without university affiliation, some of whom were not members of the university, though the figure inevitably includes a few unidentified graduates and students. The non-honours students were largely attending for elementary physics and mechanics for the Pass examination, though perhaps some did not proceed to a degree at all. Of the 149 honours students, 133 (89 per cent) were reading natural science, 13 (almost 9 per cent of the 149) were reading for non-scientific degrees, chiefly history or law, and 3 were reading mathematics. Most of the natural science honours students were attending for physics and mechanics for the Preliminary examinations; the subjects they proceeded to specialize in were predominantly chemistry (61, 41 per cent of the 149) and physiology (51, 34 per cent), only nine of them being physicists (6 per cent), together with eight zoologists and four geologists. The small number of physics finalists and the dominance of chemistry both

---

[50] MHS, MS Gunther 55; this document presents the usual difficulties in confidently identifying all the names, numbers derived from it being therefore subject to a slight margin of error. No register survives for the years 1902 to 1908.   [51] MHS, 1900 testimonials.

Millard Mechanical Laboratory, Oxford.

ELECTRICAL ENGINEERING.
Note by the Lecturer.

Since I have had the management of the Laboratory, I have noticed that nearly all our students of late years, on leaving Oxford, have taken up Electrical Engineering as a profession, or are employed in some work requiring a knowledge of electricity. As this is so, in order that the Laboratory might be of use to students of electrical science, considerable additions have been recently made of dynamo-electric machinery and apparatus used in electrical research, and a room in the Laboratory has been furnished with complete apparatus for instruction in chronographic measurements, such as are made in ballistics and physiology. I may mention here, that several of the methods which have originated in this Laboratory have now become standard methods used both by Electrical Engineers and those engaged in professional science teaching; also the methods have been adopted by some of our leading engineers. Amongst these, about fifteen in number, I may mention the names of Sir W. Armstrong (Lord Armstrong), Whitworth & Co., Elswick; also the United States Artillery have kindly given a considerable place to our work in their professional journal.

Apparatus recently made available for teaching purposes in the Laboratory.

Dynamo by Siemens, continuous current.	Chronographic apparatus for instruction in the measurements of small intervals of time.
Dynamo, Triphase.	Traversing Micrometer Microscope for use with the above, Zeiss.
Motor and Dynamometer combined.	
Ballistic Galvanometer by Crompton.	Numerous sets of resistance coils.
Potentiometer, complete.	Different forms of accumulators.
Two Meter bridges. F. J.-S. pattern.	Two Induction coils by Apps, one 10 in. spark, one 8 in. spark, with Wehnelt breaks.
Mercurial Pump for exhausting X-ray tubes.	
Three dead beat Holden D'Arsonval Galvanometers.	Two Morse telegraphic instruments by Siemens & Co., and transmitting keys and batteries.
Complete apparatus for demonstrating wireless telegraphy.	Relays by Elliott.
Electric arc lamp.	The Laboratory is supplied with electricity, direct current, at 100 volts, by the Oxford Electric Co.
Eight electric measuring instruments for Volts, Amperes, and Watts.	

F. J. J.-S.

Fig. 5.5. Printed notice issued by Jervis-Smith, c.1900, responding to the increasing demand for instruction in electrical engineering.

reflect the broader situation in Oxford; the special relevance of Jervis-Smith's precision mechanics to physiologists helped bring this subject almost to a level with chemistry.

Certain college tutors habitually sent their students to Jervis-Smith for preliminary physics. Several of these were biologists: at Magdalen College Edward Chapman and his successor R. T. Gunther (a pupil of Jervis-Smith for three terms in 1888–9), and E. B. Poulton at various colleges where he served as general science don, were vigorous supporters from the start.[52] Tutors' names occasionally noted alongside names in the register include mathematics dons C. J. C. Price at Exeter College and H. T. Gerrans at Worcester, chemists V. H. Veley, A. F. Walden, and John Watts, and Jervis-Smith's friend the physicist P. J. Kirkby. Almost every Oxford college is represented in the Millard Laboratory register. The largest patrons were Magdalen, St John's, Keble, and Exeter, with Trinity itself not surprisingly in the lead, but not by so large a margin as to dispel the impression of a facility used by the whole University. The lower fee paid by Trinity and St John's students made it easier for them to sign up; several of the non-scientists were from Trinity, as were some of those attending for an unusually high number of terms (the record being seven). Another group with a strong presence was Non-Collegiate students, a formally regulated body whose members lived in private homes in Oxford. It was a characteristic of teachers or colleges interested in liberalizing and extending university education to be specially hospitable to these students. By the same token, a small but respectable number from the new women's colleges studied at the Millard Laboratory: twelve women can be identified in the register during these years (1886–1901), the first being Laura Lester from Lady Margaret Hall (two terms, 1889–90). Another was Evelyn Hannah Berkley (Lent term, 1890), one of the earliest women to take final honours in physics (second, 1893).[53]

The most distinguished scientists to have studied with Jervis-Smith were, by chance, two close friends. These were the metallurgist Sir Henry Cort Harold Carpenter, a chemistry student at Merton College under John Watts (first, 1896), who attended for three terms in 1893–4; and the chemical physicist Frederick Soddy (first in chemistry, 1898), who attended in Trinity 1894. Soddy's memoirs contain a rare glimpse of the informal and supplementary teaching that went on unrecorded, both Carpenter and himself (in 1892 and 1893 respectively) receiving vacation coaching from Jervis-Smith before coming up to Oxford, arranged by their schoolmaster. Soddy received a bad report, but bluffed his way out of it by attributing it to having broken Jervis-Smith's 'pet galvanometer'.[54] Even so, he retained an interest in mechanics and returned to it many years later (Soddy 1956). Of physiologists, several became prominent doctors and medical teachers, while Anthony E. Mavrogordato (three terms, 1892–3), a Trinity student (second, 1896), was later J. S. Haldane's research assistant, and Edward Whitley (Lent Term, 1899), also of Trinity (third, 1902), became the founding benefactor of the university's biochemistry chair.[55] Jervis-Smith's graduate patrons included a few senior members, such as Samuel Alexander, a fellow of Lincoln College (Michaelmas Term, 1891), and Henry

---

[52] MHS, 1890 testimonials (Chapman's and Poulton's letters); for the deliberateness of Magdalen's policy see Gunther (1967: 11).
[53] Avent and Pipe (1991: 37); the women students are included in the statistics given above.
[54] Howorth (1953: 16); on Carpenter see the obituary by Edwards (1939–41); *DNB*; *DSB*.
[55] On Whitley see Morrell (1997: esp. 186–9).

Bazett, a member of Worcester (three terms, 1898–1900), neither of whom was a scientist; while George Dexter Allen, a zoology graduate and a clergyman, attending over five terms in 1898–1900, probably represents the true amateur mechanic. The small number of non-university students included a boy from St Edward's School (two terms, 1895–6), and E. H. Swain, of 'Gas Works' (Michaelmas Term 1899), who was one of those explicitly studying engineering.

Disentangling the engineers from the majority of elementary physics students would require further research. Only fourteen names – annotated by Jervis-Smith with 'Eng' or similar – can be said with certainty to have gone through his engineering course; but if H. G. Woods was correct in 1890 in saying that there had been 26 engineering students, the true number (for the register's period, 1886–1901) should be almost 100.[56] Attending for a large number of terms is a possible diagnostic; though conversely some men submitted themselves for a crash course: 'Peak' of Hertford College enrolled for 'Mechanics Practical' for Michaelmas Term, 1895, but 'Left for India' within the month; and the electrical engineer Charles Caesar Hawkins, a Magdalen graduate in classics, was a pupil during the Easter vacation of 1887. Hawkins wrote a standard textbook on *The Dynamo* (1893), which reached its sixth edition in 1922. He became superintendent of the technology department at the City and Guilds of London Institute (1920–34).[57] Others studying engineering included L. G. Hughes, of Wadham (six terms, 1892–4), Frank Nicholls, of 66 Woodstock Road (six terms, 1892–4), and George Ward William Grazebrook, of Magdalen (fourth in chemistry, 1898) (four terms, 1896–9), who later worked for Wolseley Motors, Birmingham (Gunther 1924: 238–9). Grazebrook's Magdalen contemporary, Sir Kenneth Irwin Crossley (third in chemistry, 1898), was heir to the Manchester firm of Crossley Brothers, manufacturing engineers, though his two terms at the Millard (1896) were for routine Preliminary physics (Gunther 1924: 238; *WWW*).

JERVIS-SMITH'S OXFORD RESEARCH AND CAREER

Jervis-Smith's first invention announced from Oxford – obviously representing work done earlier – was an electromagnetic siren, described in October 1885. After a pause devoted primarily to establishing the Millard Laboratory and developing his teaching courses, he burst back into creative activity in 1888. In that year he took out patents relating to mercury pumps and telephones, invented the 'tram

[56] MHS, 1890 testimonials (Woods's letter).
[57] Gunther (1924: 123, 225, 487); *WWW*; Hawkins and Wallis (1893); Hawkins (1922–5).

chronograph', and was working on 'instantaneous' photography, galvanometers, and (perhaps inspired by setting up the laboratory) improvements to various other laboratory instruments.[58] His research flourished in its new setting, generating eleven patent applications in the years 1888–99, as well as regular talks and published papers.[59] It is appropriate that the most important work of the Oxford phase of his career should have been inspired by the chemical research that he found going on in Balliol College, where H. B. Dixon was using the chronograph acquired from the physiologist Yule in measurements of the explosion rates of gases.[60] Although Dixon left Oxford in 1887, Jervis-Smith continued developing his new form of chronograph, his aim being not simply precision but a uniformity of motion superior to the existing pendulum method. The arrangement of a carriage (carrying a smoked glass) travelling on a rail, the motion along which was perfectly even once the motive force ceased acting, gave the name 'tram chronograph' to his invention of 1888. It was first shown to the Junior Scientific Club in that year and described to the Royal Society in 1889, though not patented until 1894 and 1897.

The standardized version was commercially manufactured by the instrument makers Elliott Brothers, and was still in production in 1903, when Jervis-Smith renewed the patent and published an instruction booklet for it.[61] The tram chronograph's other original feature was a highly delicate and miniaturized electromagnetic stylus. The timing was derived from a standard Koenig tuning fork, and an accuracy of one twenty-thousandth of a second was possible. One of its virtues was that it was adaptable to different purposes, including to relatively slow as well as very rapid events: those mentioned in the literature were gas explosions and other chemical reactions, physiological phenomena such as impulses along nerves, the flight of insects, the velocity of gun shot and other projectiles, and the movement of sound through gases and solids. The most unexpected application came in 1897 when Jervis-Smith employed his chronograph to record the speed of a vehicle (a horse-drawn butcher's van) in the street in Oxford, and presented the result in

---

[58] MHS, leaflet on siren (1885); patent nos. 4780, 1888 and 18715, 1888; Jervis-Smith (1888); on the chronograph see below; patent no.19024, 1889 (photography); *Journal of the Oxford University Junior Scientific Club*, 1, no. 20 (1894): 272 (list of earlier papers, including Jervis-Smith's of 4 May 1887, on the D'Arsonval galvanometer, not printed); MHS, leaflets on galvanometer (1890) and air thermometer (1890); Jervis-Smith (1889a; 1889b; 1890a).

[59] The main sources used to reconstruct Jervis-Smith's scientific work are his published papers, supplemented by a collection of archives and more ephemeral printed material in MHS, chiefly: patent documents (MS Museum 29); letters and obituaries (MS Museum 62); scrapbook of newspaper and magazine cuttings (MS Museum 89); job applications and testimonials from 1890 and 1900; bound collection of some published papers and privately issued leaflets (the latter mostly issued to accompany talks or exhibitions); and various other pamphlets and photographs; the notes that follow attempt to be representative rather than comprehensive.   [60] See above, and Simcock, Chapter 4, 123 and 133.

[61] Patent nos.16884, 1894 and 25526, 1897; *Journal of the Oxford University Junior Scientific Club*, 1, no. 21 (1894): 280 (list of earlier papers, including Jervis-Smith's of 15 June 1888, not printed); Jervis-Smith (1889c); Jervis-Smith (1890b); Anon. (1897–8); Anon. (1898); Jervis-Smith (1903); Jervis-Smith (1926: 303–4); Anon. (1911a); Boys (1913: v); Laidler (1988: 249–50); Boullin (1989).

Fig. 5.6. The earliest prototype of Jervis-Smith's tram chronograph, spread across two tables in the Millard Laboratory and photographed in about 1888; the sheets of white paper are merely added for visibility. The smoked glass plate on which the tracings are made leans behind, revealing the carriage with three wheels running within the two long rails. The weights are the motive power. The most significant innovation in Jervis-Smith's design is that a uniform velocity is achieved for a period after the weights have stopped falling (the principle of the Atwood machine); this is when the recording is made. The styli and tuning fork, which mark the event and timing respectively on the glass, are here mounted together on an ordinary laboratory stand.

Fig. 5.7. The 1890 prototype of Jervis-Smith's tram chronograph, posed on a table seemingly outside the Millard Laboratory; the instrument is exactly as described in his article in the *Philosophical Magazine*, 29 (1890): 377–83. The smoked glass plate is *in situ* on its carriage, and in front of it stand separately the tuning fork (the timing device) and the greatly improved electromagnetic styli that record the event being timed. Other improvements are the three pairs of adjustable feet for precise levelling, and a small pulley immediately above the weight, which increases the velocity achievable.

Fig. 5.8. The final production model of Jervis-Smith's tram chronograph, photographed about 1898. It was manufactured in this form by Elliott Brothers from 1897 and described by Jervis-Smith in an instruction booklet of 1903. In this version, mainly for compactness, the two rails have moved closer together and the wheels of the carriage run on the outside. The slight gradient at which the whole instrument is mounted can be clearly seen.

court in a case of 'furious driving', undoubtedly the first instance of the use of a 'speed trap' in the enforcement of traffic law. The vehicle was travelling at twelve miles-per-hour.[62] In 1895 he was appointed with several other experts to a government Commission to look into explosions of compressed gas cylinders, including how their manufacture and use should be improved to reduce the risk of explosion. They conducted experiments at Woolwich, and issued their report in 1896.[63] For the eleventh edition of the *Encyclopaedia Britannica* (1910) he wrote a new article on chronographs which amounts to a comprehensive historical and technical description of the different types (Jervis-Smith 1910b; 1926).

That he first published his chronograph in the uncharacteristic context of a paper on Dixon's subject of the velocity of gaseous explosions, communicated to the Royal Society by Harcourt, shows how he was immediately embraced by the physical chemists. He briefly pursued an aspect of Dixon's work, measuring the 'acceleration period' of explosions. But he escaped the Oxford fate of being annexed by chemistry, and concentrated his attention on developing the instrument. Its invention is a striking illustration not just of interdisciplinary collaboration, but of the benefits of the deliberate clustering of physical expertise

[62] *Oxford Times*, 17 July 1897.
[63] *The Times*, 26 June 1895; other cuttings in MHS, MS Museum 89; Anon. (1896).

on the Balliol–Trinity site. Yet there was also the danger that a mechanical laboratory staffed by a generous and modest genius would simply be seen as a means of technical support. This impression is given by several of those who wrote testimonials for Jervis-Smith in 1890, their flattery conveying less a potential Cambridge professor than a kind of glorified interdepartmental technician.[64] This inappropriate if well-meaning attitude to him diminished in time, not least because he pursued his subject, as researcher and as teacher, in a way that lent dignity and coherence to it as an academic activity in its own right. Jervis-Smith became an FRS in 1894; and moved in circles that defined him as a mainstream if minor research physicist pursuing common concerns with the likes of Kelvin and Rayleigh, as well as his closer scientific friends Sir Vernon Boys, Sir Oliver Lodge, and Silvanus P. Thompson.[65] But it remained fundamental to his nature, as his obituarist recorded, that 'He would put himself to endless trouble to help a friend in any experimental problem, and he always managed to convey the idea that one was doing him a service by asking for his help' (Anon. 1911a).

Jervis-Smith was fascinated by the possibilities of photography and related imaging techniques, both in themselves and as methods of scientific investigation. He was one of those (such as Mach and Boys) who used an electric spark to photograph high-speed events or projectiles as a serious tool in their analysis and measurement. He patented his version of instantaneous photography in 1889. Boys, in his 1892 British Association lecture to artisans, on photography of things moving at speed, described some of Jervis-Smith's work, and showed a photograph taken by him of the famous train the Flying Dutchman at full speed. He also devised a way of using spark photography in conjunction with the tram chronograph, especially to record the flight of bullets.[66] His process using electrical induction to create contact images of metal objects such as coins and engraved copper-plates was patented in 1892; he named it 'inductoscript' and publicized it widely over the next two years, convinced that it had mass potential as a copying process, though it does not seem to have achieved any popular success.[67] It consisted of two distinct processes, having in common the use of an electrical discharge instead of light: in one a latent image formed on a photographic plate, in the other a metallic oxide coating transferred from the object to a sheet of card. Inductoscript is thus one of the precursors of modern electrostatic copying (xerography). Another variation used induction to

[64] MHS, 1890 testimonials.
[65] MHS, MS Museum 62 (letters from Kelvin, Rayleigh, and Lodge); MHS, 1900 testimonials (letters from Boys, Lodge, and Thompson); on Boys's important work in Oxford in 1890–4 see Gooday, Chapter 3, 110–14.
[66] Anon. (1892a); cf. Jervis-Smith's subsequent Oxford talk in *Oxford Review*, 10 November 1892; Jervis-Smith (1896a).
[67] Patent no. 12097, 1892; MHS, inductoscript leaflets (see below); Jervis-Smith (1890–2, read 24 June 1892); Jervis-Smith (1893a); *The Electrician*, 29 (1892): 395; Anon. (1892b); *Engineering*, 55 (1893): 681; other cuttings in MHS, MS Museum 89; Jervis-Smith (1893b).

activate 'phosphorescence' in a coating and create the image on a photographic plate. This variant he later developed into a separate phosphorescent photographic technique ('phosphorograph') in an attempt to create a contact-copying method suitable for books or documents.[68]

In retrospect the inductoscript is interesting chiefly because it shows him on the brink of discovering X-rays, and possibly even radioactivity. Although he was focusing on the effects, he was not indifferent to the fundamental questions that might be asked of the phenomena. One experiment was with an exposed (spoiled) photographic plate, which still produced an image, proving 'that the electrical action on the photographic plate is different from that of ordinary light'; another experiment tried the effect of conducting the process through a vacuum. His pamphlet on inductoscript (1892) twice notes the danger of photographic plates being fogged by proximity to the induction apparatus, and by (as he speculates) 'electrically-induced phosphorescence'.[69] The story that has become legendary, dating from these experiments of 1892, was recounted by Sir Oliver Lodge in his Becquerel Memorial Lecture of 1912. Jervis-Smith, he said,

> ... told me that he missed the discovery of Röntgen rays by a trifle ... [he] noticed that boxes of photographic plates which he needed for his work were liable to be fogged if allowed to remain in the neighbourhood of active Crookes's tubes. Presumably he thought the cause to be some merely chemical effluvium, such as ozone or oxides of nitrogen getting into the box, or perhaps he did not speculate on the cause at all, but merely regarded it as a nuisance, interfering with the steady course of his work. Anyhow, he seems to have instructed the laboratory attendant to keep the boxes in a cupboard well away from the fogging influence ... (Lodge 1914: 221).

Jervis-Smith was always alert to new discoveries, and usually able to investigate or replicate them quickly. As with the phonograph and telephone in 1877, he took great interest in the new electrical phenomena of the 1890s. He was the first to make an X-ray photograph in Oxford, communicating an image of his own hand to the Junior Scientific Club on 5 February 1896, within a month of the first announcement in the English newspapers. It revived his interest in investigating and modifying Tesla induction apparatus.[70] He was early in understanding the technicalities of Hertz waves (which were not widely understood or publicly demonstrated until 1894), mentioning them in 1892 as an alternative source for the electrical effect in his inductoscript process; and he subsequently worked on several aspects of wireless equipment, including a new form of Hertz wave resonator (receiver). Two patent applications relating to such apparatus, in 1897 and 1899,

---

[68] Jervis-Smith (1901a; 1901b); information from Mr Giles Hudson.

[69] MHS, leaflet entitled *"Inductoscript"* (1892), accompanying lecture to the Ashmolean Society, esp. 3, 4; other versions of this leaflet.

[70] *Journal of the Oxford University Junior Scientific Club*, 2, no. 34 (12 February 1896): 21–2; Jervis-Smith (1896b); early X-ray tubes made by him are in MHS.

were abandoned because (he said) he had no time or facilities for developing them.[71] He demonstrated wireless telegraphy between the Millard Laboratory and his home (in Norham Gardens, about half a mile away) at about this time, 1897–9. A related series of experiments looking at the effect of environmental factors (vacuums, pressure, temperature) on electrical discharges led to the invention of a high-pressure spark gap used with Tesla inductors, based on his discovery that the discharge increases when the air-pressure is increased.[72] This was relevant to both radio transmission and X-ray work – though it could also be used for glass-cutting.

Work on dynamometry had been the main focus of his research in the years before coming to Oxford (1880–4), and the didactic use of such instruments was a constant feature of the Millard Laboratory. But with his new-found interest in the chronograph it was 1892 before he returned to his old speciality, perfecting a water-cooled brake ergometer and developing his important torsional ergometer, which could be used for testing screw propellers, spinning machinery, and other rotating mechanisms. This sophisticated dynamometer was used for testing spinning machinery at a mill in Halifax, and exhibited at the Royal Society conversazione in 1894 along with ancillary instruments allowing integration and recording of the measurements, including his mechanical integrator and optical rotostat.[73] In 1910, shortly before his death, he was working on a book on dynamometers, which was completed by his friend C. V. Boys and published posthumously in 1915. It is Jervis-Smith's only full-length book, and appropriately provides a comprehensive historical and technical description of the many different forms of dynamometers and dynamometric apparatus (Jervis-Smith 1915).

Various offshoots of his research, and incidental oddities, provide examples of his ingenuity or enthusiasm. Around 1890 he was trying to get his galvanometer to measure, and record photographically, subtle changes in temperature, magnetism, and similar phenomena, one of the results being a method for monitoring recalescence (the inherent retardation in the process of cooling of metals).[74] His knowledge of telephony and of musical acoustics combined in the invention of a way of employing organ pipes to record the rise of water level in a river, the note of the pipe heard by telephone indicating the level of water in it. It was widely reported in the technical press in 1892, though *The English Mechanic* concluded that it was 'ingenious, but rather troublesome'.[75] Work on photographing and measuring the

---

[71] Jervis-Smith (1897–8); MHS, leaflets on resonator (1898) and aerial conductors (1899); patent nos.19420, 1897 and 18574, 1899.

[72] Jervis-Smith (1902); MHS, leaflet entitled *Notes on the Electrical Discharge from Inductors of the Tesla type* (1903); *Chemical News*, 87 (1903): 245–6; other cuttings in MHS, MS Museum 89.

[73] Jervis-Smith (1893b; 1894a); *The Engineer*, 77 (1894): 380; Jervis-Smith (1896c; 1898a); Jervis-Smith (1915: 60–1, 196–202).

[74] Jervis-Smith (1891a; 1891b); Bedford Air Force Base archives, letter from Jervis-Smith to Rayleigh, 27 July 1891 (information from Dr Joanie Kennedy).

[75] Jervis-Smith (1892); *The English Mechanic*, 55 (30 July 1892); other cuttings in MHS, MS Museum 89.

Fig. 5.9. Jervis-Smith's 'integrating ergometer', probably photographed in 1894 when it was exhibited at the Royal Society. It is a composite instrument, consisting of Jervis-Smith's own design of transmission dynamometer (*centre and right*), giving the usual reading of force on one dial, combined with a version of his mechanical integrator, invented in 1882, which integrates additional variables (velocity, tension, length of drive-belt, etc.) so as to give a truer measurement of 'work' on the second dial (*to the left*).

velocity of bullets led in 1894 to experiments on their penetrative power, and discussions of bullet-proofing (Jervis-Smith 1894b). It typifies the incorrigibleness of inventors that when mains electricity came to Oxford in 1897 Jervis-Smith should have experimented with the wiring (for lights) and found that it could also be used to form a telephone circuit (Jervis-Smith 1898b). On one of his visits to the antipodes, in 1899, he devised an ingeniously simple way of measuring the rolling angle of a ship in heavy seas. He mounted a camera in such a position that the horizon provided the common horizontal against which the angle could be measured from photographs taken at the two extremities of the roll.[76]

In 1890 Jervis-Smith had applied unsuccessfully for the chair of mechanism at the university of Cambridge, where presumably someone with superior mathematical qualifications was required (such as James Alfred Ewing). More realistic expectations were attached in 1900 to Oxford's new Wykeham chair of physics, whose special role in electrical physics Jervis-Smith felt he had in some measure already been performing. Indeed it seemed to others, as Oliver Lodge wrote, 'that

[76] MHS, original prints mounted and captioned (dated 2 March 1899); Jervis-Smith (1901c).

the post was being made for you'.[77] But once more he was up against the Cambridge factor, and the job went to J. S. E. Townsend. Although disappointed, it meant at least that his focus thereafter must be upon achieving recognition for engineering in the university's syllabus, and bringing about the long-delayed establishment of that chair. A good deal of Jervis-Smith's student constituency was removed by the centralization of preliminary physics teaching in Townsend's department in 1905, so that after the end of 1904 he no longer routinely saw a portion of the natural science undergraduate intake. His elementary mechanics and physics course was replaced by lectures on basic mechanical engineering, his last course in 1908 being entitled 'The Elements of Design and Construction in connexion with Civil and Mining Engineering'.[78]

The first step to the formal subject of engineering at Oxford, and some compensation for the loss of the preliminary physics course, was the graduate Diploma in Scientific Engineering and Mining Subjects, launched in 1905, of which Jervis-Smith taught the 'Engineering Principles and Machine Drawing' component. Seven other topics formed the syllabus, taught or supervised by the appropriate professor or lecturer: mathematics, geology, mineralogy, hygiene and mine ventilation (J. S. Haldane), electricity (J. S. E. Townsend), assaying (H. B. Baker), and surveying (the university now employed a surveying instructor, H. N. Dickson). The diploma also required physics and chemistry, and French and German translation, but no special teaching of these subjects was provided. The two-year course was meant to include a four-month secondment for practical training 'in a mine or in engineering works'.[79] It was thus a prototype for a full degree course in engineering science, embodied an admirably interdisciplinary approach, and provided a channel for the more specialized offerings of Jervis-Smith, Haldane, and Dickson. But if the Diploma was not as 'bizarre' as Morrell thinks (Morrell 1997: 95), it could hardly avoid the impression of coming twenty years too late, and it was quickly overtaken by events.

It is unlikely that Jervis-Smith had any realistic hope, or even wish, at the age of 60, of being appointed to the new chair when it was finally created in 1907. Even so, he might have expected better recognition at the time (and since) of the role he and the Millard Laboratory had played in carrying the torch for an unestablished subject for twenty years, and preparing the ground for Oxford's acceptance of it. He now turned his attention to setting up a laboratory at Battramsley House, a home he had acquired several years earlier on the edge of the New Forest, to which he retired in 1908. His laboratory assistant James Mogridge moved with him, and he was also assisted by his son Eustace John Jervis-Smith, who had been one of his

---

[77] MHS, 1890 testimonials (Cambridge); 1900 testimonials (quoting Lodge's letter).
[78] OUG, 38 (1907–08): 276 and 303 (17 and 20 January 1908).
[79] OUG, 35 (1904–5): 409–10 (7 March 1905); The Times, 22 March 1905.

students (for three terms in 1895–6) when he was at Balliol College. They conducted experiments in the hope of explaining a phenomenon Jervis-Smith had noticed, that evacuated globes or similar vessels glow when moved or spun within an electrostatic field. He corresponded about it with Lodge; but although the phenomenon fascinated and perplexed him he never found an explanation for it.[80] One of his last publications was a letter to *The Electrician* in 1910, about the liquid microphone, in which he courteously reminded the public that this crucial technology for the new era of radio broadcasting was invented by him thirty years before (Jervis-Smith 1910a). He also returned to the subject of dynamometers. His health deteriorated however, and after some months of serious illness he died in 1911. Boys wrote the Royal Society obituary of 'his closest and most valued friend and counsellor', and edited together the unfinished book on *Dynamometers*, published in 1915 (Boys 1913; Jervis-Smith 1915).

THE DEPARTMENT OF ENGINEERING SCIENCE

A professorship of engineering science was formally established by Oxford University in 1907, and the Department of Engineering Science came into being with the appointment of the first professor in 1908. The defining philosophy and agenda for the new subject were laid down by Jervis-Smith, for instance in his newspaper article in December 1907.[81] It was the Kennedy philosophy he had adopted in 1886, sharpened by experience. As the chair was established without reference to an actual degree in the subject, he pressed for this logical step to be taken (which a year later it was). He also urged 'friendly cooperation' between the new professor and the engineering industry. In April 1908 his CUF lecturership expired and he retired gracefully, opening the way for the subject he had pioneered to obtain proper status in Oxford. The man appointed as professor was Charles Frewen Jenkin, who was in fact a practising professional engineer, though with a respectably academic education (Southwell 1939–1941; *DNB*; Morrell 1997: 11–12, 95–7). He was first attached to New College, but soon moved to Brasenose, which took responsibility for funding the new chair.

In his inaugural lecture on 16 October 1908, Jenkin did not neglect the custom of acknowledging one's predecessors: he laid claim to the inheritance of Rankine and Kennedy, with mention of Fleeming Jenkin (his father) and the Cambridge

---

[80] MHS, MS Museum 62, letters from Lodge to Jervis-Smith, 20 and 23 May 1908; Jervis-Smith (1908a; 1908b); Boys (1913: v).

[81] Jervis-Smith, 'Engineering at Oxford', cutting from newspaper (title unspecified) dated 11 December 1907 in MHS; and cf. *Electrical Engineering*, 11 July 1907.

professor Ewing. And he described his subject in unexceptionable terms derived from them, and with an innocence that might easily suggest that he thought Oxford had never heard of engineering before. The syllabus and teaching methods he proposed were identical in every way to Jervis-Smith's. He added a disparaging reference to the accommodation assigned to his department, which was the 'small and somewhat shabby' Millard Laboratory.[82] If Jervis-Smith was conspicuous by his absence from Jenkin's address, he was equally so from the audience hearing it. By a remarkable piece of stage-management, the University of Oxford had suddenly become so devoted to the history of science—a subject just beginning to assert itself there (Simcock 1985: 65–8)—that it had dispatched an official representative to the Torricelli tercentenary celebrations in Italy, who on the very day that Jenkin was speaking in Oxford was enjoying the speeches and exhibits in Faenza in honour of one of the pioneers of precision physics. Jervis-Smith seems to have taken pride in representing his university in this way, and on his return published a booklet on Torricelli (Jervis-Smith 1908c).

Trinity College rented the Millard Laboratory to the fledgling Department of Engineering Science, but Jenkin went in quest of alternatives to his despised premises. From 1912 all his lectures and laboratory instruction were taking place either in the University Museum (presumably in the rooms recently vacated by Townsend) or at 6 Keble Road.[83] An Engineering Laboratory was eventually erected in the Keble Road Triangle, a new outgrowth of the university's Science Area. The usual ill-informed controversies surrounded the choice of site, opponents expecting that it would be accompanied by factory chimneys, smoke, and noise.[84] It was completed and occupied in 1914. Jenkin's staff originally consisted of the instructor in surveying, now Nicol F. MacKenzie (shared with the School of Geography), and a demonstrator, initially (in 1908) the electrical physicist E. W. B. Gill, afterwards demonstrator in the Electrical Laboratory.[85] Townsend was his best ally, both in rebutting the objections to the proposed laboratory, and in collaborating in the interdisciplinary needs of the new course, allowing his demonstrator Frederick Bernard Pidduck, later reader in applied mathematics (1927–34), to begin lecture courses and practical instruction in physics specifically intended 'for Engineering Students'.[86] At the same time the physical chemist Andrea Angel developed a chemistry course for the engineers, given at Christ Church.[87]

[82] Jenkin (1908: esp. 6, 20); a curious legacy of Jenkin's is that to this day the University's annual directory, the *OU Calendar*, contains in its blurb about Engineering Science the gratuitous statement 'Originally sited in makeshift accommodation . . .' (*OU Calendar* for 2000: 156).
[83] TCA, Order Book: 241 (10 December 1907) (information from Dr Joanie Kennedy); Smith (1979: 63); *OUG*, 43 (1912–13): 23 (10 October 1912).   [84] Pamphlets for and against in MHS, MS Gunther 65.
[85] *OU Calendar* for 1909; for Gill see Simcock, Chapter 4, 143.
[86] *OUG*, 39 (1908–09): 538 (23 April 1909); *OUG*, 41 (1910–11): 25 (13 October 1910); see Lelong, Chapter 6, 220–8.   [87] *OUG*, 40 (1909–10): 304 (17 January 1910), etc.; Morrell (1997: 7).

Examinations for the engineering degree (like physics, a natural science degree with engineering as the specialism) were first taken in 1910; and it was soon producing as many honours graduates as physics, each subject having ten in 1913 and nine in 1914, before the progress of the school was halted by the War (*Historical Register First Supplement*: 239–40 and 246–7). One early candidate took both physics and engineering; but the essentially physical origins of the engineering science course, as well as the availability of physics and chemistry within it, allowed it to be free-standing. Its mathematical requirements, however, meant that (as with physics) candidates often approached it via mathematical moderations. Jenkin pursued a few specialized research topics, including his important wartime work on metal fatigue and the suitability of materials for aircraft construction, and in the 1920s refrigeration, mounting for the 1926 British Association meeting in Oxford an exhibition of 'small refrigerating machines (working), including food-safes suitable for household use'. It was hardly a coincidence that soon afterwards, in 1932, Oxford's Pressed Steel Company (an ancillary of the Morris car factory) became the first significant British manufacturer of domestic refrigerators, under the trade name of 'Prestcold' (British Association 1926; British Medical Association 1936: 214–17). While cryogenic engineering has remained a feature, the research areas for which Oxford engineering became better known, such as fluid mechanics (wind-tunnel experiments), thermodynamics, and the strength of materials, have been developed under Jenkin's successors, the first of whom (1929–42), Sir Richard Vynne Southwell, a graduate of the Mechanical Sciences Tripos at Cambridge, proved a more effective discipline builder (Morrell 1997: 97–106).

CONCLUSION

In 1894 the British Association for the Advancement of Science assembled in Oxford for the first time since 1860. The discovery of argon was announced, Hertz waves were demonstrated, flying machines were discussed, and the Brunsviga calculator was exhibited – Jervis-Smith already had one, and had lent his name to the publicity for it.[88] The high-level spokesmen for physics and engineering respectively made polite attempts to welcome their host into the modern world. Arthur Rücker assured his audience that academic physics, even at Oxford, was fully engaged with the life and work of the nation: 'Our modern universities are within earshot of the whirr of the cotton mill or the roar of Piccadilly'. Alexander Kennedy was not so sure: 'there are few places which have much less in common

---

[88] *The Electrician*, 33 (1894): 456–9; MHS, Brunsviga advertising leaflet [1894]; *Electrical Review*, 35 (1894): 207.

with the special work of Section G', he thought, than Oxford; but he went on to offer an attractive personal reflection on the teaching of 'mechanical science' and the work of an engineering laboratory (Rücker 1894: 543; Kennedy 1894: 739). He was selling a product as surely as the man from Brunsviga. Why Oxford did not buy, but kept the discipline of engineering in limbo for another decade and more, is a complicated question, the debate about which I suspect has contained as much prejudice about Oxford as Oxford itself ever had against applied science.

The usual view (as expressed recently by Morrell) is that engineering was disadvantaged, even after its establishment, by Oxford's 'characteristic suspicion of any subject which might sully the University by making positive contact with industry and by indulging in vocational training' (Morrell 1997: 95, and cf. 105). An assumption that such snobbery was endemic not just in 'Oxford' but in Oxford science derives from Sir Henry Acland's grand defiance of vocationalism in medicine – on which, incidentally, he had conceded defeat in 1890 (Acland 1890). In fact, the universities to which Oxford is usually compared unfavourably had confronted the same sensibilities about whether engineering was a fit academic subject, the solution being the one that Jervis-Smith quietly imported into Oxford in 1886. There is little evidence that he felt he must be diplomatic about the vocational nature of his subject. In setting the tone for Oxford's engineering school in 1907, he actually stressed the need to defray the self-imposed 'limitations' in university engineering by ensuring 'an equality of excellence' between the academic and industrial phases of training: 'I venture to suggest that nothing short of friendly cooperation between the two sources of instruction can bring about the best result'.[89] It sounds revolutionary. But it was a revolution that Jervis-Smith had represented all along, for it was John Percival's philosophy thirty years earlier, when he got Perry to establish mechanical teaching at Clifton College and then brought Jervis-Smith to do the same at Oxford.

So was Jervis-Smith a fifth-columnist? The common thread in much of his research – with dynamometers, integrators, galvanometers, chronographs, and even some of the uses he found for the telephone or photography – was precision measurement. His work was thus absolutely consistent with the experimental preoccupations that characterized Oxford physical science, and so was the direction in which he took his subject. Perhaps the paradox that has allowed the John Perry critique to ring true – not only as contemporary diatribe but as history – is that Oxford was always full of fifth-columnists. Its highly decentralized and non-deferential collegiate structure provided an environment in which they flourished, while having (and this is the paradox) very limited impact upon the organization that answered to the name of Oxford University in dialogues with the outside world.

[89] Jervis-Smith, 'Engineering at Oxford', cutting from newspaper (title unspecified) dated 11 December 1907 in MHS.

And as for the historical record, and our starting point in redefining the territory of 'Oxford physics', it is scarcely necessary to reiterate how inadequate a picture proceeds from assuming that the Clarendon Laboratory and Electrical Laboratory contained all of physics that Oxford had to offer. It follows that we can no longer regard Clifton's precious conservatism as a malaise shared by Oxford physicists at large, nor the wider landscape of collegiate Oxford – the true university – as a desert of barren classicism in which a Bosanquet could never blossom. It is not that the university's 'official line' is unimportant. Yet it behoves history to scratch the surface and see whether the creakings of cumbersome organizations or the pronouncements of senile professors really represented the abilities and ideals of those whose achievement was constrained by their subordinate station. Jervis-Smith provides an illuminating case-study from the middle ranks of physicists, on whom the historical spotlight is seldom turned. And whether we judge him to have been the heart and soul of Oxford physics or an idiosyncratic sideshow, his very presence must give us a different perspective on Oxford science around the turn of the century.

6

Translating Ion Physics from Cambridge to Oxford: John Townsend and the Electrical Laboratory, 1900–24

Benoit Lelong

The postgraduate career of John Sealy Edward Townsend occurred in three places: Dublin, Cambridge, and Oxford. After graduating in 1890 at Trinity College, Dublin, Townsend spent the next five years there teaching mathematics. In 1895 he obtained an 1851 Exhibition scholarship to Cambridge, where he worked as a research student in the Cavendish Laboratory under J. J. Thomson. In 1900 he was appointed the first Wykeham professor of physics in Oxford, a chair that supplemented that of Robert Bellamy Clifton, professor of experimental philosophy. Clifton's responsibility was to teach light, heat, and sound, and Townsend's electricity and magnetism (von Engel 1957: 257; *Historical Register*: 83).

I shall argue that in Oxford Townsend sought to implement practices and values inherited from the Cavendish regime. It proved a difficult enterprise, Oxford offering an environment very different from that of Cambridge. My focus will be on the making of Townsend's Electrical Laboratory, the shaping of its scientific and social identity, and the establishment of its links with various actors in the outside world, and I shall seek to relate this story both to the institutional and cultural peculiarities of Oxford and to the international development of ion physics.

A core element in what I offer as a comparative analysis is the international diaspora of the physicists trained by J. J. Thomson. Most of them left Cambridge at the turn of the century and tried, like Townsend, to transfer the 'Cavendish style' elsewhere in the world, especially to Canada, France, and the United States. Comparing their experiences helps to identify what was specific to Townsend in Oxford.

For valuable comments and criticism, I am indebted to Isobel Falconer, Robert Fox, Graeme Gooday, Janet Howarth, Roger Hutchins, Jack Morrell, Dominique Pestre, Simon Schaffer, and Andrew Warwick.

TOWNSEND IN CAMBRIDGE, 1895–1900

In 1895 a new admissions system was established by the University of Cambridge. This allowed some 1851 Exhibition scholars, even though they were not Cambridge graduates, to pursue research for one or more years as advanced students in the Cavendish Laboratory. Townsend was among the first to be accepted and he arrived in Cambridge in October 1895, together with Ernest Rutherford (from New Zealand) and J. A. McClelland (from Galway). They were followed in 1896 and 1897 by several young physicists, including Richardson, Craig-Henderson, Barkla, and H. A. Wilson from Britain; Child, Owens, and Zeleny from the United States; McLennan from Canada; and Langevin from France. These 'advanced students' constituted a new population in the laboratory. They experienced a certain sense of exclusion from students and staff trained in Cambridge, who perceived them as 'competitors for the limited facilities in the way of apparatus and the services of the workshop, and also . . . for the attention and sympathy of the Professor' (Strutt 1942: 60; quoted in Kim 1995: 215). The newcomers consequently formed a separate group, cemented by strong friendships. Townsend was particularly linked to Rutherford and Langevin. In contrast to the Cambridge students, the new scholars shared a keen enthusiasm for Thomson's research on electrical discharges, cathode rays, and conducting gases, and above all for his most cherished scientific object, gaseous ions (Sviedrys 1976; Kim 1995).

In Cambridge, Townsend imbibed the 'Cavendish style' of experimental physics, now well known to us through the work of John Heilbron, Isobel Falconer, Peter Galison, and others. Its main characteristics were the following: the charismatic leadership of J. J. Thomson; a material culture centred on quadrant electrometers, vacuum apparatus, and ionized gases; an enthusiasm for 'advanced' research on ions, electrons, and other 'fundamental' particles; a literary rhetoric based on theoretical hypotheses followed by experimental confirmation; a scientific approach based on reductionism and microphysics; a disdain for 'applied' science, engineering, and high-precision measurements; a training through laboratory collaborations and joint publications; and a social distance between undergraduates and research students (Falconer 1989; Galison and Assmus 1989).

This physics differed markedly from practices and values in the Cavendish Laboratory *before* the appointment of J. J. Thomson. Under the previous directors, Maxwell and Rayleigh, the laboratory's material and intellectual culture was that of the measurement and standardization of electrical resistances. In this period, the Cavendish was marked by its links with the electrical industry, its cult of precision, its distrust of microphysics, and the care it bestowed on teaching. As Gooday has shown, almost every British laboratory shared these features in the last third of the century, so that Thomson's new group stood somewhat apart from the practices and values of Victorian physics as a whole (Gooday 1990; Schaffer 1992).

In Cambridge, Townsend immediately adopted Thomson's research programme. He designed a new method for measuring the coefficient of recombination of ions, and another for their coefficient of diffusion (Townsend 1898b; 1900a). He examined a new case of gaseous conduction: gases recently prepared by chemical reactions (for instance hydrogen and oxygen produced by electrolysis). He learned with C. T. R. Wilson how to build cloud chambers, and used them to perform the first measurement of the elementary ionic charge (Townsend 1898a). In 1900, having published more than two papers a year since 1895, Townsend was among the three most prolific Cavendish students, along with Wilson and Vincent (Kim 1995: 222). But his enthusiasm had clearly diminished. As he wrote to Rutherford: 'I wish we had opportunities here of doing work outside the Cavendish as I must confess that eternally working at ions is beginning to be tiresome'.[1]

Nevertheless, through his work on ions, Townsend had accumulated several academic honours that distinguished him from other advanced students. He had been elected to a fellowship of Trinity College in 1899 (Kim 1995: 216). Being a mathematical expert, he was given task of delivering the course of mathematics for physicists by Thomson in 1900 (Falconer 1989: 21). In 1898, he obtained the Clerk Maxwell scholarship, a much coveted award since it was exceptionally generous (£200 a year). It had never before been offered to a non-Cambridge man and would not be so again until 1910 (Kim 1995: 220; Falconer 1989: 19). His work interested Lord Kelvin, with whom he exchanged preprints and letters and who wrote him a testimonial for a professorship at University College, Liverpool.[2]

As a rising star, Townsend came into conflict with J. J. Thomson. In 1900 he proposed a new collision theory that explained Stoletow's results on the photoelectric effect (Townsend 1899b). At the outset, Thomson preferred his own explanation, based on a layer theory, which had been published in December 1899. Later in the same year, Thomson published a paper suggesting that collisions would produce Stoletow's results, but without mentioning Townsend (Thomson 1900). Townsend was furious about the omission and doubtful about Thomson's conversion:

J. J. Thomson has been like a weathercock on all this subject during the last 9 months. You will notice that I communicated my paper myself. I don't know how much of it he agrees with, as he stuck vehemently to his theory of layers as an explanation of my results and Stoletow's. The layer theory is I believe absolute rot.[3]

The new advanced students did not remain in the Cavendish for more than a few years, their financing allowing only a temporary stay. Langevin returned to Paris in June 1898, defended his doctoral thesis there in 1902, and became *professeur suppléant*

---

[1] Townsend to Rutherford, 27 March 1900, CUL Add. MS 7653, T74.
[2] Townsend–Kelvin correspondence, CUL Add. MS 7342, T573–8.
[3] Townsend to Rutherford, 14 January 1901 (wrongly dated 14 January 1900), CUL Add. MS 7653, T73.

at the Collège de France in 1903 (Langevin 1909). In September 1898, Rutherford moved to Montreal, where he had been appointed professor of physics at McGill University (Heilbron 1979). In 1899, Zeleny returned to the University of Minnesota, where he had been assistant professor since 1896, and became associate professor there in 1900 and professor in 1909 (Fudano 1990: 445, 473). Hence, when Townsend was appointed to Oxford, most of his friends had already left Cambridge for positions in Britain, Canada, America, France, Poland, India, and elsewhere. His career is consequently just one example of the international diaspora of Cavendish physicists.

IMPLEMENTING ION RESEARCH OUTSIDE CAMBRIDGE

Townsend arrived in Oxford in January 1901. Gratified by his new professional situation, he wrote to Rutherford:

I am in great luck getting the professorship here. It is a very easy job, I have only to teach Electricity & Magnetism and Clifton does the other branches of physics. We are quite independent & I am going to have a separate lab for myself. The work will be very light and I will have plenty of time for research. If I like I need only lecture two days a week for two terms in a year, but of course I will do more than that.[4]

Townsend's enthusiasm owed much to Oxford's reputation in late Victorian Britain. In the 1860s, the superiority of Oxford to Cambridge with respect to facilities for science teaching had been widely acknowledged. Yet in the 1880s the situation changed; in Janet Howarth's words, 'Oxford acquired a reputation that dogged it well into the twentieth century as a university particularly ill-adapted, even hostile, to science' (Howarth 1987: 335). The famous dictum 'Oxford for arts, Cambridge for science' was firmly established at the turn of the century (Howarth 2000: 457). Oxford was known for its grandiose gothic buildings, for the glamour of college life, and for the domination exerted by professors of law, ancient languages and literature, and philosophy. It is easy to understand, therefore, why Townsend was charmed (he wrote to Rutherford that 'Oxford is a very jolly place') and why he was also surprised to discover that 'most of the classical men stick up for science'.[5]

Townsend's salary was about £500 per year. He was paid and accommodated by New College. Merton College provided £700 for 'fitting up the laboratory' and £250 per year for the first two years 'for assistance and maintenance'.

---

[4] Townsend to Rutherford, 14 January 1901 (wrongly dated 14 January 1900), CUL Add. MS 7653, T73.
[5] Ibid.

Townsend used these grants to buy electrical instruments, mainly for teaching. In 1901, a temporary classroom and a workshop were allocated by the University, followed in February 1902 by three more rooms on the first floor of the University Museum (von Engel 1957: 259–60). At the end of 1901, the delegates of the Common University Fund instituted a demonstratorship, to which Townsend appointed the Revd Paul Jerome Kirkby, like him a fellow of New College. Kirkby had graduated in 1891 in mathematics and held a fellowship at the college since 1895.[6]

From April 1901, Townsend lectured on electricity and magnetism, while practical instruction was given by Kirkby from October 1901 onwards (Townsend 1901–2). Meanwhile, Townsend campaigned successfully for a modification of the university syllabus and added to the practical examination the use of electroscopes, magnetometers, ammeters, voltmeters, and ballistic galvanometers; hitherto the syllabus had required the mastery of only *one* electrical instrument, the tangent galvanometer. Reforming the syllabus and examination enabled Townsend to transform his laboratory into an indispensable facility for students. It allowed him to create, so to speak, a new clientele, a new market for his laboratory classes.[7]

Townsend developed not only teaching but also research. He accumulated the *instruments* and the *skills* necessary to reproduce the Cavendish Laboratory's material culture. G. A. Bennett, whom he engaged as laboratory assistant, was 'a skilled mechanic' and, most importantly, a glassblower able to make vacuum apparatus. S. W. Bush was engaged as instrument maker; he had been employed by the Cambridge Scientific Instrument Company and was able to make quadrant electrometers. Townsend bought the necessary instruments to produce and measure ionic phenomena: electrical cells, X-ray tubes, and quadrant electrometers (Townsend 1901–2; 1902–3). He also tried to inculcate the methods and values of ionic physics. As part of this programme, he undertook a long laboratory study with Kirkby (a work on ionic collisions in hydrogen), leading to a joint paper published in June 1901 (Townsend and Kirkby 1901). Another paper was published by Kirkby alone in February 1902. At the time, Kirkby was not yet an 'ion physicist' in the Cambridge sense of the term. Whereas Cambridge papers and textbooks usually presented formalized schemes of electrical connections, Kirkby published a three-dimensional drawing of his apparatus, showing its external physical shape but not its internal electrical structure. His vocabulary was sometimes different too. For instance, Kirkby explained that ions were 'disintegrating' molecules (instead of 'dissociating' them), or that there was an 'extraordinary' difference between positive and negative ions (such enthusiasm would never surface in the dry, objective tone of Cambridge papers). These singularities disappeared in Kirkby's later publications (Kirkby 1902).

---

[6] *OU Calendar* for 1900, 485.
[7] OUA, Natural Science Faculty Board Minutes 1892–1912, FA4/13/1/2, 92, 94, 107; Natural Science Reports 1892–1912, FA4/13/2/1, 27 April 1903.

In this work of accumulation and reproduction, Townsend's network and allies played a crucial role. Lord Kelvin helped him to obtain a £100 grant from the Royal Society to buy electrical cells,[8] while Townsend visited Langevin in Paris and bought there the high-quality X-ray tubes produced by Victor Chabaud.[9] The 'ion diaspora' generated similar episodes elsewhere in the world. Most of the Cavendish physicists who were appointed to other laboratories beyond Cambridge tried to perform and publish new experiments on gaseous ions, to transmit their know-how, and to create research schools on the Cavendish model. In Montreal, Rutherford immediately undertook new experiments on ions and taught the young Canadian McClung how to measure their rate of recombination (Heilbron 1979: 50). In Paris, from 1902, Langevin taught Cavendish Laboratory methods; his demonstrator Eugène Bloch used them to detect ions in phosphorus emanation, to measure their speed, and to establish their condensation properties. Similar episodes occurred with McLennan in Toronto (Gingras 1991: 30–1), Zeleny at the University of Minnesota (Fudano 1990: 472), and Cunningham in Calcutta.[10] Their influence helped to fashion an international network of ion specialists who frequently exchanged letters and visited one another. Most of their scientific publications referred to one another's work.

TOWNSEND VERSUS CLIFTON

Although formally created in 1900, the Wykeham chair of physics already had a long history. In 1877, Clifton had explained to the University of Oxford Commission that his teaching duties were so heavy that he had had to give up practical instruction in electricity and magnetism in order adequately to teach the other subjects: heat, light, and sound. In 1878, the Commissioners recommended the foundation of the Wykeham chair by New College to supplement Clifton's chair of experimental philosophy. But for more than twenty years New College delayed the creation of the chair, giving priority to its own building projects whilst suffering from a loss of college income due to the agricultural depression.[11]

Over this period, Clifton's situation deteriorated, as Gooday shows in Chapter 3. Faced with an increasing number of students in the 1880s, Clifton asked time and time again for more rooms, posts, and grants, but his requests were systematically

---

[8] Kelvin to Townsend, 9 December 1901, CUL Add. MS 7342, T575; Townsend to Kelvin, 13 December 1901, T577; Kelvin to Townsend, 14 December 1901, T578.

[9] Townsend to Langevin, 5 March 1901, Langevin papers, Ecole supérieure de Physique et de Chimie industrielles, Paris, L76/72.

[10] Cunningham to Langevin, 10 November 1903, Langevin papers, L74/36.

[11] Gooday (1989), Chapter 6. On Clifton's laboratory management and scientific style see Gooday, Chapter 3, 107–17.

refused. Significantly, his constant argument was the need to restart practical instruction in electricity: 'it is much desired by students who are anxious to take advantage of the openings presented for employment in connection with scientific work by the recent development of electrical engineering'.[12] In 1894 the University tried to obtain funds from the Drapers' Company for the construction of an electrical laboratory.[13] Later, the Vice-Chancellor, Thomas Fowler, deplored the fact that 'students who require advanced teaching in [electricity and magnetism] are obliged to have recourse to foreign or other British Universities, or to some of the University Colleges', which was a 'notable deficiency in our teaching'.[14] This suggests that, because of the growing opportunities for employment in the electrical industry, the University recognized the need to resume practical instruction in electricity and magnetism; it also suggests, however, that Clifton was no longer seen as the right man for such an expanded commitment.

From 1901, therefore, Clifton and Townsend taught the same students, preparing them for the same degree (though not for the same subjects). This meant that they competed for the same resources, and the result was bitter rivalry. In 1901, when the newly appointed Townsend still had no laboratory, Clifton did not allow him to experiment in the Clarendon. Townsend and Kirkby had to pay rent for rooms in the University Observatory and later in the Physiological Laboratory. When space on the first floor of the University Museum became available, Clifton requested it for his teaching to Preliminary Honour candidates. He was furious to see it given to Townsend. In 1903, Townsend timetabled his lectures to clash with Clifton's: suddenly deprived of an audience, Clifton had to cancel his lectures![15]

Townsend's strategy was to discredit Clifton in order to capture his clientele and his resources. In 1903, he proposed that he himself should teach all the branches of physics to the Preliminary students, i.e. not only electricity and magnetism but also Clifton's subjects (heat, light, and sound). This was accepted in 1904 by the Hebdomadal Council. To support the additional teaching, Townsend was given a sum of £600 to buy new instruments, and an annual grant of £200, which allowed him to recruit two more demonstrators, Craig and Lattey. Clifton used his annual report for 1904 to complain bitterly about the heavy loss caused to his own income by the new arrangement, and he soon lost two of his four demonstrators, being unable to pay all four of them any more (Clifton 1904–5; Townsend 1904–5; Heilbron 1974*a*: 36, n. 19).

[12] Clifton, 'Proposed extension for the Clarendon Laboratory', 3 February 1887, Bodl. G.A. Oxon b.139 (89).
[13] Drapers' Company Archives, Court of Assistants minute-books, 6 December 1894.
[14] Thomas Fowler, Letter to the Warden and Fellows of Merton College, 19 February 1900, Bodl. G.A. Oxon c.281 (17).
[15] The unfolding of this episode is conveyed vividly in Townsend's and Clifton's annual reports for 1901, 1902, and 1903 (Clifton 1901–2; 1902–3; 1903–4; Townsend 1901–2; 1902–3; 1903–4).

Fig. 6.1. The Revd Paul Jerome Kirkby. As a fellow of New College (from 1895) and Townsend's first demonstrator (from 1901), Kirkby worked mainly on gaseous explosions. He had a distinguished academic record in both mathematics (in which he took a first in 1891) and physics (first, in 1893) and published a number of significant papers. But he abandoned science in 1910, when at the age of 41 he left the University to pursue a clerical career, first in Norfolk and finally, until his death in 1931, as rector of Chorley in Lancashire.

A comparison between Townsend's and Clifton's incomes between 1901 and 1914 shows the consequences for student fees (see below). Between 1904 and 1905 Clifton's fees diminished from £314 to £121, while Townsend's rose from £125 to £583. The figures are evidence of a transfer of credibility on the part of the University. The sums that the University gave to Clifton had initially been higher than those given to Townsend, but while they remained constant, Townsend's increased, so that the ratio changed.[16]

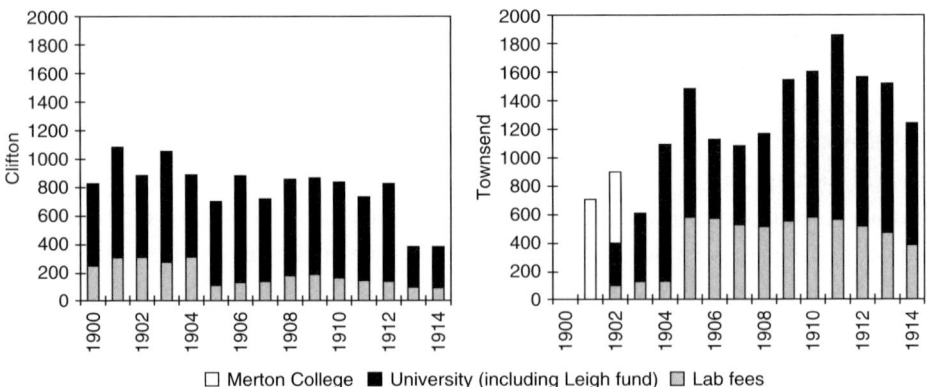

☐ Merton College ■ University (including Leigh fund) ☐ Lab fees

Departmental incomes £, comparing Townsend's and Clifton's laboratories. Balances from previous accounts are not included

Source: University's accounts published annually in *OUG*.

[16] Information from Townsend's and Clifton's departmental accounts, in 'University of Oxford. Abstracts of accounts for the year ending December 31, 1904', published as a separately paginated, undated appendix to *OUG*, 35 (1904–5): 22 and 24; and the corresponding accounts for 1905 in *OUG*, 36 (1905–6): 22 and 24.

There were many differences between Clifton's and Townsend's management styles and institutional rhetoric. If only because of the financial implications for the Clarendon, Clifton gave priority to the teaching of Preliminary students at the expense of Honour students and original research; then in his annual reports he repeatedly asked for more money and space to improve the teaching of Honour students and to allow his staff to do research. Conversely, Townsend's reports showed that he had little personal interest in candidates working for the Preliminary examination, often paying advanced students to teach them. It was typical of his priorities that he encouraged his demonstrators to perform experiments and to publish in scientific journals as an overriding duty, while he asked for additional resources to support teaching at Preliminary level.

The laboratories also had very different scientific ideals. This has been reported by Hurst, who learnt physics in both settings between 1901 and 1903:

The laboratory course at the Clarendon was very well organized, as each experiment had a card with a clearly-written description. A great point was made of accuracy in results, and a student could not pass on from an experiment until his numerical result was within certain limits of accuracy.

Instruments were perfectly tuned, and then treated as objects of distant worship:

Clifton . . . was describing to us how to use a certain instrument. He pointed out that the zero of its scale was in error, and a small correction must be added to its reading. One of the group said 'Would it not be a good thing to make a scratch on the scale to indicate the true zero?' Clifton said 'Don't talk to me about putting a scratch on a scale; it sends a shiver down my spine' (Hurst 1969–70: 59).

Such a respect for precision and instruments did not exist in Townsend's laboratory. There, Hurst learnt that an agreement to the first order between gaseous phenomena and ion theory was more than enough (Hurst 1906; Townsend and Hurst 1904).

This contrast is partly due to the fact that Clifton and Townsend belonged to different generations of physicist. Clifton was 64 in 1900. To a large extent he was marked by the Victorian tradition described by Gooday, in which teaching was ranked before research as a professional duty, and precision was a technical and moral imperative of scientific conduct (Gooday 2004). Townsend, by contrast, was 27 in 1900. He had been trained by J. J. Thomson in the Cavendish Laboratory. Like his master, he perceived himself primarily as a researcher, considered teaching as a mere financial necessity, and viewed excessive precision as an obstacle to the advancement of microphysics.

Most ion physicists had a strong research ethos. That inclination generally set them in conflict with their universities, which paid them to *teach* physics and seldom gave them resources for research. The tension took different forms in each context. Back in the University of Minnesota, where he was now associate pro-fessor in the physics department, John Zeleny's research orientation was obstructed by his colleagues' indifference, by instruments convenient for instruction but not for research, and by heavy teaching duties, which were for him 'an unavoidable cruelty', according to his biographer Erikson (Fudano 1990: 471–2). In the University of Toronto in 1901, McLennan experienced similar difficulties (Gingras 1991: 51). He wrote to Langevin: 'I have been very very busy since coming back to Canada and have had but little time for any research, and so feel rather disappointed'.[17] Rutherford's situation in Montreal looked better: 'I suppose I should consider myself very lucky as the laboratory there is about the finest equipped in the world as regards scientific instruments. I have 3 demonstrators . . . so I should have a little time for research'.[18] In Paris, Langevin had little space and frequently performed experiments on the table of his lecture-room. Preparing his lectures required 'a large amount of bibliographical work'. Yet, as a professor at the Collège de France, he was not responsible for undergraduate teaching. Consequently he had time to give to his research students and to his own scientific work (Langevin 1931).

Another aspect was that most ion physicists encountered resistance from local communities that shared different intellectual and professional values. In Paris, Langevin met the traditional French hostility towards atomism. In 1900, 1902, and 1903, he published detailed epistemological justifications of the existence of ions, and of their accessibility to experimental investigation (Langevin 1900a; 1900b; 1903; Lelong 2001). In Montreal, Rutherford confronted a research tradition, in the high-precision measurement of thermodynamic quantities, that was represented by the assistant demonstrator, Barnes, by two other demonstrators, and by three graduate students. Heilbron observes that Rutherford did not like their 'old fashioned physics, with its tedious exactness and proximity to engineering' (Heilbron 1979: 45). Conversely, his demonstrators scorned him for the low precision of his ionic measurements; in fact, he could only train ion physicists among undergraduates. In 1900 two distinct research groups coexisted within the Montreal laboratory: Barnes's on calorimetry and Rutherford's on ions and radioactivity.

The opposition between Rutherford and Barnes parallels the conflict between Townsend and Clifton. Each pair of men embodied opposing cultures with different instrumental bases and different attitudes to precision, teaching, and engineering. In Montreal as in Oxford, the contrast was due to the fact that the physicists concerned belonged to different generations. High-precision calorimetry had been introduced there by the previous director, H. L. Callendar, who had been trained in the Cavendish Laboratory in the pre-Thomsonian tradition of electrical resistances. His experimental approach, like Clifton's, was rooted in Victorian precision physics (Gingras 1987). But the conflicts had different endings. In Montreal, from 1903, collaborations began between calorimetry and radioactivity, a difference that

---

[17] McLennan to Langevin, 11 December 1901, Langevin papers, L76/01.
[18] Rutherford to Langevin, 3 September 1898, Langevin papers, L76/40.

owes much to the institutional conditions. Rutherford was the head of his department, and the calorimetry group was under his administrative and financial control. In contrast, the Oxford situation was characterized by two directors independent of each other and competing for the same resources, which led perforce to the domination of one by the other.

OXFORD CHEMISTS AND COLLEGE FELLOWS

Apart from Clifton's group, Townsend had to deal with another scientific community in Oxford: the chemists. His first research student was H. E. Hurst, who finished his B.Sc. dissertation, a study of collisions between ions in hydrogen, in 1905. Two examiners were appointed: Townsend and Herbert B. Baker.[19] Baker was Dr Lee's reader in chemistry at Christ Church. He was a chemist, not a physicist, so that his appointment as an examiner warrants comment. First, Baker belonged to a research tradition that Oxford chemists had pursued since the 1880s: the study of chemical reactions between gases, especially the combination of hydrogen and oxygen by electric sparks. More than any physicist in Oxford, Baker was an expert on the electrical properties of gases (Thorpe 1932–5; Baker 1910; Howarth 1987: 338). Secondly, chemistry in Oxford was far more developed than physics. Of all the sciences, it was the best catered for in college laboratories, and by the last two decades of the nineteenth century it was far more popular among undergraduates than was physics. Oxford chemical research covered some territories, such as electricity in gases, that in Cambridge were appropriated by physicists.[20] That Baker appeared in Oxford as the most competent examiner for Hurst is therefore not surprising.

Baker, however, severely criticized Hurst's work. Hydrogen had been prepared by the electrolysis of a solution of potassium hydroxide, i.e. not by Baker's method, which required the electrolysis of barium hydroxide. Consequently, Baker considered that Hurst's hydrogen was not pure enough and that impurities had certainly affected his numerical results. He refused to approve the award of the degree,[21] and Hurst was required to perform new experiments using samples of barium hydroxide provided by Baker (Hurst 1969–70). This new method soon became assimilated in Townsend's laboratory, where Kirkby and Haselfoot started using barium hydroxide 'kindly supplied' by Baker (Haselfoot and Kirkby 1904: 481).

Kirkby's career deserves attention, because it sheds light on the role played by Oxford chemistry, and by the collegiate structure of the university, in the development of Townsend's laboratory. From 1904 to 1907 Kirkby's experiments showed a close intellectual and technological proximity to Baker's line of research. In 1904 he

---

[19] OUA, Natural Science Faculty Board Minutes 1892–1912, FA4/13/1/2: 123, 125, 127.
[20] Howarth 1987: 337–9; Howarth 2000: 462–5; Laidler 1988. On the practice of physics by Oxford chemists see Simcock, Chapter 4, 121–45.
[21] OUA, Natural Science Faculty Board Minutes 1892–1912, FA4/13/2/1, 39, 41.

examined one of Baker's favourite reactions, the explosion of hydrogen and oxygen, and he tried to explain it with Townsend's concepts, i.e. by collisions between ions (Kirkby 1904). In 1905, with Baker's help, he designed new apparatus for studying the reaction at low pressures (Kirkby 1905). His paper was reviewed by the *Journal of the Chemical Society* as a contribution to 'Inorganic Chemistry', and not to 'General and Physical Chemistry', as was usually the case for papers on ion physics. Meanwhile, Kirkby and Baker collaborated within Oxford, creating a 'Committee for Engineering and Mining'.[22]

Kirkby, therefore, borrowed his research goals, theoretical tools, and material technologies from two different fields: Cambridge ion physics and Oxford gaseous chemistry. This disciplinary hybridization was very much a function of the local environment. In Paris, Langevin's students, for instance Bloch in 1905, explicitly rejected any interest in the chemical properties of gaseous ions (Bloch 1904). Each ion laboratory drew upon new instrumentation and concepts from local scientific communities. In Montreal, Rutherford collaborated with Barnes, using Barnes's expertise and his thermometers to measure the energy required to produce an ion (Heilbron 1979: 58). In Paris, Maurice de Broglie used the ultramicroscope designed by Jean Perrin to observe Brownian motion: he transformed it in order to see macroscopic ions and to photograph their trajectories (de Broglie 1909). In Cambridge, Wilson's cloud chamber established a link between ion physics and meteorology: the device could be used either to investigate cloud formation or to photograph the tracks and collisions of electrified particles (Galison and Assmus 1989). These exchanges owed much to locally available expertise, but also to social networks that allowed the transfers or even required them. Examples are, in Paris, the informal circles of Langevin, Perrin, and Curie; or, in Oxford, the Natural Science Faculty Board or Amabel Moseley's dinner parties (Heilbron 1974a: 41).

Such comparisons also show that Kirkby performed very few transfers. Conversely, de Broglie in Paris had the skills required to use an ultramicroscope: he taught them to other ion physicists in Langevin's laboratory, for instance to Brizard, Reboul, and Léon Bloch. In Cambridge, C. T. R. Wilson did the same with his cloud chamber in the Cavendish Laboratory. In contrast, Kirkby did not transmit his new chemical technical know-how to other workers in the Electrical Laboratory. This seems to reflect his independence and his growing isolation in the department (de Broglie and Brizard 1910).

Kirkby's independence was of, above all, a financial and institutional kind. As a fellow of New College, which provided him with a salary and accommodation, Kirkby enjoyed an income unrelated to his demonstratorship. Moreover, among Townsend's demonstrators, he alone was a college fellow, paid not by Townsend

---

[22] OUA, Natural Science Faculty Board Minutes 1892–1912, FA4/13/2/1, 39, 149.

but directly by the Secretary of the University Chest, W. B. Gamlen. In 1910, Kirkby resigned his demonstratorship to become the Rector of Saham Toney, Norfolk (von Engel 1957: 263), whereupon Townsend had to negotiate with Gamlen to have Kirkby's salary paid to his successor, F. B. Pidduck.[23] In this way, Townsend experienced what Jack Morrell has called 'the peculiarities of the Oxford system, which gave a professor charge of a university laboratory but which in part staffed it with college fellows who were statutorily independent of him' (Morrell 1993: 122). We have seen that the independence gradually became a scient-ific one too. Kirkby's experiments were far more distant from Townsend's research than were those of other demonstrators, such as Brown, Gill, Hurst, or Smith. After 1910, Kirkby continued experimenting in Oxford, but not in Townsend's laboratory. Instead, he

Fig. 6.2. Henry Gwyn Jeffreys Moseley in the main teaching laboratory of the Balliol–Trinity Laboratories. The photograph, taken during the brief period (in the summer of 1910) between his graduation and his departure for Manchester to work with Rutherford, has often been reproduced. However, its significance has never been discussed. It is one of a number of portraits (drawn on for this volume) of those who were actively associated with the Balliol–Trinity Laboratories at the time. It shows Moseley, a Trinity College physics graduate, as part of a community that, though largely chemical in its interests, made little distinction between chemistry and physics. The photograph calls into question the impression given in many studies of Moseley (and in some of his own comments) that he considered the college teaching of his subject of poor quality and that he held himself somewhat aloof from it.

---

[23] Kirkby to Gamlen, 16 October 1910; Townsend to Gamlen, 19 October 1910; Townsend to Gamlen, 24 February 1911. All in OUA, 'Electrical Laboratory', UC/FF/77/1.

worked in the chemistry department with one of the demonstrators there, James E. Marsh. In 1913, he and Marsh published a joint study of the explosion of azoimide (Kirkby and Marsh 1913).

Kirkby's scientific and social behaviour is easier to grasp through a comparison with Harry Moseley. Although Moseley graduated in Oxford, he felt alien to the place. As an undergraduate at Trinity College, he worked in the Balliol–Trinity Laboratory, where he built a primitive cloud chamber. Then, 'partly from tension over his course of study', he obtained only a second class in the 'bookish' Final Honour School (Heilbron 1974b: 542). Soon after taking his degree in 1910 he moved to Rutherford's laboratory in Manchester, where he enthusiastically experimented on radioactivity, sometimes for sixteen hours a day. In 1912, as soon as news of von Laue's discovery of X-ray diffraction reached him from Germany, he moved to Bragg's laboratory in Leeds to learn the new technique. Back in Oxford in 1913, in a private capacity and with research space in the Electrical Laboratory, he observed: 'Things seem to move slowly here compared to Manchester'.[24] But by the following year his experiments on X-ray spectra were already known, praised, and used on an international level.

The differences between Moseley and Kirkby are striking. Kirkby worked for thirteen years on the same subject and quietly published less than one paper every two years. His work on the chemical action of gaseous ions was not mentioned by the three major treatises on ion physics (Abraham and Langevin 1905; Thomson 1906; Townsend 1915). To the best of my knowledge, in fact, nobody outside Oxford paid any attention to it at all. Predictably, Moseley held Kirkby in contempt:

Firstly he is by nature rather lazy himself, and therefore his ideal is on a less strenuous plane than my own – he would probably think an Oxford fellowship would be the full stretch of my ambition. Secondly, as an Oxford man he looks down on all things outside . . . (Heilbron 1974a: 169–70).

More charitably, it should be said that Kirkby's college was his favourite environment and that he was ideally suited to that world. To quote Morrell again:

. . . the tradition of connoisseurship at high table in fellowship divine sat uneasily with the notion of specialist and allegedly narrow scientific publication. From the college enclave, research could be seen as an ungentlemanly and boorish Germanic notion, and postgraduate supervision as a Yankee device for inserting plebeians into a patrician university (Morrell 1994: 141).

In contrast with Moseley, Kirkby was professionally established, well integrated and at ease in Oxford. For him, research was neither a vital priority nor a route to international fame, but rather a stimulating hobby and a way of socializing with the Oxford chemists.

[24] Moseley to Rutherford, 7 December 1913, quoted in Heilbron (1974a: 217).

THE DRAPERS' COMPANY AND THE TEACHING OF ENGINEERING

Ion research was sometimes financed by industrial companies or private foundations, such as McDonald or the Carnegie Foundation in Canada (Gingras 1991: 64; Heilbron 1979: 57). In Townsend's case, this role was played by the Drapers' Company, a medieval guild turned into a charitable institution. From 1877 onwards, under political pressures mediated by the City and Guilds of London Institute, the Drapers' Company gave substantial support for the development of technical education in Britain (Girtin 1964: 336–54; Johnson 1914–1922, vol. 3: 487). Between 1887 and 1922 the Company donated a total of £532 000 to various universities and colleges, funding physiological or electrical laboratories, and mining or engineering departments in London, Cambridge, Sheffield, Leeds, Oxford, and elsewhere (Johnson 1914–1922, vol. 3: 488–9).

At the end of the nineteenth century, close links were already established between Oxford and the Drapers' Company. The Master of the Company, the Revd Henry Boyd, was also the Principal of Hertford College (Croft 1986: 53). In 1894, the Drapers decided to finance a new building, the Radcliffe Science Library, in order to support 'the educational work of the University, especially in those subjects which in more recent times it has been found necessary to include in the

Fig. 6.3. The Electrical Laboratory, built to the design of T. G. Jackson and inaugurated in 1910.

Fig. 6.4. The plaque commemorating the role of the Drapers' Company in the construction of the Electrical Laboratory. The company donated £23 000 'for the promotion of the Study of Electrical Science'. The plaque is on an outer wall of the laboratory, next to the main entrance. The plaque reads:

'The Worshipful Company of DRAPERS of the City of LONDON erected this building for the promotion of the Study of Electrical Science and presented it to the Chancellor Masters and Scholars of the University of Oxford on the 21st day of June A.D. 1910.

> Keddey May Fletcher (Master)
> John Barrow, Bernard Francis Harris, Webster Glynes, Gerald Walton Williams (Wardens)
> Ernest Henry Pooley (Clerk)'

course of study'. Three years later, T. G. Jackson was appointed as the architect, to the Company's evident pleasure:

We may feel confident that the Library in his hands will become not merely a suitable home for the literary studies of our scientific men here, but will be a permanent and attractive memorial of the splendid and thoughtful generosity of the Drapers' Company'.[25]

Townsend was supportive of these local proponents of education in the sciences and soon he became a beneficiary of the Drapers' Company's generosity himself. On 1 June 1908, after three years of negotiation, the company took the momentous decision that 'Oxford stands in great need of a new laboratory for the teaching of Physics and Electrical Science'.[26] It offered £23,000 for the purpose and again, predictably, appointed Jackson as the architect. The new Electrical Laboratory, built close to the Clarendon and inaugurated on 21 June 1910,[27] transformed the teaching of electricity. With ample space at his disposal, Townsend was able to offer instruction not only to physicists but also to undergraduates reading engineering science, which was added as one of the subjects for the Final Honour School of Natural Science by a decision of Hebdomadal Council in December 1908. Townsend, who had formally proposed the move in Congregation in June 1908, was happy to assume overall responsibility for the

---

[25] Drapers' Company Archives, Court of Assistants minute-books, 6 December 1894 and 15 July 1897.
[26] Drapers' Company Archives, Court of Assistants minute-books, 1 June 1908.
[27] *Nature*, 73 (1910) 510–11.

new teaching, and soon his laboratory, initially in the University Museum and then in the new building, was offering courses at two levels: one in general electrical engineering (in parallel with a course in applied chemistry taught by his colleagues in chemistry), the other in 'Advanced applied electricity and magnetism', which included alternating currents, electromagnetic waves, telegraphy, and telephony.[28]

The new teaching, for a degree scheme that never attracted more than ten Honours candidates a year before the first world war, was not unduly burdensome, and engineering students were always outnumbered by physicists. From 1909 to 1912, for example, the attendance at practical classes in electrical engineering varied between two and fourteen each term, compared with between twelve and twenty-four in the classes for undergraduates specializing in physics for the Final Honour School.(Townsend 1909–10, 1910–11, 1911–12, 1912–13, 1913–14).[29] Michael Sanderson has stated that 'the work there [in the Electrical Laboratory] was not evidently concerned with technical problems of electrical engineering' (Sanderson 1972: 39). And it is conceivable that the teaching did reflect some residual influence of the values that Townsend had imbibed in the Cavendish Laboratory. But too much should not be made of this point. Townsend was unreservedly keen to contribute to the success of the new chair of engineering science, to which Charles Frewen Jenkin was appointed in 1908 (Jenkin 1911–12; see also Simcock, Chapter 5, 204–6). Even if he delegated the task of teaching to his demonstrators (as the annual reports in the *Oxford University Gazette* between 1909 and 1914 indicate), he did not treat his involvement in Jenkin's work lightly, and his readiness to promote instruction in the applied aspects of electricity is beyond question.

SCIENTIFIC WORK ON COLLISIONS, 1901–10

In order to assess Townsend and his laboratory in the international context of ion physics, I begin with the decade from 1901 to 1910. A first point is that Townsend's laboratory focused on the specific subject of collisions between ions. Physicists there ionized gases with different agents and measured collisions in different gases and in different conditions of temperature and pressure. The subject was controversial, and they talked of the 'theory of collisions', presenting their results as 'confirmations of Professor Townsend's theory' (Hurst 1906; Brown 1906; Gill and Pidduck 1908). Other ion laboratories specialized too. Langevin's group studied slow-moving ions and recombination. McLennan's examined spontaneous ionization in closed vessels. Rutherford's investigated ions produced by radium rays (and increasingly radioactivity itself rather than its ionization effects). By comparison, the

---

[28] OUA, Natural Science Faculty Board Minutes 1892–1912, FA4/13/1/2, 16 June 1908; 17 October 1908; 1 December 1908; Natural Science Reports 1892–1912, FA4/13/2/1, 64.
[29] On the teaching of engineering in Oxford see Howarth (1987: 342; 2000: 17 and 23–7).

Oxford group seemed unusually homogeneous in its research topics, which confirms Townsend's reputation as a very authoritarian director (von Engel 1957: 269).

How were Oxonian results perceived and used internationally? Here again, a crucial role was played by Townsend's network, at the heart of which were his friends trained in the Cavendish. In 1906, Langevin praised Townsend's theory of ionization by collision for its all-embracing explanatory power, ranking it as the most important advance in discharge physics since 1900 (Langevin 1906). Moreover, ionic collisions multiplied charges and allowed the detection of single particles. When, in 1908, Rutherford and Geiger used the multiplication coefficients measured by the Oxford physicists to design their new particle detector (Trenn 1986), Townsend himself seems to have taught them how to produce and stabilize a steady multiplication effect.[30] In his papers, Rutherford frequently mentioned 'Professor Townsend's principle of collision', cited his publications, and publicly gave him credit for the new counting method (Rutherford and Geiger 1908: 161).

As a research subject, collisions were less important in Cambridge. In 1903, J. J. Thomson published a new treatise entitled *Conduction of Electricity through Gases*. In this he praised Townsend's 'valuable' experiments on collisions but he described his own layer theory of Stoletow's results while ignoring Townsend's alternative explanation (Thomson 1903: 341). In 1904, Townsend wrote: 'I have lots of things to be done in my particular job on collisions. I am very glad they are not working at it at the Cavendish. I expect J. J. keeps them off it as much as possible'.[31] In 1906, in the second edition of his treatise, Thomson once again ignored Townsend's explanation of Stoletow's results, also arguing that Langevin's collision concept was inconsistent and that it should be dropped (Thomson 1906: 270–1). Townsend responded in 1910 with a small book entitled *The Theory of Ionization of Gases by Collision*. In this, he defended his team's results and angrily undermined Thomson's layer theory (Townsend 1910; Townsend 1915: 263). The positions of both men remained unchanged until the First World War.

AFTER 1910: INTERNATIONAL MARGINALIZATION AND LOCAL STRENGTH

After 1910, Townsend gradually became marginalized within the new context of international physics. The 1910s were marked by the consolidation of a new transnational network of physicists, a process initiated by the first two international physics congresses (Paris 1900 and Saint Louis 1904) and continued by the Solvay congresses for the discipline in 1911 and 1913. A key role was played by the new generation, especially

---

[30] Townsend to Rutherford, 10 March 1908, CUL Add. MS 7653, T76.
[31] Townsend to Rutherford, 9 July 1904, CUL Add. MS 7653, T75.

by Langevin, Perrin, and Rutherford. Townsend did not attend these congresses. At the 1911 Solvay meeting on 'Radiation and quanta', he was cited three times, though briefly. At the 1913 Solvay meeting on 'The structure of matter', only one of Townsend's scientific results was mentioned (his measurement of the ionization energy of hydrogen in 1910), and there was no mention of any of his students. Instead, discussions focused on specific heats, radioactivity, electronic theories of magnetism, or Brownian motion; none addressed collisions between ions. After 1910, a social and intellectual distance gradually separated Townsend's laboratory from the new 'modern physics' and its networks (Mehra 1975: 13–72, 75, 78, 81).

The distance was also a technological one. Experimental results examined at the Solvay congresses were produced with equipment very different from Townsend's: X-ray diffraction apparatus, ionization chambers, photoelectric cells, spectroscopes, mass-spectrographs. Access to this new 'symbolic market' required new techniques.

This is clearly illustrated by the attitudes of ion physicists in Toronto and Paris. In 1912–13, McLennan's team in Toronto gradually gave up counting ions in atmospheric air and closed vessels and started investigating the spectrum of mercury (McLennan 1912; McLennan and Henderson 1915; Gingras 1991: 71). In Paris, in 1912–13, the two most prolific ion physicists also abandoned ions; Maurice de Broglie entered the field of X-ray diffraction, and Eugène Bloch turned to ultraviolet spectroscopy (Bloch E. 1912; Bloch L. and E. 1914; de Broglie 1913a and 1913b). Townsend's attitude was different. In the 1910s he bought new thermogalvanometers, ammeters, voltmeters, and Geryk pumps.[32] His most expensive new instrument was a liquid-air plant for use in producing a high vacuum.[33] In this way, he sought to improve his technical system, therefore, but in a manner very different from his Canadian and French counterparts, who built and used a new one.

This technological gap is illustrated by Moseley's experience in Oxford, when he came back to Oxford from Manchester to work in Townsend's laboratory in November 1913. Learning that he wanted to work on X-ray spectroscopy, Townsend required all his 'patience and tact'. Moseley did not find in the Electrical Laboratory the instruments he needed (a Gaede pump and an X-ray spectrometer), and he complained that Townsend's assistant lacked the skills that were necessary to help him. The instrument-sellers linked to Townsend responded to Moseley's demands, but slowly and inadequately (Heilbron 1974a: 70–2, 95–6). This reveals a marked contrast with Townsend's technical system, with its complex of instruments, phenomena, skills, and routine collaborations. Equally significant is the fact

---

[32] Records of these purchases are in the folder 'History. Archaic technical literature and drawings', CLA (filing cabinet 2, drawer 1). For further details see the inventories in Watson (1994): 16–17, and the website http://www.physics.ox.ac.uk/history.asp.

[33] OUA, 'Electrical Laboratory', UC/FF/77/1, Townsend to Gamlen, 21 November 1911; Townsend to Gamlen, 17 December 1913.

that nobody in Oxford took the opportunity of learning the new technology with Moseley (whereas Georges Urbain came from Paris to do so).

These unflattering perceptions need to be qualified, however. Townsend was still known and respected internationally. In 1919, for example, when looking for a new president for the Solvay congresses, Marcel Brillouin suggested, in a letter to Emile Tassel, that Lorentz's successor be chosen from among the 'new generation', which meant for him 'Rutherford, Langevin, Perrin, Townsend, etc.' (Marage and Wallenborn 1995: 212). But he seriously considered only the first three and finally recommended Langevin. In 1910 the German publisher Marx decided to put out a new treatise on 'modern physics' in six volumes. Townsend was responsible for the volume on gaseous ions. Other contributors included Rutherford and Marie Curie, but not J. J. Thomson.[34] This reflected not only Townsend's growing prominence but also the fall of J. J. Thomson as the international expert in discharge physics. The manuscript of the third edition of Thomson's treatise was finished in 1913.[35] But it was not published until 1928, and then in a much revised form (Thomson and Thomson 1928–33). At the 1913 Solvay congress, when Thomson presented his new atomic model, he was severely criticized by Rutherford and Langevin. In these years, Thomson opposed the new views on electrons, even refusing to use the word and keeping his own term of 'corpuscle' (Falconer 1987: 242–3 and 273–4). In summary, it is clear that Townsend, whose treatise was published in Oxford by the Clarendon Press in 1915 (Townsend 1915), successfully built a disciplinary niche, gaseous ions, in which he secured his leadership. This speciality was still important for the design of detectors and experimental set-ups, but it was no longer at the forefront of research on the structure of matter. Townsend's leadership in gas ionization, therefore, resulted not just from his scientific excellence but equally from Thomson's increasing obsolescence and from the abandonment of the field by most of its practitioners.

The First World War dispersed Townsend's group. Gill, Tizard, and Edmunds left for military service in 1914. At the end of 1915, Lattey obtained a commission in the Royal Flying Corps and Pidduck one for special duties in the Ministry of Munitions (Townsend 1915–16: 545). Moseley had obtained a commission in the Royal Engineers in 1914 and after eight months' training he was shipped to the Dardanelles as a signals officer (Heilbron 1974a: 544). In contrast with the eight physicists provided by the Electrical Laboratory, the Clarendon contributed only I. O. Griffith, an assistant demonstrator, to the war effort (Morrell 1992: 266). Moseley's death in August 1915 in an offensive led by Kemal Ataturk was a shock to Townsend. It led him to write to Rutherford (a member of the Board of Inventions and Research since July 1915), asking him to 'move the Inventions board to try &

---

[34] Townsend to Langevin, 11 July 1911, Langevin papers, L76/74.
[35] Thomson's manuscript is in CUL, Add. MS 7654, BD7.

get some of the very clever physicists, who are now in the army, back to England', and suggesting Bragg among them.[36]

Aged 46 in 1914, Townsend was under no obligation to join the services. He seems nevertheless to have volunteered for a wireless unit that was to be sent to Russia. In the event, the plan was never implemented, and he stayed in Oxford (von Engel 1957: 264; Croft 1986: 57). The Electrical Laboratory being almost empty, most of the rooms were given over to the Royal Flying Corps in April 1916 to assist them in establishing the School of Military Aeronautics, and were used by them until December 1918 (Townsend 1916–17; 1917–18). The few remaining Preliminary students were taught by Baynes. Having no more advanced students, Townsend offered his services to the Royal Naval Air Service.[37] By September 1915 he had enrolled, had shaved his moustache, and was wearing a blue uniform with brass buttons: 'I feel quite a fraud in naval uniform as I would get sea sick if I went on the sea for 10 minutes!! However the Admiralty have not discovered my weakness yet in this respect as up to this I have been working on land.'[38] Townsend undertook wireless research for the Navy at Oxford and Woolwich. In 1916 he developed small, robust wavemeters that measured wavelengths of 800–3200 m and 3000–12 000 m , the sort of frequencies used by radio transmitters during the war. Eighteen of them were to be supplied to the Admiralty in 1918.[39] Townsend continued to improve the wavemeter after the war, patenting it several times between 1919 and 1921.[40]

The break with international physics became more marked after the war. Townsend first ignored and then rejected the emerging quantum theories. In 1915, Niels Bohr reinterpreted Franck and Hertz's experiment of 1914 as an important confirmation of his atomic theory. The work of Franck and Hertz led eventually to their receiving the Nobel Prize in 1925. Between 1923 and 1926, however, the Oxford group measured new ionization potentials, which Townsend published as contradicting Franck and Hertz's results and refuting the quantum theory of electron collisions (Hon 1989: 87–8, 98–101). Meanwhile, the Ramsauer effect of slow electrons came increasingly to be viewed as a quantum phenomenon, which led Max Born to apply wave mechanics to collision processes from 1926 onwards. To very different ends, slow electrons were studied in the 1920s by Townsend and his collaborators (Bailey, Skinner, and White), who constructed a 'Townsend effect' as a non-quantum alternative to the Ramsauer effect (Im 1995: 280, 282–7).

[36] Townsend to Rutherford, 22 September 1915, CUL, Add. MS 7653, T88. Rutherford was one of the twelve scientists of the Consulting Panel of the Bureau; see Hackmann (1988: 98).
[37] Townsend to Rutherford, 12 February 1915, CUL, Add. MS 7653, T86.
[38] Townsend to Langevin, 11 November 1915, Langevin papers, L76/75.
[39] Instrument no. 30, CLA; contract with the Admiralty to supply eighteen wavemeters, 1918, in Townsend Papers, Bodleian Library, IV/11; diagrams for wavemeter IV (7 sheets, nd), ibid., IV/16.
[40] Numerous patents are in the Townsend Papers, Bodleian Library. For instance: 'Improvements in electric wave meters', patent number 129579, 4 February 1919 (IV/13); 'Improved method of and means for measuring the length of electric waves', with James Herbert Morrell, patent number 177938, 15 February 1921 (IV/18). See also Townsend (1920) and Townsend and Morrell (1921).

Fig. 6.5. John Sealy Edward Townsend, Wykeham professor of physics from 1900 until his retirement in 1941.

All this occurred in a context of increasing detachment between Oxford and the wider world of physics. Between 1924 and 1953, Townsend attended no international scientific gathering. His laboratory has been described as 'an isolated island' whose inhabitants 'were rather impatient in explaining unfamiliar concepts', only 'occasionally carrying the flag into a far land' (von Engel 1957: 269). In 1938, with his lectures being ever more out of touch, the University tried to convert his chair into one for theoretical physics, which happened at the end of the Second World War (Morrell 1992: 304). Townsend withdrew into Oxford in his personal life also. After 1918, he apparently stopped visiting and exchanging letters with Rutherford and Langevin. And home became his most important focus. Here, his marriage, in 1911, to Mary Georgiana Lambert, who became 'a delightful hostess' and an active worker in municipal affairs, an alderman, twice mayor, and an honorary freeman of the City of Oxford, did much to fashion his activities (von Engel 1971: 985).

By his lifestyle, despite having neither independent means nor a country house, Townsend resembled other Oxford physicists, such as Dobson, Merton, and Jackson, the wealthy 'Clarendonian gentlemen' depicted by Morrell (1992: 275–8). His favourite hobbies were tennis, riding, shooting, and fox-hunting. He was often seen riding on his horse to the laboratory, although his house in Banbury Road was less than ten minutes' walk from it. The practice must have appeared eccentric in streets gradually invaded by cars from the Morris factories, but Townsend's social manners were of a piece with the college life of an Oxford character. According to his friend and obituarist Hans von Engel, 'he was a charming conversationalist and his stories about people he had met and other anecdotes of an amusing nature were always well received in common-rooms and at private parties' (von Engel 1957: 268–9). Yet Townsend rarely socialized with the physicists of the Clarendon Laboratory. The only exception was Derek Jackson, whom he frequently visited in the laboratory to discuss fox-hunting (Bleaney 1994: 260). Jackson was a very wealthy man; he bought himself expensive spectrographs, moved in artistic and aristocratic circles, had a succession of six wives, and owned horses, which he rode when fox-hunting and competing in important races, including the Grand National (Kurti 1991; Morrell 1992: 278). His friendship with Townsend was favoured by their common passion for aristocratic hobbies, sarcastic jokes, eccentricity, and country life.[41]

After several decades in Oxford, Townsend's scientific and social identity had changed dramatically. I suggest that this owes much to the cultural idiosyncracies of Oxford science, to its patterns of socialization, and to its model of the gentlemanly scientist. Analysing William Perkin's later career in Oxford, Morrell has observed that 'Oxford's structures and interests' were 'inimical' to industrial research. Hence Perkin's integration there, after his move from Manchester, might well have turned out to be a 'lamentable failure'. But it was in fact a success, an 'Indian summer' due partly to 'his character, which was amiable and endearing', and partly to 'his wide interests in music, horticulture, hospitality and travel' (Morrell 1993: 122–3). By contrast, Frederick Soddy's settlement in Oxford was a failure, one that owed much to the incompatibility of his personality and social manners with Oxonian codes (Cruickshank 1979). Oxford was clearly a place that required a considerable degree of adaptation from scientists arriving from outside.[42]

The contrast between Townsend's evolution and that of other ion physicists from the Cavendish Laboratory is striking. Between the wars, Langevin became a French *intellectuel*, a communist fighting fascism, a reformer of university education, an ardent proponent of relativity and quantum physics, and a divorcee (Bensaude-Vincent 1987). McLennan was involved in the institutional making of the Canadian physics community and became 'the best known and most influential physicist in Canada' (Gingras 1991: 122). Like Townsend in Oxford, each ion

---

[41] On the relations between Townsend and Lindemann's group, see Morrell, Chapter 7, 243–4 and 262.
[42] See also Howarth (1987: 353).

physicist encountered a specific national and cultural context; each one adapted to it and shaped accordingly his scientific and social practices and his public and private identity.

CONCLUSION

Townsend's original intention was clearly to reproduce the Cambridge style of experimental physics in Oxford. From 1901 to 1914, he planned and built the new Electrical Laboratory on the Cavendish model and strengthened its position within the University. He reformed the curriculum to create and expand a student clientele for the teaching that was offered in his laboratory. He accelerated the discredit into which Clifton had fallen, in order to capture part of the Clarendon's resources, and by allying himself with the local proponents of technical education, he secured the support of the Drapers' Company for the construction of his laboratory. As a result of his encouragement for demonstrators and students to experiment and publish in scientific journals, the laboratory soon became a centre for research on gaseous ions, with instrumental equipment, precision norms, organizational structure, and professional ideals that were clearly inherited from the Cavendish Laboratory. In the international context of physics, Townsend built a disciplinary niche in which he secured a position of leadership.

But Townsend had to deal with several peculiarities of the Oxford system, such as the institutional and scientific domination of physics by chemistry, the high degree of independence of college fellows in a collegiate university, and the Oxford model of the gentlemanly scientist. These long-term structural constraints and cultural legacies (as well as a number of external influences, such as that of the Drapers' Company) led him to depart from his self-imposed brief. At the end of the 1920s, the Electrical Laboratory had a scientific and social identity clearly distinct from its original model. So too did its director.

7

The Lindemann Era

Jack Morrell

In 1919 Frederick Lindemann assumed the chair of experimental philosophy and responsibility for the Clarendon Laboratory. He had high ambitions: to launch and nurture high-quality research; to raise the status of science at Oxford; and to promote low-temperature physics as a distinguished feature of the Clarendon. By 1933 the first and second aims, but not the third, had begun to be realized. Lindemann himself faded as a researcher in the 1920s, but he did recruit some talented physicists, three of whom (Merton, Dobson, and Jackson) were affluent gentlemen who pursued research mainly solo. Only Egerton, a specialist in combustion, ran a research group.

Lindemann saw the advent of the Third Reich as an opportunity as well as a tragedy. Between 1933 and 1935 he invited to Oxford two mathematical physicists (Schrödinger and Fritz London), an atomic physicist (Szilard), a spectroscopist (H. G. Kuhn), and four low-temperature physicists (Simon, Mendelssohn, Kurti, and Heinz London, all from Breslau), most of whom were supported by ICI via a temporary scheme that Lindemann had instigated. Though his attempts to promote mathematical physics and atomic physics met difficulties, low-temperature physics and spectroscopy were so successful that Lindemann managed to keep Simon, Mendelssohn, Kurti, and Kuhn in Oxford. The Breslau cohort enabled Lindemann to fulfil his long-cherished aims. Their success in establishing productive research groups, in introducing research-training, and in giving the Clarendon a prominent research identity enabled him to secure a new laboratory completed in 1939 notwithstanding the small number of finalists in physics – about twenty per annum (see Table 7.1).

For permission to cite the Hebdomadal Council Papers (HCP), the files of the University Registrar (UR/SF), the minutes of the Natural Sciences Faculty Board (NS/M), and the reports of the Physical Sciences Faculty Board (PS/R), I am grateful to Mr D. Vaisey, formerly Keeper of the Archives, University of Oxford. For permission to cite manuscripts I am grateful to the Bodleian Library, Oxford; Lincoln College, Oxford; Nuffield College, Oxford; the Royal Society of London; and the Society for the Protection of Science and Learning (SPSL). For valuable responses to my queries I am indebted to Professor B. Bleaney, the late A. H. Cooke, the late C. H. Collie, E. H. Cooke-Yarborough, the late H. G. Kuhn, the late Professor R. V. Jones, H. D. Megaw, the late A. R. Ubbelohde, and especially the late Professor N. Kurti.

Table 7.1. Number of finalists in physics 1920–39

	Women	Men	Total
1919–20	0	5	5
1920–21	0	17	17
1921–22	1	20	21
1922–23	1	21	22
1923–24	2	18	20
1924–25	0	13	13
1925–26	2	13	15
1926–27	1	23	24
1927–28	0	12	12
1928–29	0	19	19
1929–30	0	11	11
1930–31	1	23	24
1931–32	2	15	17
1932–33	3	18	21
1933–34	1	27	28
1934–35	3	19	22
1935–36	1	18	19
1936–37	1	20	21
1937–38	1	19	20
1938–39	0	22	22

Source: Class lists published annually in *OUG*

Simultaneously Lindemann became a public and frequently reviled figure as a Conservative politician and as Churchill's scientific adviser about air defence.

In this chapter I support the traditional view that Lindemann was the reviver of the Clarendon, for which he became responsible after an interregnum of four years after Clifton's resignation in 1915. But the mode of resuscitation was neither widespread in British physics nor in the University of Oxford. Lindemann did not shine as a discoverer; indeed, he relinquished research at the bench in the mid-1920s. As a research supervisor he gave no training in research. His lectures verged on the inaudible; and his personality, though formidable, was often seen as austere. But he comprehended physics as a whole and on that basis became the impresario of the Clarendon, promoting and publicizing it whenever and wherever he could. He delegated daily management of the laboratory to Keeley and, as its guardian, tried to secure adequate rewards and facilities for those he had appointed as researchers and demonstrators. In this chapter, I adduce new evidence to characterize not only Lindemann's activities as an impresario but also the performances of the Clarendon troupe that he recruited and nurtured.[1]

---

[1] Bleaney *et al*. (1986) is a perceptive 'traditional' vignette. For earlier versions of this chapter see Morrell (1992; 1997).

Fig. 7.1. Group photograph of the participants in the second Solvay physics conference, held in Brussels in 1913. *Seated (left to right)*: Nernst, Rutherford, Wien, Thomson, Warburg, Lorentz, Brillouin, Barlow, Kamerlingh Onnes, Wood, Gouy, Weiss; *Standing*: Hasenohrl, Verschaffelt, Jeans, Bragg, Laue, Rubens, Mme Curie, Goldschmidt, Sommerfeld, Herzen, Einstein, Lindemann, de Broglie, Pope, Gruneisen, Knudsen, Hostelet, Langevin. Lindemann was co-secretary for the conference, as he had been for the first conference two years earlier. By the time of Lindemann's death in 1957, a further eight Solvay physics conferences had been held, but he attended none of them.

THE PROF: LINDEMANN IN THE TWENTIES

Lindemann's career before 1919 was unusual for a future Oxford professor. Though English, he was born in Germany and his higher scientific training was obtained there and not at Oxford, Cambridge, or London universities. His greatest debt was to Walther Nernst, with whom he worked for almost five years in Berlin, obtaining his doctorate in 1910. They collaborated in research on specific heats at low temperatures attained with liquid hydrogen. From Nernst, Lindemann derived a deep interest in low-temperature physics pursued in the contexts of the quantum theory and of Nernst's heat theorem enunciated in 1906. He admired the way in which Nernst swung the whole laboratory on to cryogenics, thus identifying the laboratory with it. He learnt that Nernst's late rising was no bar to eminence, and he copied it. Nernst ensured that Lindemann was co-secretary of the first Solvay physics conference held in Brussels in 1911. Through Nernst he made useful contacts with

German physicists and friends with two of Nernst's young English researchers, Henry Tizard and Alfred Egerton. For the rest of his life Lindemann was consumed with nostalgia for his heady days in Berlin with Nernst. They remained in contact, their last meeting occurring in 1937 when Lindemann arranged for an honorary D.Sc. to be conferred on Nernst by the University of Oxford. After Nernst's death Lindemann paid his last tribute in an affectionate obituary.[2]

When he left Berlin, Lindemann took time in deciding on his future career because he was very wealthy: for him science was not a livelihood but a lifetime's choice. In 1915 he joined the Royal Aircraft Factory, Farnborough, where he took flying courses and as a pilot undertook his celebrated experiments on the spin of aircraft. At Farnborough one of his laboratory assistants was Thomas Keeley, a recent graduate in physics from Cambridge. At the Central Flying School, Upavon, Wiltshire, Lindemann met Gordon Dobson, a Cambridge physicist in charge of research there, and I. O. Griffith, an Oxford don who had demonstrated in the Clarendon Laboratory.

When Lindemann assumed his Oxford chair in May 1919, aged 33 and at a salary of £900 per annum, he had high ambitions. He longed to establish a research school. He insisted that good men [sic] were more important than big apparatus in showy buildings. He knew that some Oxonians wanted the Clarendon to end its eclipse by the Cavendish Laboratory at Cambridge. That was more easily said than done: Ernest Rutherford had just been appointed director of the Cavendish; and at Oxford the Clarendon competed for staff and facilities with the Electrical Laboratory run by John Townsend, the Wykeham professor of physics. After a short honeymoon, this structural division of physics led to bitter rivalry between Lindemann and Townsend with little co-operation between their laboratories until 1945, when (four years after Townsend's resignation) the University merged them.[3]

Lindemann arrived in Oxford at an unpropitious time. The University was in such a desperate plight financially that it was saved from financial collapse by government who gave an annual emergency grant (£30 000) that in 1922 became a regular grant distributed by the University Grants Committee (established in 1919). The University had little money to lay out on research in the Clarendon: it preferred to spend limited funds on renovating the Clarendon for teaching. The financial constraints were such that in 1922 and 1923 Lindemann himself contributed £100 to his department's income (see Table 7.2).[4] Even so, Lindemann launched a small research group in his favourite field of low-temperature physics. By 1922 it was a

[2] The main biographies are Birkenhead (1961); Thomson (1958); Bleaney et al. (1987); Jones (1986–7); Berman (1987); Keeley (1958); Harrod (1959). On Hermann Walther Nernst see Mendelssohn (1973); Lindemann and Simon (1942); Mehra (1975: 13, 52–3, 67–8).

[3] Lindemann to R. A. Millikan, 8 March 1919, LP, A21; J. Wells, Warden of Wadham, to Lindemann, 23 April 1919, LP, A22; Jones (1987: 113–26); on John Sealy Edward Townsend, Wykeham professor of physics 1900–41, see von Engel (1957); Llewellyn Jones (1957).

[4] Lindemann to D. Vickers, 8 November 1919, LP, B4; Lindemann, memorandum on Clarendon, n.d., [1920], Bodleian Library, Asquith Commission papers, MS Top Oxon b107, ff 122–5; Birkenhead (1961: 85–90).

Table 7.2. Income of Clarendon Laboratory 1919–39 (£) (to pay for demonstrators, assistants, apparatus, etc.)

Year	University contribution	Laboratory fees	Leigh Fund	Lindemann donation	Christ Church grant	Lecture fees	Rockefeller grant	Duke of Westminster gift	Higher Studies Fund	Total
1919	500	126	–	–	–	–	–	–	–	1767
1920	3041	333	–	–	–	–	–	–	–	3704
1921	2850	436	200	–	–	–	–	–	–	3513
1922	2467	643	–	100	–	–	–	–	–	3407
1923	2350	428	300	100	–	–	–	–	–	3275
1924	2367	468	200	–	–	–	–	–	–	3126
1925 (Jan–Jul)	1600	299	–	–	–	–	–	–	–	1941
1925–6	2300	384	300	–	90	50	–	–	–	3353
1926–7	2650	463	–	–	30	60	–	–	–	3465
1927–8	1880	398	–	–	50	81	–	–	–	2822
1928–9	1950	337	200	–	50	59	–	–	–	3085
1929–30	1950	440	–	–	20	89	–	–	–	2758
1930–1	1800	557	–	–	20	113	–	–	–	2600
1931–2	2200	516	–	–	20	98	–	–	–	3050
1932–3	1824	764	400	–	110	–	–	–	–	3561
1933–4	1824	787	250	–	50	–	500	–	–	4566
1934–5	1824	671	300	–	50	–	500	500	–	4109
1935–6	2324	616	300	–	50	–	–	–	–	3507
1936–7	2762	622	300	–	50	–	–	–	–	3750
1937–8	2762	682	200	–	50	–	–	–	300	4070
1938–9	2762	693	200	–	50	–	–	–	300	4318

Note: The categories of income are not exhaustive

Source: University accounts published annually in *OUG*. From 1932–3, the laboratory fees included lecture fees.

patent failure: nobody at the Clarendon could work the hydrogen liquefier, designed by Nernst and built by Hoenow, Nernst's head mechanic, which Lindemann had bought, and the instrument was never seriously used. Perhaps Lindemann realized that a research assistant or lieutenant, familiar with the temperamental intricacies of liquefiers, would be necessary for future success.[5]

After this fiasco Lindemann's own experimental research soon petered out, his last original paper being published in 1924. From the mid-twenties Lindemann did not lead by example from the laboratory bench; he was too proud to develop a lieutenant system of supervising research vicariously, and he failed to build up a research group focused on a topic with which he was publicly associated. Instead he became a Socratic oracle, stimulating by argument the research of Dobson on the ozone layer and of Cyril Hinshelwood on chemical kinetics.[6] Simultaneously, he lost interest in current trends in theoretical physics: he referred contemptuously to recent mathematical physics as mere squiggles on paper and deprecated Schrödinger's wave mechanics as arid mathematical formalism (Lindemann 1932: vi; 1933: 29; Mendelssohn 1973: 173; Mott 1932).

As a research supervisor, Lindemann did not pursue a main line of research. He launched his postgraduates on a series of isolated topics that often were not consolidated. Lacking any research training from him, his men were thrown in at the deep end and sank or swam. He attracted few research pupils from abroad or from other British universities. Although he produced a steady stream of D.Phil. students in the 1920s, none of them became disciples who modelled themselves on him: his interests were too scattered and his temperament too restless for him to be a renowned practitioner whose reputation was based on a continuing mastery of a particular field (Jones 1957; 1987: 120–2; Calvert 1957; Thomson 1958: 55–6; Birkenhead 1961:114). Unlike Rutherford he did not enjoy a far-reaching influence on physics in Britain through his research pupils.

EARLY RECRUITS

For senior research workers, Lindemann had no cohort of college tutorial fellows in physics upon which he could call and rely in 1919. He inherited just one such

---

[5] *OUG*, 50 (1919–20): 688–9 (27 May 1920); *OUG*, 52 (1921–2): 75–6 (19 October 1921); *OUG* 53 (1922–3): 671 (13 June 1923); LP, B4; Mendelssohn (1973: 70–71, 164–5). Lindemann's researchers were: Ivan George Evans, who relinquished a research scholarship in March 1920 to take up a post at the Patent Office; Henry Herman Leopold Adolf Bröse, whose 1925 D.Phil. taken under Townsend was on the motion of electrons in oxygen; Tielman François Tertius Malherbe returned to South Africa where he pursued a career in secondary school education; John Gustave Pilley, first in chemistry 1922 and later professor of education, University of Edinburgh, 1951–66. The liquefier is now in the Museum of the History of Science, Oxford.

[6] Dobson et al. (1926: 5), Dobson (1926; 1968), Laidler (1988: 256–60). On Cyril Norman Hinshelwood tutorial fellow in chemistry, Trinity, 1921–37, Dr Lee's professor of chemistry, Oxford, 1937–64, Nobel prizewinner chemistry 1956, see Thompson (1973).

fellow, Griffith, who acted as senior demonstrator until 1931. Griffith played golf every day in term and was not a noted researcher: his forte was university politics where he was an emollient ally of Lindemann. Given the dearth of available college physicists, Lindemann recruited two sorts of workers who were mainly non-Oxonians. He appointed three chums, Egerton, Keeley, and Dobson, from his Berlin or Farnborough days. Egerton was a promising researcher (FRS 1925) who specialized in high-temperature thermodynamics, which complemented Lindemann's interest in the low-temperature variety. Keeley continued as Lindemann's research assistant and administrative director of the laboratory.[7] Dobson was another promising researcher (FRS 1927) who, like Lindemann, was interested in air science and was affluent. Lindemann also gave facilities to two wealthy spectroscopists, Thomas Merton and Derek Jackson. The latter was a promising yearling (FRS 1947) but the former an established figure who, like Lindemann, was elected FRS in 1920. Through these men and their different lines of research, Lindemann hoped to achieve his first and second aims.

Three of Lindemann's recruits, Dobson, Merton, and Jackson, possessed private means. For Lindemann that was a desirable qualification because it guaranteed commitment to science as a lifetime's vocation and not as a mere livelihood. As a rich man, Lindemann saw scientific gentlemen as superior to players – before 1933. Wealthy researchers enabled him to build up research at low cost to the University. On the other hand, they had the independence to withdraw from the Clarendon. Dobson and Merton were cases in point. When Dobson became a demonstrator in the Clarendon in 1920 he enjoyed a private income through his wife; in 1924 he himself inherited land from his father. He was financially able to pursue research on ozone at his country home on Boars Hill, Oxford. Though he was made reader in meteorology in 1927, no college fellowship came his way until 1937, when Merton College elected him to one. Mainly a solo worker, his private meteorological laboratory became by 1927 the centre of an international network of research on atmospheric ozone, aided by grants from the Royal Society and the Department of Scientific and Industrial Research (DSIR). In 1937 he improved his private facilities when he moved to a 10-acre site on Shotover Hill, Oxford, where he built a new laboratory and an observing platform superior to its predecessor. Though Dobson was the only person who demonstrated continuously in the Clarendon from 1920 to 1939, his research life was based on his home, his wealth, and grants from the Royal Society and the DSIR for apparatus and assistance.[8]

[7] On Idwal Owen Griffith, fellow of St John's 1904–11, 1915–20, tutorial fellow in mathematics and physics, Brasenose 1920–41, see *The Brazen Nose*, 7 (1931–44): 165–9, and *Nature* 148 (1941): 589; on Thomas Clews Keeley, tutorial fellow in physics, Wadham, 1924–61, see *The Times*, 27 December 1988, and *The Independent*, 2 January 1989.

[8] On Gordon Miller Bourne Dobson see Houghton and Walshaw (1977); Dobson (1968); Dobson memorandum, 26 November 1938, PS/R/I/7. Dobson's main collaborators were Alfred Roger Meetham, who ended as a principal scientific officer at the National Physical Laboratory, and Douglas Neill Harrison as

Fig. 7.2. Gordon Miller Bourne Dobson. After meeting Dobson at the Royal Aircraft Factory, Farnborough, during the First World War, Lindemann brought him to Oxford as demonstrator in 1920. Chiefly through his important work on the measurement of atmospheric ozone, he did much to establish Oxford's reputation in atmospheric physics.

Merton was an Oxford graduate who was appointed stipendless reader in spectroscopy in October 1919 and a research fellow of Balliol before Lindemann was elected to his chair. Initially Merton researched in the inorganic chemistry laboratory run by Frederick Soddy; but the irascible Soddy ejected him in 1920 and Lindemann appropriated him for the Clarendon, where in cramped conditions he ran a small but productive research group using students from his own college. In 1924 he left Oxford for a rural home in Herefordshire, where he researched in his private laboratory, took out patents, collected Botticellis, and enjoyed fishing for salmon. He went on to be Treasurer of the Royal Society for seventeen years and to be knighted for his inventions used in the Second World War.[9]

In 1927 spectroscopy was resuscitated by Jackson, who had graduated in physics at Cambridge that year. As a rich man, Jackson defied Rutherford's invitation to

head of upper air instruments at Meteorological Office Development branch, Harrow. Dobson's Merton fellowship, tenable from 1937 to 1944 in the first instance, brought him £300 per year.

[9] On Thomas Ralph Merton see Hartley and Gabor (1970); OUG 49 (1918–19): 288, 368 (26 February and 24 April 1919); OUG, 50 (1919–20): 717 (2 June 1920); OUG 52 (1921–2): 75, 679 (19 October 1921 and 14 June 1922); 4 February 1919, 9 March 1920, NS/M/I/3; HCP 116 (1920: 109–10); HCP 119 (1921: 155–6). Merton's research collaborators were: Sidney Barratt, chairman of Albright and Wilson 1952–67; Raynor Carey Johnson, lecturer in physics at Belfast and King's College, London, then Master, Queen's College, Melbourne, Australia, 1934–64; and D. N. Harrison. Frederick Soddy, Dr Lee's professor of chemistry 1919–36 and Nobel prizewinner 1921, ran the inorganic chemistry laboratory.

Fig. 7.3. Thomas Ralph Merton in the Balliol–Trinity Laboratories, 1910. Although he was professor of spectroscopy from 1923 to 1947, Merton performed most of his research from 1924 in his own private laboratory.

work in the Cavendish but turned up at the Clarendon with all his apparatus bought by himself. Though fearless, wild-tempered, and wickedly amusing, he was by 1933 the most prolific solo publisher in the Clarendon since 1930. Neither his temperament nor style of life permitted him to take on research students: he moved in aristocratic and artistic circles and rode his own horses at important meetings (Kuhn and Hartley 1983; Bleaney *et al.* 1986: 262).

Until 1933 the most productive research group was run by Egerton, who had been attracted to Oxford in 1919 by Lindemann and Tizard. Initially Egerton researched in Soddy's department supported by a grant from the DSIR. Early in 1921 he succeeded Tizard as reader in thermodynamics at a salary of £300 per annum. Later that year Soddy ejected him. Egerton then began a fifteen-year stint in improvised accommodation in the attics and cellars of the Clarendon. His main research was on 'knocking' in internal combustion engines. As a supervisor he was patient and helpful with his researchers, both chemists and physicists by training. Though not a college fellow, Egerton attracted a new researcher almost every

Fig. 7.4. Derek Ainslie Jackson. Educated at Trinity College, Cambridge, Jackson entered the Clarendon Laboratory in 1927 as a spectroscopist. Later he worked closely with H. G. Kuhn.

Fig. 7.5. Alfred Charles Glyn Egerton, reader in thermodynamics from 1921 until his departure to the chair of chemical technology at Imperial College, London, in 1936. Egerton was knighted in 1943, chiefly for his advisory work on internal combustion engines and alternative fuel supplies.

year, initially from Oxford but increasingly from elsewhere. Such was the eminence of his group that from 1927–8 it was processed separately as a subdepartment of the Clarendon in the accounts of the University, from which it received between £250 and £288 p.a. In 1931 the problem of space was partly solved by decanting some of Egerton's research into the Engineering Science Laboratory. But the experience of supervising research in split locations and of using consultancy fees to buy apparatus led him to propose in 1934 the establishment of a separate new laboratory for thermodynamics. Nothing came of this proposal. The University was undecided about the relation of Egerton's applied thermodynamics to physics, and Franz Simon and Kurt Mendelssohn had shown by then that Egerton's combustion research was not the only way of pursuing thermodynamics profitably. In 1936 Tizard, as rector of Imperial College, London, invited Egerton to assume the vacant chair of chemical technology there. Eager to be in sole charge of a department with excellent facilities for research, Egerton moved to London where his career prospered: he became a secretary of the Royal Society in 1938 and was knighted in 1943.[10]

While Dobson, Jackson, and Egerton were pursuing productive research in the late 1920s, Lindemann found it difficult to promote the Clarendon. When the new position of university demonstrator was established in 1927, Townsend secured five such posts at a cost to the University of £2450 per annum, whereas Lindemann had just two (Griffith and Dobson) at a cost of £1100 p.a. That year, Lindemann enjoyed the services of two college fellows (Griffith and Keeley), as did Townsend; but Lindemann subsidized Keeley's salary as science fellow of Wadham College for ten years from his election in 1924.[11] From 1927 to 1931 the annual income of the Clarendon dropped from £3465 to £2600, the decrease being almost totally due to a reduction in the University's grant (see Table 7.2). Though the Clarendon's income overhauled that of the Electrical Laboratory for the first time in 1927–8, the former did not begin to outstrip the latter until 1931–2 (Tables 7.2 and 7.3). Not surprisingly,

---

[10] On Alfred Charles Glyn Egerton see Newitt (1960); Egerton (1963); Ubbelohde (1960); *OUG*, 50 (1919–20): 689, 691 (27 May 1920 (both)); *OUG*, 51 (1920–21): 225, 337 (24 November 1920 and 19 January 1921); *OUG*, 52 (1921–2): 78, 681 (19 October 1921 and 14 June 1922); HCP, 119 (1921: 155); Egerton memorandum, November 1931, PS/R/1/3; HCP, 159 (1934: 47, 173–4); *OUG*, 62 (1931–2): 664 (15 June 1932); *OUG*, 64 (1933–4): 203 (8 December 1933); *OUG*, 65 (1934–5): 198 (5 December 1934); *OUG*, 66 (1935–6): 211 (6 December 1935); *OUG*, 67 (1936–7): 225 (11 December 1936). Egerton's chief postgraduate collaborators from Oxford were: William Bell Lee, fourth in physics 1922; Stanley Frederick Gates, second in chemistry 1921, later ICI research, plant, and works manager, Lancashire; William Edmondson, second in chemistry 1925, later research chemist, Coxeter Ltd; Alfred Rene Jean Paul Ubbelohde, first in chemistry 1930, on whom see Weinberg (1990); John Mylne Mullaly, second in chemistry 1924; Michael Milford, second in physics 1928, left the Clarendon in 1935 to be physics and senior master, Repton School, England; and Frank Allan Cunnold, first in physics 1933, D.Phil. 1935, became a civil servant in the war, defence, supply, and air ministries. Those from other universities were Frederick Llewellyn Smith, later chairman of Rolls Royce, obituary in *The Times*, 24 August 1988; John Wilson Drinkwater (both firsts in engineering from Manchester), F. F. Coleman, R. J. Bracey, and L. M. Pidgeon, later a researcher, National Research Council, Canada. On Henry Thomas Tizard, tutorial fellow in science, Oriel, 1911–20, reader in thermodynamics 1920, rector of Imperial College, London, 1929–42, President of Magdalen 1942–7, see Farren (1961) and Clark (1965).

[11] Wells to Lindemann, 1 December 1924, LP, B121; Birkenhead (1961: 95–9).

Table 7.3. Income of Electrical Laboratory 1919–39 (£)

Year	University contributions	Laboratory and lecture fees	Laboratory fees	Lecture fees	Leigh Fund	Hulme Fund	Total
1919	1850	1174	–	–	150	397	3933
1920	3050	1479	–	–	–	–	5272
1921	3050	1504	–	–	–	–	5450
1922	2917	–	1131	335	–	–	5049
1923	2850	–	999	273	150	–	5156
1924	3818	–	966	266	200	–	5386
1925							
Jan–Jul	2185	–	530	154	–	–	2869
1925–6	3170	–	822	278	–	–	4612
1926–7	3170	–	819	214	200	–	4820
1927–8	1120	–	868	217	150	–	2659
1928–9	950	859	–	–	–	–	2227
1929–30	950	924	–	–	100	–	2345
1930–1	950	1132	–	–	–	–	2628
1931–2	931	1097	–	–	–	–	2684
1932–3	912	960	–	–	100	–	2852
1933–4	800	873	–	–	–	–	2537
1934–5	1100	861	–	–	100	–	2270
1935–6	1655	817	–	–	–	–	2483
1936–7	1300	747	–	–	–	–	2313
1937–8	1300	917	–	–	–	–	2588
1938–9	1300	1052	–	–	–	–	2806

Note: The categories of income are not exhaustive
Source: University accounts published annually in *OUG*

in the late twenties Lindemann became more polemical about the status of science at Oxford. Desperate not to lose any resource, he led a vain campaign from 1930 to 1934 to stop the Radcliffe trustees moving their observatory, hitherto in Oxford, to South Africa (Morrell 1997: 248–55). Outside Oxford he pursued the life of a bachelor socialite, moving in aristocratic society and becoming scientific adviser to his friend Winston Churchill (Birkenhead 1961: 121, 124, 127–31; Jones 1966: 66–9).

THE NEW DIASPORA: EXPERIMENTALISTS

In 1930 Lindemann decided to make a second attempt to promote cryogenics in the Clarendon and turned to Franz Simon at the University of Berlin for help in producing liquid hydrogen. Simon was well known as a designer of liquefiers and, like Lindemann, a pupil of Nernst. Wishing to avoid repeating the fiasco of 1922, Lindemann sent Keeley to Berlin in late 1930 and accompanied him in spring 1931 to see Simon's latest hydrogen liquefier with a view to buying one for the Clarendon. They did not meet Simon: he had left for Breslau where he had been appointed full professor at the Technische Hochschule. It fell to Mendelssohn, a Ph.D. pupil of Simon and his chief research associate from 1929, to demonstrate the apparatus. In late spring 1931 a Simon liquefier was working in the Clarendon to the delight of Lindemann, who tried to lure Mendelssohn to come to Oxford in 1931–2. Mendelssohn declined because in June 1931 he was moving to Breslau, where he would be in charge of research during Simon's leave of absence in the USA from January to June 1932.[12]

In early spring 1932 Lindemann decided to extend his low-temperature resources by securing a small helium liquefier like the one he had seen working at Berlin in Simon's department. As Simon was away, Mendelssohn was the natural contact and the man Lindemann wanted to come to Oxford to install a helium liquefier and launch experimental work with it in session 1932–3. With Simon's approval, Lindemann applied to the Rockefeller Foundation for a fellowship for Mendelssohn tenable that session, but the application was too late to be processed. When Mendelssohn told Lindemann about Simon's new helium liquefier (which expanded helium gas under pressure at the temperature of liquid hydrogen), Lindemann quickly arranged for Mendelssohn to visit Oxford early in 1933 for a few days with all expenses paid to install the new type of liquefier in the Clarendon.[13] Within a week, Mendelssohn became the first in Britain to produce liquid helium. Lindemann was cock-a-hoop: through Mendelssohn he had at last upstaged Cambridge and wiped the eye of the Cavendish, where the expensive Mond Laboratory for cryogenics built for Kapitza was nearing completion.[14] Mendelssohn returned to Breslau, but as it was afflicted by Nazi violence in April 1933 he decided to seek refuge in Oxford. Lindemann was so keen to capture Mendelssohn's expertise that he induced ICI to make Mendelssohn a provisional offer of about £400 per annum for two years. By July 1933 Mendelssohn had signed a contract with ICI backdated to 1 May 1933.[15]

---

[12] On Franz Eugen Simon see Kurti (1958); Arms (1966). On Kurt Alfred Georg Mendelssohn see Shoenberg (1983); Keeley to Simon, 2 December 1930, 7 March 1931, 12 March 1931, SP, 14/1/2; Mendelssohn to Simon, 22 October 1932, SP/14/1/1; Mendelssohn (1966: 129–31); Mendelssohn to Lindemann, 22 May 1931, MP, B8.

[13] Keeley to Hoenow, 19 February 1932, Hoenow to Mendelssohn, 23 February 1932, MP, B10; Mendelssohn to Lindemann, 15 and 28 June 1932, 16 September 1932, 6 October 1932, 21 and 25 November 1932, 8 December 1932, MP, B8; Lindemann to Mendelssohn, 23 June 1932, 3 October 1932, 14 and 29 November 1932, MP, B8; Mendelssohn to Simon, 22 October 1932, SP, 14/1/1; Simon to H. M. Miller, 21 July 1932, SP, 14/1/13; Simon to W. E. Tisdale, 19 and 26 November 1932, SP, 14/1/13; Miller to Simon, 6 April 1933, SP, 14/1/13. Keesom (1942: 150–80) gives a good account of helium liquefiers and liquid helium techniques. For low-temperature research in general see Scurlock (1992).

[14] *The Times*, 28 January 1933; Lindemann and Keeley (1933); Jones (1987: 122); on Piotr Leonidovich Kapitza see Shoenberg (1985: 341–2).

[15] Mendelssohn (1966); Keeley to Mendelssohn, 22 March 1933, 2 April 1933, MP, B12; Mendelssohn file, SPSL, 335/2; Mendelssohn to J. M. King, ICI, 28 July 1933, MP, B14; Mendelssohn (1973: 166).

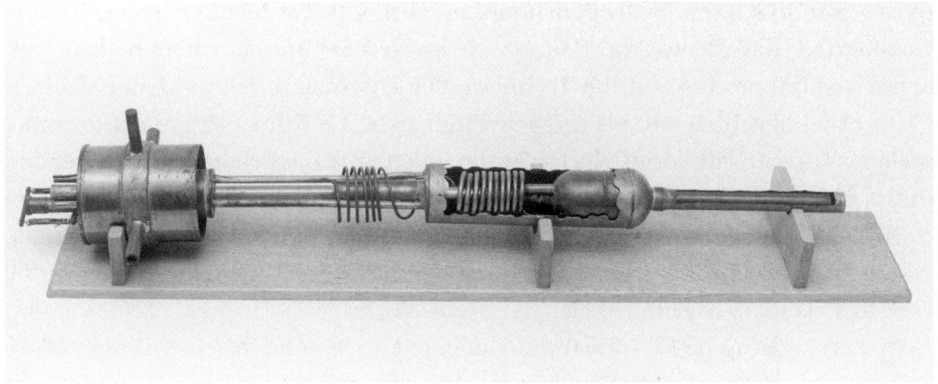

Fig. 7.6. Helium liquefier and cryostat, designed by Simon and used in the early Oxford experiments on liquid helium, 1933. The apparatus, 60 cm in length, is now displayed, with other historic apparatus, in the entrance to the Clarendon Laboratory's Lindemann Building.

Mendelssohn's coup induced Lindemann to contemplate importing other low-temperature physicists from Breslau to Oxford, where they would be paid neither by the University nor by the colleges but by ICI. Through them he could outsmart the Cavendish, put the Clarendon on the international map, and within Oxford elevate the status of science. Moreover, there were personal resonances: Lindemann and Simon were Nernst's cryogenic sons and Mendelssohn his grandson. Through Mendelssohn's Breslau colleagues Lindemann could sponsor research in which he had been involved as a young man. This scheme satisfied Lindemann's enduring loyalty to Nernst. In implementing it he was lucky: Kapitza was detained in Russia in 1934, thus removing permanently the Clarendon's chief British rival in cryogenics.

It was the combination of Nernstian loyalties and Mendelssohn's success that led Lindemann to take the lead in helping German physicists, especially Jews who were threatened by the Nazis (Sherman 1973; Beyerchen 1977; Bentwich 1953; Beveridge 1959; Rider 1984; Hoch 1983, 1991). With car and chauffeur he toured Germany in spring 1933, hoping to provide refuge for German physicists, albeit temporarily, in British universities including his own. Exploiting his friendship with Harry McGowan, chairman of ICI, he persuaded the company to finance German refugee scientists for two years. He co-operated with the Academic Assistance Council (founded late May 1933) that helped all German refugee scholars, but he did not sign its famous first circular. He preferred to act independently by turning to ICI to help chosen German physical scientists.[16] Mainly through ICI he gained

---

[16] Birkenhead (1961: 101–4); Rider (1984: 146–51); Hoch (1983: 221–31); Lindemann to McGowan, 24 May 1933, LP, D95; Lindemann to F. Haber, 9 January 1934, LP, D86; Lindemann and W. Rintoul to McGowan, 24 July 1934, LP, D96. Harry Duncan McGowan, chairman of ICI 1930–50, was rewarded with an honorary DCL in 1935.

for the Clarendon six experimentalists, namely four low-temperature physicists from Breslau (Mendelssohn, Simon, Nicholas Kurti, and Heinz London); Heinrich Kuhn, a spectroscopist from Göttingen; and Leo Szilard, a nuclear physicist from Berlin. Of these Mendelssohn, Simon, Kurti, and Kuhn were retained in Oxford until 1939 and spent the rest of their working lives there. Lindemann also recruited from Berlin two distinguished theoretical physicists, Erwin Schrödinger and Fritz London, neither of whom stayed long.

Simon was an even more desirable catch than Mendelssohn. He was the senior man with an international reputation that attracted non-German researchers to his laboratory. His principal fields of research, specific heats at low temperatures and Nernst's heat theorem, were dear to Lindemann. As much of Simon's work had sought to reformulate and elevate the theorem into the third law of thermodynamics, it was an extension of Lindemann's own pre-war research. Lastly, Mendelssohn was a solo worker who had not had the time or opportunity to attract researchers, whereas from 1931 Simon had enjoyed the services of Kurti, a Ph.D. pupil who was his collaborator and right-hand man. Lindemann's top priority was to acquire Simon and Kurti via Mendelssohn as a helpful intermediary. In July 1933 Simon accepted a post in the Clarendon, in

Fig. 7.7. Franz Eugen (later Francis) Simon, with low-temperature apparatus, either in Berlin or (more probably) in his laboratory at the Technische Hochschule of Breslau, where he became professor in 1931.

Fig. 7.8. Franz Simon in his laboratory in Breslau.

Fig. 7.9. 'Tug of war' between Bosch and Simon, probably in the Breslau laboratory.

Fig. 7.10. Heinrich Gerhard Kuhn. Before his arrival at the Clarendon Laboratory in 1933, Kuhn had lectured at the distinguished physics institute of the University of Göttingen. He made major contributions to high-resolution spectroscopy, some with D. A. Jackson.

August he arrived in England bringing with him two helium liquefiers, and by October had accepted ICI's offer of £800 per annum for two years from 1 September. Kurti arrived in September and Heinz London, yet another PhD. pupil of Simon, joined the Breslau colony in Oxford early in 1934, both being supported by ICI.[17]

The acquisition of Kuhn was both opportune and opportunist. During a recruiting tour in Germany Lindemann met an old friend in Göttingen, James Franck. One of his most promising pupils was Kuhn, who had been dismissed from his post in Franck's department. In early June Kuhn met Lindemann for the first time and was invited to Oxford to work on spectroscopy with Jackson. By 18 June Kuhn had accepted an ICI grant of £400 per annum, initially for two years, to enable him to work in the Clarendon. By mid-August Kuhn and his wife had arrived in Oxford and met Jackson for the first time. In ten weeks, Kuhn's career had been drastically changed.[18]

---

[17] On Nicholas Kurti see Sanders (2000); Kurti (1987: 86–90). On Heinz London see Shoenberg (1971). Lindemann to Simon, 24 April, 18 May, 18 June, 24 and 29 July 1933; Simon to Lindemann, 12 May, 3 July 1933; Simon to A. Black, ICI, 5 October 1933, all SP, 14/3/L; Black to Simon, 18 September 1933, SP, 14/1/3; Kurti, transcript of interview for Archive for History of Quantum Physics; H. London file SPSL, 334/5.

[18] On Heinrich Gerhard Kuhn see Bleaney (1996); Kuhn to author, 23 January 1987; Lukes (1987); Kuhn file, SPSL, 333/3. James Franck was professor of experimental physics, Göttingen University, 1921–33.

THE NEW DIASPORA: THEORETICIANS

Lindemann was also keen to recruit theoretical physicists. This was partly the result of his frustration with Oxford's mathematicians who, as researchers and teachers, were generally indifferent to experimental physics and pursued a pure version of their subject. Their attitude was not helpful to those undergraduates who spent their first year studying for the mathematics moderations examination before specialising for two years as honours physicists. Lindemann had high hopes of Edward Milne who had become the first occupant of the newly created Rouse Ball chair of mathematics in 1929. But even Milne's inaugural lecture was disappointing to experimentalists. He stressed that for him mathematical physics had higher tasks than merely seeking to account for specific empirical data.[19] In contrast to Milne, Einstein was Lindemann's beau ideal of a theoretical physicist. Through Lindemann's initiative Einstein first appeared in Oxford in spring 1931 as Rhodes Memorial Lecturer. He soon accepted a research fellowship at Christ Church for five years at £400 p.a., making a summer visit in 1932. His last appearance in Oxford was in June 1933 as Herbert Spencer Lecturer.[20] Einstein's visits presumably convinced Lindemann that a couple of distinguished German theoretical physicists might have their uses for all the experimentalists in the Clarendon. He had in mind Fritz London and Hans Bethe. In the event, London came but not Bethe. London had been dismissed from his post at Berlin as assistant to Schrödinger. With Mendelssohn's help he arrived in summer 1933 in Oxford where ICI had agreed to support him initially for two years at £600 p.a.[21] The second theoretical catch was Schrödinger himself. Though Lindemann was deeply suspicious of Schrödinger's wave mechanics, he accepted Einstein's high opinion of Schrödinger *qua* theoretical physicist. He was supported by a grant of £1000 per annum from ICI, initially for two years, and his base was Magdalen College, where he was a non-stipendiary supernumerary fellow. In November 1933 Schrödinger arrived in Oxford and soon heard that he had been awarded a Nobel prize. At this time, with two eminent theoretical physicists installed in Oxford, it seemed possible that theoretical physics would blossom to complement and help the experimentalism pursued in the Clarendon.

---

[19] On Edward Arthur Milne, Rouse Ball professor of mathematics 1929–50, see McCrea (1950–1); Whitrow (1970); Milne (1929); Harrod (1959: 66); Lindemann memorandum, 19 May 1938, OUA, UR/SF/PHE/7.

[20] Birkenhead, (1961: 51–2, 159–62); Harrod (1959: 47–50, 91); Elsa Einstein to Lindemann, 11 May 1931, Lindemann to Einstein, 29 June 1931, both LP, D56; *OUG*, 61 (1930–31): 224 (23 December 1930); *OUG*, 62 (1931–2): 131 (28 October 1931); *OUG*, 63 (1932–3): 582 (8 June 1933); Einstein (1931; 1933).

[21] Lindemann to Einstein, 4 May 1933, LP, D57. On Fritz Wolfgang London see Gavroglu (1995: esp. 96–138); F. London file, SPSL 334/4; Hans Albrecht Bethe, Nobel prizewinner in 1967 for his theory about the source of energy in stars, spent 1933–5 at the universities of Manchester and Bristol before leaving for Cornell University, USA.

That did not happen with Schrödinger. He lectured on elementary wave mechanics but had little contact with the Clarendonians. Though married, his domestic life was unusual. His latest mistress was Hildegunde March, wife of Arthur March, Schrödinger's assistant. All four came to Oxford. It soon became apparent that Mrs March was pregnant – by Schrödinger, who sometimes lived openly with his wife and mistress in a *ménage à trois*. In May 1934 Schrödinger's child by Mrs March was born: for Lindemann the Nobel prizewinner became a bounder. After two and a half years in Oxford, where he did no important research, Schrödinger left for Graz.[22] Fritz London was a married man of impeccable bourgeois morality who was stimulated by the experimental work on superconductivity done in the Clarendon by his younger brother, Heinz. Their fruitful partnership produced important results about the electrodynamics of superconductors but it ended early in 1936 when Heinz moved to Bristol to be a research assistant. Fritz then endured an unhappy period. He knew his ICI grant would expire irrevocably in August 1936, he broke his ribs in an accident, and his spoken English was poor. When his ICI grant ended, the Academic Assistance Council supported him for two months before he went to Paris for a post at the Institut Henri Poincaré. Lindemann wanted to keep him in Oxford but could not find the appropriate funding.[23]

FUNDING AND FACILITIES

The experimentalists in the Clarendon also faced funding problems. Generally they were initially on two-year ICI grants from 1933 to 1935. Though ICI extended the grants by a year, it wished to end its scheme in summer 1936. Heinz London took the hint and was prepared to go to Bristol on £200 per annum. Kuhn managed to stay in Oxford from 1936 because two colleges, St John's and Queen's, between them gave him £300 per annum. Simon and Kurti not only remained at the Clarendon, but exceptionally their ICI grants were renewed in 1936 for five years.[24] In October 1936 Simon became the only refugee physicist at Oxford to gain a University post, when he succeeded Egerton as reader in thermodynamics at £500 per annum. Simultaneously his ICI grant, augmented to £1000 per annum in autumn 1935 for three years, was reduced to £500 but guaranteed until 1941.

[22] Hoch and Yoxen (1987); Moore (1989: 267–71, 273, 280, 293–8, 317); Schrödinger to Sidgwick, 1 March 1936, Sidgwick papers, Lincoln College, Oxford, V.75.
[23] SPSL 334/4; A.H. Cooke, interview with author, September 1986; K. J. Spalding to Lindemann, 8 August 1936, LP, D148; London and London (1935).
[24] H. Molson to Lindemann, 1 May 1933, LP, D161; H. London file, SPSL, 334/5; H. G. Kuhn file, SPSL, 333/3; Mendelssohn (1964: 8); A. M. Tyndall to Simon, 21 January, 26 June 1936, SP, 14/3/T-Z; Simon to D.C. Thomson, 13 June 1939, SP, 14/3/K.

ICI saw that Simon's partnership with Kurti was fruitful and gave him in 1936 a five-year grant of £300 per annum. Procuring emolument for Mendelssohn involved protracted negotiations. His first contract was renewed in May 1935 for three years at £400 p.a.[25] Late in 1936 Lindemann persuaded ICI to spread out its grant until September 1939. At the same time Lindemann induced Sir Robert Mond to transfer to Mendelssohn for three years a grant of £300 p.a. previously given to a refugee mathematician who had acquired a permanent post. Thus from December 1936 Mendelssohn received £200 per annum from ICI and £300 per annum from Mond.[26]

The migrants who were to remain in the Clarendon were sometimes anxious about their situation. In the mid-thirties Kuhn and Mendelssohn applied unsuccessfully for posts abroad. Having rejected offers of chairs at Istanbul and Jerusalem, Simon applied unsuccessfully for the professorship of physics at Birmingham in 1936. He wanted more income, the power to run his own department, and once again to have the status of a professor. He was also unhappy about the lack of a 'physikalische Atmosphäre' in Oxford. Until he was naturalized in 1938, he could not apply directly for grants to British institutions but had to work through Lindemann. In order to continue his interest in the properties of fluids at high pressure he had to travel to Amsterdam to use Michels's apparatus. To develop his work on magnetic cooling he and Kurti went seven times for a month between 1935 and 1938 to use the powerful magnet in the Laboratoire de l'électroaimant, Bellevue, Paris, run by Aimé Cotton. In neither case was the appropriate apparatus available in the Clarendon.[27]

Once the migrants had settled in, there were contrasting developments in spectroscopy and cryogenics. In the former, Jackson had never worked under or with anyone but he soon established a fruitful partnership with Kuhn, working on the sharpening of spectral lines. When Jackson was horse-riding and socializing away from Oxford, Kuhn continued his established interest in the pressure-broadening of

---

[25] Lindemann and Rintoul to McGowan, 24 July 1934, LP, D96; Lindemann to Slade, ICI, 4 October 1936, LP, D156; Lindemann to C.G. Robinson, 5 March 1936, LP, D228; Lindemann memorandum, 6 June 1936, PS/R/1/5; Simon, application for Birmingham physics chair, 29 February 1936, SP, 14/2/67; Lindemann to A. D Lindsay, 23 September 1938, LP, D154.

[26] Mendelssohn to J. M. King, ICI, 28 July 1933; Mendelssohn to L. A. Inglis, ICI, 25 April 1935; Inglis to Mendelssohn 14 August 1935, all MP, B14; H. L. Nathan to Lindemann, 17 November 1936; R. E. Slade, ICI, to Lindemann, 20 November 1936; Lindemann to Nathan, 24 November 1936; Lindemann to Slade, 24 November 1936; Lindemann to R. Mond, 27 November 1936; Mendelssohn to Mond, 27 November 1936; Lindemann to Veale, 19 May 1939, all MP, B15; OUG, 70 (1939–40): 51 (13 October 1939); on Robert Ludwig Mond see Thorpe (1936–8).

[27] Kurti, AHQP tape; Kuhn to author, 23 January 1987; Kuhn file, SPSL 333/3; Mendelssohn file, SPSL, 335/2; CM. Skepper to Mendelssohn, 1 November 1934, MP, A4; Arms, (1966: 69–73, 86); Simon application for Birmingham chair, 29 February 1936, SP, 14/2/67; LP, B15; Kurti, interview with author, December 1986; Kurti (1958: 236–40); Antonius Mathias Johannes Friedrich Michels, director of the van der Waals physics laboratory, University of Amsterdam, 1929–60; Bellevue Laboratory, Paris, correspondence, SP/1/8E; Aimé Auge Cotton, professor of physics at the Sorbonne and director of physics laboratories, 1920–41, on whom see Shinn (1993).

spectral lines. The nature of their partnership and Jackson's personality precluded their supervising any research students. For Lindemann, spectroscopy was a side-show, albeit distinguished, compared with low-temperature and atomic physics. Hence Kuhn's financial position was difficult from 1936; but Jackson poured £4000 by 1939 into their laboratory for apparatus. It was not the University or the colleges but Jackson who paid for the two lines of experimental spectroscopy pursued in the Clarendon.[28]

CRYOGENICS

In cryogenics, where there was no indigenous researcher, a tense situation quickly developed. Mendelssohn had launched cryogenics at Oxford, he was the first of the Breslau cohort to settle in Oxford, and he had helped Simon, Kurti, and Fritz London to migrate to Oxford. He then found himself number two to Simon, a situation he resented. In order to avoid abrasive conflict, Simon and Kurti focused on

Fig. 7.11. Kurt Alfred Georg Mendelssohn (*centre*) with his Oxford collaborator John Gilbert Daunt (*left*) and Rex Bush Pontius. Daunt and Pontius (who was an American Rhodes Scholar) both graduated in physics in 1935 and went on to doctoral research in Mendelssohn's areas of interest, taking their D.Phils. in 1937 and 1938 respectively. Bodleian Library, Mendelssohn Papers, MS Eng. Misc. b.388, J.1.

[28] Kuhn and Hartley (1983: 274–80); Kuhn to author, 23 January 1987; *OUG*, 64 (1933–4): 335 (17 January 1934). *OUG*, 66 (1935–6): 655 (27 May 1936); Kuhn (1934); Lindemann memorandum 17 February 1939, LP, B25, ff.3-4; OUA, Spectroscopy. Reader and lecturer in, UR/SF/Spec/1.

thermodynamic properties, leaving superconductivity to Mendelssohn. It was fortunate for the Clarendon that Simon did not retaliate against Mendelssohn's jealousy so that it was possible for each to build up a research group.[29] Mendelssohn attracted fewer researchers than Simon but in collaboration with four Oxonian postgraduates, including one woman, he made important discoveries about superconductivity and superfluidity in liquid helium. He was a heavy publisher and a driving supervisor without a research lieutenant; but none of his D.Phil. pupils made an academic career in Oxford.[30]

The larger cryogenic group was that of Simon and Kurti, who focused on the new field of magnetic cooling and the established one of specific heats in relation to Nernst's heat theorem. By the late 1930s the Clarendon had overtaken Leiden and Berkeley as the world's leading centre for magnetic cooling research; and Simon had so effectively reformulated the theorem that he had elevated it to the third law of thermodynamics (Kurti 1958: 233–5, 238–40; Simon 1956; Ruhemann and Ruhemann 1937: 239–40). Simon soon attracted research students from Oxford

Fig. 7.12. Judith Rachel Moore graduated in physics at Lady Margaret Hall in 1933 and took a D.Phil. in 1937. Working with Kurt Mendelssohn, she was the first woman to undertake postgraduate research in the Clarendon. Later married to Richard Hull (*see Fig. 9.6*), she did not go on to an academic career. Bodleian Library, Mendelssohn Papers, MS Eng. Misc. b.388, J.3.

---

[29] Shoenberg (1983: 381–2); Simon to Bragg, 1 November 1946, SP, 14/3/M; Simon to Rollin, 12 December 1938, SP, 14/3/R.

[30] Shoenberg (1983: 371–6); Daunt and Mendelssohn (1938*a*; 1938*b*); John Gilbert Daunt, first in physics 1935, D.Phil. 1937, ended his career as professor of physics, Queen's University, Kingston, Canada; Judith Rachel Moore, second in physics 1933, D.Phil. 1937, see caption to Fig. 7.12. Rex Bush Pontius, second in physics 1935, D.Phil. 1937, a Rhodes Scholar from Idaho, spent most of his career in the Kodak research laboratory, USA; John David Babbitt, a Canadian Rhodes Scholar, first in physics 1932, D.Phil. 1934, became a physicist with the National Research Council, Ottawa.

Fig. 7.13. Bernard Vincent Rollin. After reading physics at Wadham, where he graduated in 1933, Rollin joined Simon's low-temperature group. During the Second World War, he worked in the Clarendon Laboratory on microwave devices for use in radar. Subsequently, he was concerned with a variety of subjects involving very low temperatures and with the new field of nuclear magnetic resonance.

(on average one per year), Cambridge, and the European continent. His research pupils made successful academic careers in Oxford: Cooke, Bleaney, and Hull became college tutorial fellows in the 1940s, with Rollin becoming a university lecturer.[31] By the late 1930s the Belgian government recognized Simon's eminence by paying for two physicists from Liège to be trained by him. Nearer home, Helen Megaw came from Cambridge in 1935 to work with Simon for a year. She was a Cambridge graduate who had taken her Ph.D. there in X-ray crystallography under J. D. Bernal who had advised her to study under Simon. She became the first woman to be given a staff appointment at the Cavendish Laboratory and ended her

---

[31] Arthur Hafford Cooke, first in physics 1935, D.Phil. 1938, was fellow (1946–76) and Warden of New College (1976–85), obituaries in *The Independent*, 1 August 1987, and *The Times*, 1 August 1987; Brebis Bleaney, first in physics 1937, D.Phil. 1939, fellow of St John's 1947–57, Dr Lee's professor of experimental philosophy 1957–77; Richard Albert Hull, first in physics 1932, D.Phil. 1936, fellow of Brasenose 1944–9, obituary in *The Brazen Nose*, 9 (1949–54), 53–4; Bernard Vincent Rollin, first in physics 1933, D.Phil. 1935, Commonwealth fellow at Berkeley, California 1937–9, university lecturer 1945–69 and fellow of Wolfson College 1965–9, obituary in *The Times*, 27 June 1969. Simon's other Oxonian researchers were: George Lawson Pickard, first in physics 1935, D.Phil. 1937, ended his career as professor and director of the Institute for Oceanography, University of British Columbia, Vancouver, Canada; Ralph Trevor Kerslake, first in physics 1936, did not complete his D. Phil. and ended his career as a senior brewer, Guinness, London; Henry Shull Arms, second in physics 1938, D. Phil. 1949, a Rhodes Scholar from Idaho, became chief engineer, Atomic Power Division, English Electric Company, Rugby.

Fig. 7.14. Nicholas Kurti in 1926, at about the time he left Hungary to begin his undergraduate studies at the Sorbonne. The photograph appeared on his student record card at the Sorbonne.

career as tutorial fellow at Girton College. Before the arrival of Simon it would have been inconceivable for a Cambridge postdoctoral physicist to make a rewarding pilgrimage to the Clarendon.[32] A key element in Simon's success was the work of Kurti as research collaborator and experienced research lieutenant. Mendelssohn taught his own students how to use the apparatus, whereas in Simon's group Kurti often taught these essential practical skills. The small expansion helium liquefiers took three or four weeks to build, and, once they were working, up to eight hours could elapse before the investigation proper could be begun. The successful use of these liquefiers required tacit skills as well as caution about the dangers of liquid hydrogen leaking and exploding and of releasing helium gas at 150 atmospheres pressure via a valve into glassware. Thus the techniques brought to Oxford from Breslau were neither easily learned in Oxford nor easily copied elsewhere: they could not be acquired from textbooks or even exhibitions. For the Breslau trio small-scale cryogenic engineering was an essential part of low-temperature physics and continued to be so until the mid-1950s. That conjunction exacerbated the difficulty for other laboratories of copying the Clarendon's cryogenic techniques and programmes quickly and easily.[33]

[32] D'Or to Simon, 8 April 1938; Simon to D'Or, 27 April 1938, both SP, 14/3/NO; Helen Dick Megaw, Cambridge Ph.D. 1934, later fellow of Girton 1947–70, worked in Oxford from Easter 1935 to summer 1936; Megaw to author, 30 November 1988, 26 January 1989.   [33] Contrast Birkenhead (1961: 105).

NUCLEAR PHYSICS

In his plans for the new Clarendon, Lindemann gave as much prominence to nuclear as to low-temperature physics (Lindemann 1938: 55). Perhaps stimulated by the discovery of the neutron in 1932, he began to give more attention to nuclear physics though his own successful experience in it was meagre. The results obtained by 1939 made a telling contrast with those in cryogenics: nuclear physics did not take off before the war but cryophysics did. No major discoveries in nuclear physics were made, the publication rate was not high, and the Clarendon did not begin to rival established centres. Blithely ignoring the difficulties of starting nuclear physics from scratch, Lindemann installed in the Clarendon in 1933 a high-voltage apparatus. He had available to use it three Oxford graduates and recently elected college fellows (Carl Collie, Claude Hurst, and James Griffiths) whom he hoped would be as effective a trio as Simon, Kurti, and Mendelssohn. None of them had strayed outside Oxford to learn about nuclear physics and only one of them (Collie, radioactivity) was an expert on a related subject, so Lindemann turned

Fig. 7.15. Claude Hurst in 1932, the year of his election as a Research Student (i.e. research fellow) of Christ Church.

Fig. 7.16. Group photograph taken in the 1930s: (*left to right*) D. A. Jackson, T. C. Keeley, C. Hurst, and C. H. Collie.

again to the device of importing a Continental expert or cohort.[34] By summer 1935 he had secured, on Simon's recommendation, Leo Szilard and an ICI fellowship for him for three years.

Szilard had left Berlin in 1933 and, having failed to secure facilities at Cambridge and Imperial College, London, persuaded Lindemann to employ him in the Clarendon to build up nuclear physics by studying chain reactions with Collie and Griffiths. For two years they worked on the absorption of slow neutrons but Szilard was restless and worried about the political situation in Europe. In summer 1937 he told Lindemann that henceforth he wished to spend at least half his time in the USA. Though Lindemann deplored this plan, it was formalized in early 1938, Szilard's ICI fellowship was not extended, and the brilliant, mercurial, and exotic bird of passage lived from autumn 1938 in the USA. Szilard was not interested in taking research pupils and, as a quintessential maverick, he was not an easy colleague. Even the charitable Simon told him in 1938 that he had exhausted the patience of his Oxford friends and ICI.[35]

[34] *OUG*, 64 (1933–4): 200 (8 December 1933); *OUG*, 65 (1934–5): 194 (5 December 1934); Carl Howard Collie, Dr Lee's reader in physics and tutorial fellow in physics, Christ Church, 1930–71, obituary in *The Times*, 28 August 1991; Claude Hurst, tutorial fellow in physics, Jesus, 1934–75, on whom see Houghton (1988); James Howard Eagle Griffiths was undergraduate, postgraduate, fellow in 1934, Vice-President and finally President (1968–79) of Magdalen (ter Haar 1982).

[35] Szilard (1978: 18–21, 41–2, 48–52;); Lanouette (1992: 139–72); Szilard file, SPSL, 167/2; Szilard to Lindemann, 30 March 1935, LP, D237; Lindemann to Szilard, 30 July 1937, LP, D237; Simon to Szilard, 23 August 1938, SP, 14/3/S; Mendelssohn (1960).

Though Lindemann knew that Szilard was yearning to leave Britain, in autumn 1937 he secured a grant from the University of £300 per annum for five years for apparatus for accelerating electrons to be developed by Szilard. As an insurance measure, Lindemann lured to the Clarendon, as a collaborator with Szilard and possible substitute, James Tuck, a rich physical chemist from Manchester University. Supported by a fellowship from the Salters' Institute of Industrial Chemistry, Tuck's research on electron acceleration done with Morgan, an Oxford graduate, was promising but unpublished when interrupted by the war. Tuck's technical astuteness led to his being Lindemann's personal assistant during the war and to his involvement in the atomic bomb project. Thus two imported nuclear physicists, Szilard and Tuck, enjoyed only limited success in the Clarendon.[36] A third projected import, Lise Meitner, soon to be the co-discoverer of nuclear fission, could well have settled as Szilard's successor and promoted nuclear physics; but Lindemann could not raise facilities for her and felt that keeping Simon and Mendelssohn was his overriding duty.[37]

The indigenous nuclear physicists in the Clarendon pursued different paths. Hurst was put in charge of an expensive Cockcroft–Watson accelerator of which he had no experience. There is no record of anyone from Cambridge teaching him its use. By 1938 a caustic critic described the accelerator as derelict. Unlike Hurst, who had one D.Phil. pupil, Griffiths had none, but his research on neutrons was not negligible.[38] Collie was the busiest supervisor in nuclear physics with six D.Phil. students under his care in the 1930s. He researched in the Clarendon on counting nuclear particles; and in the Christ Church chemistry laboratory continued his research on radioactivity, using from 1935 for three years materials secured on loan from the Czech government by Lindemann. Collie was also interested in electronics, which he promoted with two of his Christ Church pupils, E. H. Cooke-Yarborough and Martin Ryle, later a Nobel prizewinner. With three research interests, two laboratories, his D.Phil. students, and his Christ Church undergraduates occupying his busy days, the urbane Collie had neither the time nor temperament to build up a research school focused on a particular aspect of nuclear physics.[39]

---

[36] Lindemann to Veale, 29 October 1937, UR/SF/PHE/4; Lindemann memorandum, 6 May 1938, PS/R/1/6; HCP, 168 (1937: 170–1); Kerst (1946); James Leslie Tuck, Manchester B.Sc. 1930 and M.Sc. 1931, ended his career as associate leader, physics division, Los Alamos Laboratory, 1954–73; William Clifford Morgan, second in physics 1938, ended as a director of EMI Electronics.

[37] Lindemann to A.D. Lindsay, 23 September 1938, LP, D154.

[38] E. W. B. Gill to Veale, 1 March 1938, OUA, UR/SF/PHE/7; Hurst's research pupil was William Douglas Allen, an Adelaide graduate 1936, D.Phil. 1940, ended his career as associate divisional head, Rutherford High Energy Laboratory, Didcot, England.

[39] Collie's research pupils were: Orvald Arthur Gratias, Canadian Rhodes Scholar, D.Phil. 1932, spent most of his career in the Canadian branch of J. C. P. Coats; Douglas Roaf, second in physics 1932, D.Phil. 1936, became demonstrator and lecturer in physics in 1946; Arthur Gerald Touch, first in physics 1933, D.Phil. 1937, left Oxford in 1936 to work on radar under Watson-Watt and ended as chief scientist, British Government Communications Headquarters 1961–71; Eugene Theodore Booth, a Rhodes Scholar from

Fig. 7.17. James Howard Eagle Griffiths, successively an undergraduate and postgraduate at Magdalen College, worked on neutron cross-sections during the 1930s. He became a fellow of Magdalen in 1945 and was President of the college from 1968 to 1979. Griffiths was the first physics graduate to be elected head of an Oxford college.

The modesty of Oxford's nuclear physics was confirmed in 1940 when the British atomic bomb project was launched, the main research being done at Birmingham, Cambridge, and Liverpool. The Clarendon did, however, make a significant contribution through the work done by Simon, Kurti, Kuhn, and Arms (a D.Phil. student of Simon) on separating the isotopes of uranium. In contrast, only one senior Oxford nuclear physicist was involved: Tuck went to Los Alamos but others, such as Collie and Griffiths, devoted themselves to radar research as part of the war effort (Collie 1946–7; Gowing 1964: 57–8, 68, 119–20, 219–20; Sanderson 1972: 342–4).

AIR DEFENCE

In the 1930s Lindemann was not closely involved in the daily running of the Clarendon partly because he was deeply concerned about national defence, which by 1934 had become an obsession he shared with Churchill. They worried about

Georgia, D.Phil. 1937, became professor of physics at Columbia University after the war; Ivor Rhys Jones, first in physics 1937, made his early career in the Indian Civil Service; John Clifford Duckworth, first in physics 1938, was managing director, National Research and Development Corporation 1959–70 and was chairman, Lintott Control Equipment Limited from 1980. See Birkenhead (1961: 110); *OUG*, 66 (1935–6): 17 (10 October 1935). Edmund Harry Cooke-Yarborough, wartime honours physics 1940, ended as head of electronics and applied physics, Atomic Energy Research Establishment, Harwell, England; Martin Ryle, first in physics 1939, Nobel laureate 1974; Smith (1986: 499–500); E. H. Cooke-Yarborough to author, 18 September, 15 October 1990.

Fig. 7.18. Frederick Alexander Lindemann, professor of experimental philosophy from 1919 until his retirement in 1956. Lindemann's chair was renamed the Dr Lee's professorship of experimental philosophy in 1922, following the death of Robert Baynes. In recognition of his service as an influential scientific adviser to Winston Churchill during the Second World War, he was created Baron in 1941 and Viscount in 1956. Photograph by A. H. Bodle.

Hitler's Luftwaffe and opposed the view that the bomber would always get through. Suspicious of the Air Ministry, which they regarded as defeatist, they persuaded Ramsay MacDonald, the prime minister, to set up in early 1935 an air defence research subcommittee of the Committee of Imperial Defence. By reporting directly to him this would bypass the Air Ministry that in 1934 had established its own committee for the scientific survey of air defence with Tizard as its chairman. In summer 1935 Lindemann, as Churchill's scientific arm, agreed to serve on Tizard's committee and began a long and famous feud with him (Birkenhead 1961: 146–56, 172–210; Harrod 1959: 163–5, 177–8; Clark 1965: 105–148; Snow 1961).

One cause of disagreement was the priority given by Tizard and his colleagues to radar, which Lindemann thought was of dubious value. Instead he promoted his own pet schemes, one of which was to establish the position of an approaching aircraft not with searchlights but by detecting the infra-red radiation emitted by its engines. In Oxford Lindemann had at hand a ranking expert, R. V. Jones, an Oxford graduate who had completed his D.Phil. in the Clarendon on infra-red detection. Though Jones was supposed to be working in the University Observatory, he spent most of 1935 in the Clarendon working on infra-red aircraft detection. Through Lindemann's lobbying of Harry Wimperis, director of research in the Air Ministry, Jones continued this research at the Clarendon as an employee of the Ministry from 1936 to 1938. In 1937 George Pickard, an Oxonian and D.Phil. pupil of Simon, joined him for a year. Jones and Pickard continued to apply science to national defence in 1938, when the former went to the Ministry headquarters and the latter to Farnborough. Lindemann did not confine his concern with air defence to the Clarendon. In 1937 he stood unsuccessfully as the unofficial Conservative candidate in a University by-election, stressing his expertise in the scientific aspects of air rearmament and defence. Lindemann was prepared to split the Tory vote, which allowed the Independent candidate to win a previously safe Tory seat, because of his obsession with air defence.[40]

THE NEW CLARENDON

The spurt of research triggered by the *émigrés* provided good reason for the erection of a new Clarendon. In this expensive enterprise Lindemann was unwittingly helped by Townsend because by the mid-1930s it was clear to the University that Townsend was declining and that his Electrical Laboratory was outclassed in research and teaching by the Clarendon. From 1931–2 the Clarendon's income outstripped that of the Electrical laboratory, owing to a larger contribution from the University and the Leigh Trust Fund for physics and to grants specifically to it from the Rockefeller Foundation, the Duke of Westminster, and the Higher Studies Fund that the University Appeal created in 1937 (see Tables 7.2 and 7.3). In undergraduate teaching the Clarendon, like the Electrical laboratory, provided bread-and-butter elementary courses but also arranged for more specialist lecture courses given by the appropriate experts in research. The Electrical Laboratory had the edge on the Clarendon in only one respect – income from fees for lectures and practicals where it was heavily dependent on first-year teaching. In 1938 the University set up a working party to consider the relations between the two laboratories and their heads. William Lawrence Bragg, the external adviser, deprecated duplication of apparatus and leaned towards converting Townsend's chair into one for theoretical physics; but the problems in doing this were so formidable that the University shelved the issue until Townsend died or retired.[41]

Lindemann's campaign for a new Clarendon began in 1934, when Griffith and Richard Southwell, professor of engineering science, produced a plan for what became known as the Science Area. Initially Lindemann argued that an extension

[40] On Reginald Victor Jones, first in physics 1932, D.Phil. 1934, professor of natural philosophy, Aberdeen, 1946–81, see Cook (1999); Jones (1978, 9–44); Lindemann to Wimperis, 11 December 1935, 25 November 1936; Jones to Lindemann, 18 and 20 September 1936; Wimperis to Lindemann, 5 October 1936, all LP, D123; Harry Egerton Wimperis, director of scientific research, Air Ministry, 1925–37.

[41] Cooke-Yarborough to author, 18 September 1990; OUA, UR/SF/PHE/7, covered 'Physics. Question of future of department of experimental philosophy and of electrical laboratory'; Townsend inquiry, OUA, MR/7/1/7.

or a new Clarendon costing £80 000–£100 000 was needed. By 1936 the University had conceded that the Clarendon was wholly obsolete and that a new Clarendon was among the three most needed science buildings. In spring 1937 the University agreed to pay for a new Clarendon costing £77 000. This was not just to promote physics but also because it would enable the most urgent changes to be made in the planned development of the congested Science Area, the first being the move of geology into the existing Clarendon (see Figure 7.19). This argument did not apply to a big magnet, like that at Paris, an atom smasher, and a cyclotron, each costing about £5000. Unsure about the future of physics, the University would not pay for these three major items. Lindemann was furious with the University for what he regarded as its niggardliness.[42] To his chagrin, both Lord Nuffield and ICI also frustrated his ambitions for expensive apparatus for the new Clarendon.

Nuffield, who first met Lindemann as late as 1927, saw physics at the Clarendon as abstract and remote from practical application. By the end of 1937 he had donated almost £4 000 000 to the University and some colleges but not one of his pleasant financial shocks galvanized the Clarendon. His preferred beneficiaries were medicine, his new college, poor colleges, the Higher Studies Fund set up by the University Appeal (1937), and to Lindemann's irritation £100 000 for a new University physical chemistry laboratory, with £70 000 allocated to the building and £30 000 for apparatus. Nuffield's largesse reduced Lindemann's chances of fund-raising elsewhere because it gave the impression that Oxford was so well endowed that it did not need more benefactors; and senior University officials discouraged him from making a personal approach to Nuffield. In 1938 Nuffield rubbed salt into Lindemann's wounds when he gave £60 000 to the physics department at the University of Birmingham. How Lindemann must have longed for Nuffield to have emulated Herbert Austin, his chief British rival in motor car manufacturing, who gave £250 000 to the Cavendish in 1936. As for ICI, even though Lindemann was a close friend of its chairman he failed to persuade it to earmark for physics its donation to the University Appeal: it gave £10 000 specifically for the new physical chemistry laboratory. The Clarendon's chief industrial sponsor for materials was, in fact, the British Oxygen Company that in the late thirties gave it liquid air worth £1500.[43]

The new Clarendon was well designed, partly because of inspections made of the laboratories at Eindhoven (Philips), Amsterdam, and Leiden; but Lindemann felt that, like its predecessor, it would lack money for permanent posts and expensive apparatus. The University's policy was to spend its funds on urgent non-repeatable

---

[42] HCP, 159 (1934: 171–2, 174–5); OUG, 67 (1936–7): 25 (8 December 1936); HCP, 167 (1937: 31–4, 57); HCP, 170 (1938: 33–5); HCP, 173 (1939: 15–16); Lindemann (1938: 62); Lindemann to Veale, 15 March 1937, OUA, UR/SF/PHE/4; Richard Vynne Southwell, professor of engineering science 1929–42.

[43] Crowther (1974: 230–1); Andrews and Brunner (1955: 259–60, 309–12); Lindemann to A. D. Lindsay, 13 October 1937, 26 March 1938, UR/SF/PHE/4; Lindemann to H. McGowan, ICI, 18 March 1937; McGowan to Lindemann, 27 April 1937, LP, D99; Birkenhead to Morris, 21 November 1927, LP, B12.

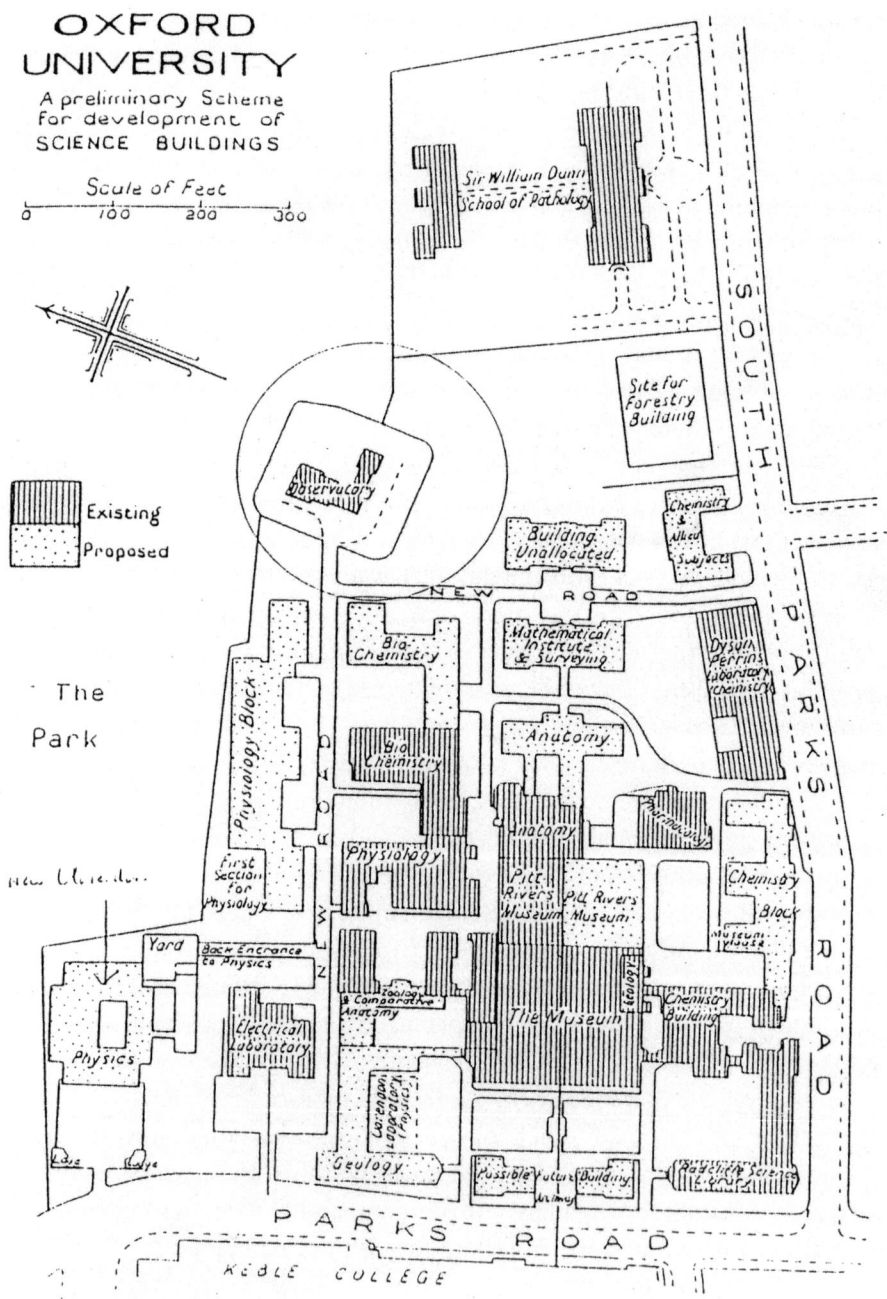

Fig. 7.19. Plan of the Science Area, 1937, showing projected new buildings. From *Oxford. Special Number*, February 1937, p. 59. Note the plan (never implemented) to build on the lawn in front of the University Museum.

Fig. 7.20. The new Clarendon Laboratory, built at a cost of £77 000 and inaugurated in 1940.

capital projects, such as the new Clarendon, leaving heads of department to secure external funding for costly equipment; and to avoid if possible new recurrent commitments such as posts with an incremental salary scale. In exchange for a consistently bigger annual grant from the University to the Clarendon than to the Electrical Laboratory (see Tables 7.2 and 7.3), the former enjoyed the services of just two university demonstrators and the latter four in the 1930s.[44]

In such circumstances Lindemann sought external funding, not always successfully, for apparatus and posts.[45] By 1939 he was so desperate that he raised with the University what was then a novel issue at Oxford, namely, payment of researchers as opposed to teaching staff. He revealed that most of the 23 researchers in the Clarendon (excluding Jackson) were supported by temporary external grants, many of which would soon expire. Lindemann feared that he would occupy the

---

[44] *OUG*, 62 (1931–2): 22 (8 October 1931); *OUG*, 67 (1936–7): 23 (8 October 1936); Lindemann memo about demonstratorship for Hull, 4 May 1939, PS/R/1/7.

[45] Lindemann (1938–9: 496–7); Royal Society Council minutes, 15 June 1939; Lindemann to Duke of Westminster, 21 March 1937, LP, B17; Lindemann to A. D. Lindsay, 23 June 1937, 18 October 1937, LP, B18; Lindemann to Sir Lionel Faudel-Phillips, 12 November 1937, LP, B19; Lindemann to Leverhulme Trust, March 1938; Lindemann to Lindsay, 11 May 1938, both LP, B22; Lindemann to Veale, 15 March 1937, UR/SF/PHE/4.

new Clarendon with plenty of space but few men. Of the total annual salary bill of £7230 about £1800 (25 per cent) was contributed by the colleges, £1400 (19 per cent) by the University, £1120 (15 per cent) by the Department of Science and Industrial Research, £1000 (14 per cent) by ICI, and £1910 (26 per cent) by various outside sources. Of the nine recipients of college funding, only Keeley, Collie, Hurst, Dobson, and Griffith were college fellows, each of them receiving an average of about £300 from his college. The University contribution of £1400 was spent mainly on Keeley as a demonstrator (£350) and Simon as a reader (£500), with £350 going into departmental funds to pay for three part-time demonstrators. About two-thirds of the Clarendon staff were dependent on external funding. ICI provided for the trio of Simon, Mendelssohn, and Kurti. The DSIR supported two post-doctoral researchers (Cooke, Hull) and four postgraduates. Outside sources, other than the DSIR and ICI, included the Salters' Company (Tuck, £300), the 1851 Exhibition Commission (Holbourn, £400), the Duke of Westminster (Roaf, £400), Sir Robert Mond (Mendelssohn, £300) and the Rhodes Trust (Arms, Allen). At the end of the academic session or soon afterwards about £1850 of the outside benefactions would end, none of them being renewable. The five most senior cryophysicists (Simon, Mendelssohn, Kurti, Cooke, and Hull) were supported by temporary grants, three of which were due to end in 1939 and two in 1941. Thus the low-temperature groups that made up almost half of the Clarendon's population and dominated its publications were likely to decline or expire.[46]

This appalling prospect dismayed the University, which in principle acknowledged the justness of Lindemann's case. But it offered no immediate aid; it confined itself to agreeing that additional financial support for the Clarendon's researchers was necessary and that the University Grants Committee should be apprised of the situation. The University had accepted that in the Clarendon an undue proportion of the research done was financed by temporary grants from outside sources, with its attendant dangers for all concerned. The desired peaceful and planned remedies were more college fellowships, more university demonstratorships, and an increased government grant; but, in the event, many of the Clarendon's researchers were saved for physics and for Oxford by the outbreak of the Second World War and their employment as boffins.

---

[46] Lindemann memorandum, HCP, 173 (1939: 225–35), draft in LP, B25, ignored Dobson's £300 per annum from Merton from 1937. My figures ignore Dobson's college income because the University responded to the figures as presented by Lindemann. Lindemann's memorandum also ignored Hans Epstein, who had arrived in England from Germany in early 1939 and, through Simon's influence, had been given accommodation in the Clarendon in May. The destitute Epstein was funded from June by a temporary grant from the SPSL. Athelstan Hylas Stoughton Holbourn, D.Phil. 1936, was laird of Foula, Shetland Islands, Scotland; on Hans Georg Epstein, SPSL 327/1.

8

Redefining the Context: Oxford and the Wider World of British Physics, 1900–1940

Jeff Hughes

Surveying fifty years of British physics at the centenary meeting of the British Association for the Advancement of Science in 1931, the octogenarian J. J. Thomson picked out five key developments that had advanced the science to its present state:

'the increase in opportunities for teaching and research in Physics caused by the foundation of many new laboratories, the increase in attention paid to the teaching of Physics in our schools, the endowment of research workers and the increase in the opportunities for these to obtain remunerative employment, the increased recognition of the importance of research in industry, and . . . the improvements made in instruments used in research and the increase in the magnitude of the forces, mechanical, electric and magnetic, which are now at our disposal' (Thomson 1932: 19).

Professor of experimental physics and director of the Cavendish Laboratory at Cambridge from 1884 to 1918, Thomson recalled that in the early 1870s, when he began his studies of physics at Owens College, Manchester, there were only six physics laboratories in England: those at the Royal Institution, King's and University Colleges, and the Royal School of Mines in London, Owens College, Manchester, and the Clarendon Laboratory at Oxford. By 1931, he observed, the number of physics laboratories at the various universities, colleges, technical schools, and institutes of technology in Britain was over 300. The demand for science teachers after the educational reforms of the 1870s, the emergence of industrial laboratories from the 1890s, the establishment of the National Physical Laboratory, and the rapid proliferation of physical laboratories in academia had led to an enormously increased demand for physicists. With increasing numbers had come increasing diversification and increasing specialization. Almost with alarm, Thomson noted that *Science Abstracts* for 1930 contained abstracts of 4165 papers on physics, or almost a dozen a day, of which it was 'obvious that no one can read more than a small fraction' (Thomson 1932: 27).

As Thomson's remarks indicate, British physics changed radically between 1900 and 1930. The establishment of many new institutions in colleges and universities, industry, and the various branches of government and the military, the development of numerous new support mechanisms for research, the elaboration of entirely new kinds of physics with consequent changes in the disciplinary boundaries of the subject, and a rapid increase in student numbers all made physics a different entity in the 1930s from what it had been in the 1890s. In this chapter, my concern is not so much with the buildings or staffs of Oxford physics that are surveyed elsewhere in this volume, but rather with the place of Oxford physics within the broader *disciplinary* context of British physics. I therefore examine some of the key elements of the changing intellectual and institutional landscape of British physics in this period, and attempt to situate the development of Oxford physics within this wider framework.

An immediate historiographical difficulty faces such an undertaking, however. Given its enormous diversity, it is paradoxical that the history of modern British physics has long been dominated by the institution with which Thomson was associated for most of his working life: the Cavendish Laboratory in Cambridge. The reasons for this dominance are complex, reflecting both the power of its own publicity machine and a post-war historical narrative in which the history of nuclear and particle physics has come to stand for the history of physics *tout court* (Hughes 2003). One effect of this overemphasis on microphysics – the part of physics devoted to the exploration of subatomic particles and forces – has been that physicists and their historians have tended to exercise a heavily retrospective judgement about which physics institutions and which physicists in the past are to be accorded significance: those that made 'significant' contributions to present-day understandings. Thus, because of its contributions to subatomic physics and nuclear science, the Cavendish Laboratory has dominated the history of British physics to the exclusion of almost every other institution; explicitly or implicitly, it has acted as the 'standard' against which other institutions are to be judged – Oxford physics included. By normalizing the values of 'pure' research that were idealized – at least in their public pronouncements – by Cambridge physicists during and after the inter-war years, this historiography has also acted as a significant constraint on our understanding of the relations between academic and industrial, governmental, and military physics in the first half of the twentieth century. Clearly, then, in addition to contextualizing microphysics, we need an even more thoroughgoing historical appreciation of the 'pure research' imperative itself.

The resources for such an approach are widely distributed, but have not been brought together before. In the 1970s, Russell Moseley (1976; 1977) explored the institutionalization of British physics in the early twentieth century, particularly the roles of establishments like the National Physical Laboratory and professional associations like the Physical Society and the Institute of Physics in promoting the discipline. At the same time, Michael Sanderson (1972) offered a comprehensive, if subsequently undervalued, account of the relationship between British universities and industry in the period 1850–1970. More recently, David Edgerton (1991; 1996) has drawn attention to the military and industrial dimensions of British science throughout the twentieth century, and has offered a compellingly critical historical and political frame for the reinterpretation of the history of the physical sciences. Several other strands of recent work contribute to such a revisionist view. As Graeme Gooday (1990; 2004) has shown, for example, late-Victorian physics laboratories must be understood in the context of teaching for the operation and maintenance of imperial communications networks as much as for the promotion of research *per se*. Robert Fox and Anna Guagnini (1999) have drawn attention to the interplay between industry, government, and education in the development of physics in the late nineteenth and early twentieth centuries, and particularly to the practice of physics in its relationship to electrical engineering. In a similar vein, the late Paul Hoch and others (Hoch 1983; Keith and Hoch 1986; Hoddeson *et al.* 1992) have explored the development and importance of solid state physics in the context of the relationship between science and industry in interwar Britain.

Our understanding of Cambridge physics has been changing too. Simon Schaffer (1992) has demonstrated the constitutive role of imperial and industrial concerns in the metrological research of the Cavendish Laboratory under James Clerk Maxwell and Lord Rayleigh. Andrew Warwick (1992; 1993; 2003) has explored the significant differences in the contexts of Cambridge mathematical and experimental physics, while Isobel Falconer (1987), Dong-Won Kim (1995; 2002), and Geoffrey Cantor (1994) have begun to prise open the black-box of the Cavendish Laboratory to demonstrate the tensions that lay within what often seemed from outside to be a homogeneous institution. In a more extensive study sweeping across the twentieth century, Peter Galison (1997) has argued for the importance of material culture and traditions of instrumentation and theoretical and experimental practice in creating and sustaining the microphysics that came to dominate physics in the 1900s. Elsewhere, I have argued for a non-essentialist and non-retrospective reading of the history of radioactivity and nuclear physics that situates them and their practitioners in their social, cultural and material contexts (Hughes 1998*a*; 1998*b*; 2003). Together, these studies have substantially changed our appreciation of the Cavendish Laboratory and begun to make possible a broader understanding of British physics. In this chapter, I extend the process of contextualization to explore more broadly the construction of institutional credibility and intradisciplinary hierarchies within early-twentieth-century British physics. Through an essentially chronological survey of physics research and teaching in the university sector, as well as in industrial, governmental, military, and medical institutions, I attempt a preliminary reassessment of the significance of the many and various kinds of activities constituting physics in Britain from 1900 to 1940. In so doing, my goal is to try to understand the development of Oxford physics not by comparing it directly with its Cambridge counterpart, but by situating both Oxford and Cambridge within the development of British physics as a whole.

1900–1907: RETHINKING THE CAVENDISH TRIUMPHANT?

What would a graduate hoping to undertake advanced work in physics have found in Britain in 1900? Consider the case of Vincent James Blyth. Having studied physics as an undergraduate at the University of Glasgow, Blyth was awarded an 1851 Exhibition scholarship in 1900 for two years of postgraduate work. When these awards were established in 1891, most '1851' physics scholars had chosen Göttingen, Berlin, Glasgow, Manchester, or their 'home' institution for advanced study. Awarded a scholarship in 1892, for example, Bangor-educated Edward Taylor Jones studied at Bangor and Berlin; three years later, Bristol-educated Samuel Roslington Milner took up his award at Bristol and Göttingen (Phillips 2001). By 1900, however, things had changed. Short of staying on at Glasgow to work with newly appointed Andrew Gray, who had been Lord Kelvin's private secretary and assistant, Blyth might have briefly considered a number of options: Arthur Schuster's new state-of-the-art physics laboratory at Manchester, home to research in optics, spectroscopy, and various aspects of thermodynamics; John Henry Poynting's department at Mason College, Birmingham, with its facilities for photometric and photographic work; University College, Bristol, where Arthur Chattock ran a small but busy department specializing in studies of conductivity; research towards a London degree with Arthur Rücker at the Royal College of Science or with Hugh Callendar at University College, where precision measurement reigned supreme; or even the possibility of entry to the Davy-Faraday Laboratory at the Royal Institution, opened only four years earlier and making a reputation for itself for the diversity of its researchers, the independence it offered them, and the fractiousness of its Director, James Dewar (Brock 2002; Watson 2002). Liverpool, a possibility in previous years, would be out of the question because of the departure of Oliver Lodge to the Vice-Chancellorship of Mason College. As the chapters in this volume imply, the Clarendon Laboratory at Oxford, too, would have been ruled out for its lack of any research tradition, except perhaps in optics. While Blyth might conceivably have considered taking his chances as a first-comer at the new Electrical Laboratory in Oxford, where John Townsend, fresh from the Cavendish, was newly installed as Wykeham professor of physics, the prospect was risky. Other, much smaller departments, such as that at Nottingham, where Edwin Henry Barton did research on acoustic and mechanical vibrations, would scarcely have crossed his mind. No, in 1900 it was the Cavendish Laboratory, Cambridge, which was Britain's foremost physical laboratory and which Blyth would have selected almost without hesitation (Board of Education 1900; Forman, Heilbron, and Weart 1975; Kim 2002).

Under J. J. Thomson the Cavendish had for several years been pursuing the experimental and theoretical physics of gas discharges and a form of ionic physics related to physical chemistry. Of all British physics laboratories, it was there that several discoveries of the 1890s – X-rays in 1895, radioactivity and the Zeeman effect in 1896, the corpuscle of negative electricity in 1897, radium and polonium in 1898 – had been taken most seriously and integrated into research. More conventional kinds of research flourished too, particularly after the change of regulations in 1895 that allowed non-Cambridge graduates to register for advanced degrees and achieve formal recognition for original research. As Isobel Falconer (1989) and Dong-Won Kim (1995; 2002) have shown, out-of-Cambridge recruitment contributed significantly to the development of the Cavendish 'school' after 1895, so that by 1900 the laboratory was becoming a powerful vehicle both for the production and reproduction of the electrical theory of matter and for professional mobility, as its graduates began to be appointed to chairs of physics in Britain and the Empire.

Yet there was more to the Cavendish than research on microphysics, the new world of subatomic particles and forces opening up as a result of the experimental discoveries of the 1890s. As Falconer has pointed out, a fundamental split between teaching and research began to emerge at the Cavendish circa 1900. Although historians have tended to focus on research, it was most often on the grounds of their teaching and administrative credentials that men were appointed to senior university positions, and it was the problems of teaching that occupied their time and energy. Take the very different cases of two senior Cavendish men going to posts elsewhere in 1900: Lionel Wilberforce to the Lyon Jones professorship of physics at Liverpool (in succession to Lodge) and John Townsend to the new Wykeham chair of physics at Oxford. These appointments illustrate the very different meanings of 'Cavendish physics' to men of different generations at this time. Wilberforce had been a successful and popular demonstrator at the Cavendish since the 1880s, and had been instrumental in reorganizing the teaching work of the laboratory. Like another Cavendish demonstrator, G. F. C. Searle, Wilberforce did not participate in

Thomson's gas-discharge research programme or subscribe to the ideology of 'original research' that motivated and sustained it. He was appointed at Liverpool on the strength of his teaching experience at the Cavendish, and spent his first few years there organizing teaching to large classes with only a small staff (Wilberforce 1910; Roberts 1944). As an 1895 entrant to the Cavendish, by contrast, Townsend was one of Thomson's ablest research students and had become a fellow of Trinity College in 1899 (Jones, F. L. 1957; von Engel 1957). Specifically appointed to a new chair at Oxford on the basis of his research experience and potential and in a very different context of expectation, he took significant elements of Cavendish research practice with him when he left Cambridge, establishing a programme of research in ionic physics for himself and the research students he hoped to attract.

The contrast between Townsend and Wilberforce illustrates the tension between physics teaching and research in universities around 1900, and the different demands that could be made on new appointees according to local circumstances. In many ways, Wilberforce was the more typical of the pair, and his experience helps us to understand the values and conditions that shaped the experiences of perhaps the majority of physicists at the time. Caution is necessary here, however. Despite being one of those whom Falconer identifies as the Cambridge 'old guard' who did little research of their own, Wilberforce devoted his first years at Liverpool to planning the new George Holt Physics Laboratory. Opened in 1904, the laboratory was one of the best-equipped in Britain, designed 'to allow teaching and research to flourish side by side, not hampering but supporting each other' (Anon. 1904: 64). Wilberforce himself took little advantage of the facilities of his new laboratory for what the canonical historiography would recognise as 'research'. But, like Clifton at Oxford, he did substantive work on instrumentation for teaching purposes. He also exploited the endowment of an Oliver Lodge fellowship to appoint Charles Glover Barkla, whom he had known from the Cavendish Laboratory and whose work on X-ray scattering gave Liverpool a place in national and international debates on the emerging microphysics (Allen 1945–8; Wheaton 1983: 44–5, 71–103). On the other hand, Townsend may have taken the Cavendish research imperative west, but, as Lelong shows, he faced considerable institutional difficulties in his attempts to transplant ion physics to Oxford. The triumph of 'modern' research was far from being as straightforward as many have assumed.

Clearly, the complexities of individual cases must be examined carefully. Nevertheless, the experiences of Wilberforce and Townsend allow us better to appreciate the wider context in which Oxford physics should be placed in this period. Although the Clarendon Laboratory of 1900 was not producing a 'modern' research-oriented physics cadre comparable with that of the Cavendish, the same has to be said of most other physics laboratories. At Manchester, Glasgow, Birmingham, the various London colleges, and elsewhere, where there were original investigations, they were for the most part in the 'traditional' physics of optics, spectroscopy, thermodynamics, standards, precision measurement, meteorology, acoustics, and so on (Anon. 1906). At London's Royal Institution, a different form of 'modern' physics – low-temperature work – was under development (Brock 2002). At the Royal College of Science, Hugh Callendar even condemned studies of conduction in gases and radioactivity as the 'playground of dilettante physicists'; for him and his students, 'classical physics was the only physics that mattered' (Thomas 1949: 583; Smith 1932). Similarly, the physicists obtaining academic posts were typically more like Wilberforce, hired for their skills in teaching and administration, rather than in original work. For example, Arthur Laidlaw Selby was assistant demonstrator at the Clarendon from 1884, and then assistant professor (1890–7) and professor (1897–1926) of physics at Cardiff University College, where he distinguished himself 'as a leader in committee and as an inspiring lecturer in the class-room' (Dunbar 1942).

More broadly, the economy of physics was beginning to expand in significant ways in 1900. The establishment of the National Physical Laboratory under the directorship of a Cambridge contemporary of Wilberforce, Richard Glazebrook, offered a new source of employment for talented young physicists (employees were expected to hold first-class honours) that grew steadily over the next few decades. With fewer than twenty physicists employed in 1900, the numbers rose to 70 in 1914, 105 in 1920, and 162 in 1938, and it is noteworthy that the first recruits at the NPL were largely from outside Oxbridge (Moseley 1976; Pyatt 1983; Magnello 2000). The NPL's role was shaped not simply by metrological concerns but by its connections with aeronautics and naval research, complemented by an emphasis on work bearing on industrial problems, on which a great deal more historical research is needed. While most physics as yet had little institutional presence in British industry, its close links with electrical engineering and instrumentation and the mobility between academic and military establishments and the NPL in this period also remain to be explored in depth.

Despite the growth of the NPL and the breadth and diversity of academic physics, it is undeniable that the Cavendish Laboratory dominated the discipline in Britain in 1905, not least because of its vibrant research ethos and its concentration of researchers and resources. The directors of physical laboratories outside Cambridge, many of them former Cavendish men (gender used advisedly; see Gould 1997), were beginning to push their students towards the Cavendish Laboratory to allow them to experience life in a large research laboratory as well as to further their research experience. When the young Australian physicist Thomas Laby took his '1851' award to Birmingham in 1905, for example, Poynting advised him to transfer to the Cavendish (Picken 1948: 736). This became a self-reproducing cycle, as Laby was one of the many Cavendish men who later pushed their own students towards Cambridge and the Cavendish (Dean 2003). And, of course, American and European academics continued to come for a year at a time to partake of Cambridge scientific and social life. Though we lack a sufficiently detailed study of the Cavendish researchers in this period, what united them was undoubtedly a desire to learn more about the new microphysics of matter being developed and publicised at the Cavendish by Thomson and his students. Yet what these students learned at the Cavendish was more than simply ion physics, the electronic theory of matter, and a reductionist approach to physics and chemistry based on the premise that all atomic and molecular behaviour could be understood in terms of the behaviour of the 'ultimate' constituents of matter (Kim 1995, 2002; Chayut 1991; Buchwald and Warwick 2001). As Falconer and Kim have argued, they learned a way of scientific life, based on intensive discussion of physics formally and informally, within and without the laboratory (the Cavendish Physical Society, established in 1893, was the first and most formal of many discussion groups). In this context, Thomson's Nobel Prize for physics, awarded in 1906 for his work on the conduction of electricity by gases, seemed only to seal the success of the Cavendish Laboratory and its placemen over the previous few years (Kim 2002).

1907–1918: REDUCTIONISM RAMPANT OR DIVERSITY AND DIVISION?

A significant moment in the development of British physics came in 1907 when one of those placemen, Ernest Rutherford, accepted the chair of physics at the University of Manchester in succession to Arthur Schuster. Schuster had directed a large, well-resourced, and active research laboratory, working mainly in the fields of optics, spectroscopy, and meteorology (Anon. 1906; Kargon 1977: 214–37). One of the most energetic products of Thomson's Cavendish, Rutherford turned the laboratory over to radioactivity, the speciality he had developed in Cambridge in 1897 and made his own at McGill University from 1898. Working with existing staff, among them Hans Geiger and Walter Makower, Rutherford reorganized physics teaching at Manchester to include a significant element of 'modern' physics and original research in the undergraduate physics course. He also established a training course in research and exploited the one-year M.Sc research degree to promote research and publication within his carefully defined programme. As well as work on the properties of the various radioactive radiations, Rutherford and his co-workers studied the interactions of these radiations with matter. It was out of these experiments that Rutherford's speculative nuclear model of the atom emerged in 1910 (Makower and Geiger 1912; Heilbron 1968; Wilson 1983: 216–338).

Rutherford forms an interesting and illuminating contrast with another 1907 appointee, William Peddie, who took up the Harris chair of physics at the University College of St Andrews at Dundee in that year and held it until 1942. Educated at Edinburgh where he had studied under P. G. Tait, Peddie's academic output was mostly in the form of textbooks, suggesting again the importance of teaching as against his own research interests in thermodynamics and oscillations (Allen 1946). At the same time, however, he taught his assistant Robert Watson-Watt the fundamentals of radiotelegraphy, laying the foundations for the latter's career in meteorological physics and military science, of which more later (Watson-Watt 1957; Ratcliffe 1975). In a larger and better-resourced department imbued with the Thomsonian ideology of research, however, Rutherford enthused everyone with his scientific exuberance and the intense social and intellectual atmosphere he and his lieutenants created. Manchester quickly became a key node in the international network of radioactivity research and a magnet for those interested in atomic physics – and a supplement, or even an alternative, to Cambridge. The visibility and reputation of the Manchester school were clearly enhanced by the award of the 1908 Nobel Prize for chemistry to Rutherford, and by his participation in the 1911 and 1913 Solvay Congresses, which put him in an elite international group of physicists explicitly devoted to the emerging physics of atoms and quanta (Wilson 1983; Hughes 1998c). At the same time, it is worth noting that not all those in Rutherford's department worked in radioactivity: a fair proportion of them taught or researched in 'electrotechnics' (still part of the physics department at Manchester, as in many other places) and in optics and spectroscopy, which continued to be a strong component of the department's research profile. Even in departments where 'modern' physics flourished, traditional subjects were still important (Broadbent 1998).

As the subject-specification of Rutherford's Nobel Prize indicates, we should also remember radioactivity's unusual disciplinary position between physics and chemistry. Contrary to later accounts that have retrospectively identified substantial parts of radioactivity as belonging unproblematically to nuclear physics, it is worth noting that the other British nodes in the small international radioactivity research network were in chemistry, rather than physics, departments; the cases of Rutherford's erstwhile collaborator Frederick Soddy at Glasgow and of William Ramsay at University College, London, illustrate the point (Merricks 1996; Travers 1956). While the issue of the boundary between chemistry and physics in this and other areas remains to be explored in detail, it is clear that by 1910 Rutherford had firmly established himself as a leading player in the politics of British physics. His part in British negotiations over the international radium standard gave him a minor role within the counsels of state as an official government representative, while his institutional authority was augmented by his frequently being asked to advise on appointments elsewhere in British and Empire university physics (Wilson 1983, 249ff.). He used this authority to promote 'modern', reductionist physics and the research ethos wherever possible. He played an important role, for example, in the appointment of William Bragg as successor to William Stroud in the chair of physics at Leeds in 1908. Bragg brought from Adelaide an active research interest in radioactivity and X-rays, though the nascent research programme underway at Leeds by 1911 gave way in 1912 to research on the recently discovered technique of X-ray crystallography, to which Bragg and his son William Lawrence Bragg were both to devote the rest of their careers and for which they were to win the Nobel Prize for physics in 1915 (Andrade 1942–4; Caroe 1978; Phillips 1979).

'Modern' physics was beginning to flourish elsewhere, too, and very little of it had to do with radioactivity or nuclei. At King's College, London, for example, a succession of former pupils of J. J. Thomson began to turn out research in gas discharge and electron physics as a supplement to their teaching, among them C. G. Barkla (professor, 1909–13, Nobel Prize for physics, 1917) on the properties and effects of X-rays, and Owen Richardson (appointed professor in 1914; Nobel Prize for physics, 1928) on the emission of electrons from hot bodies (Allen 1945–8; Wilson 1959). At the NPL, G. W. C. Kaye arrived from the Cavendish and was soon to head up radiation standards and research (Griffiths 1941; Smith 1975). Outside London, Barkla's appointment to the prestigious chair of natural philosophy at Edinburgh in 1913 took radiation research north of the border. At Bristol, Arthur Tyndall began research on ionic mobility, while Liverpool took Rutherford's advice and appointed James Rice, a specialist on relativity and 'modern' theoretical physics, in 1914 (Donnan 1936). And, most significantly for our purposes, the Drapers' Company's gift of £23 000 for the new Electrical Laboratory for Townsend at Oxford only helped to confirm the Electrical Laboratory's reputation for being rather 'more lively and productive than the Clarendon' by 1914 (Morrell 1992: 266).

From around 1910, then, the Cavendish Laboratory was being partially eclipsed by the development of Rutherford's research school at Manchester and the growth of 'modern' physics at Leeds, London, Oxford, and elsewhere. In some ways it was a victim of its own success. On the one hand, a string of appointments of senior and experienced figures out of the Cavendish had depleted its upper ranks and helped to establish ion and electron physics elsewhere, in competition with it. On the other, this efflux of talent had serious repercussions for the organization of research in the laboratory and for morale, with 'too many students chasing too few ideas for research and too little apparatus' (Bragg 1965: 117). Thomson's own research on positive rays and his atomic theory were in the doldrums, and the temper of the Cavendish seemed to have changed: it had lost the cohesiveness, excitement, and tightness of direction it had possessed at the turn of the century

(Falconer 1988; Kim 2002: 169–174). Symptomatically, the young Niels Bohr was piqued by Thomson's lack of interest in his postdoctoral project on electron theory, and he left the Cavendish in 1911 for Manchester. There, he was impressed by Rutherford's energy and enthusiasm and quickly became enrolled in the Manchester programme to map atomic structure (Heilbron and Kuhn 1969; Pais 1991: 117–29).

Like Bohr, Harry Moseley offers a revealing window into the world of British physics at this moment. Eton-educated, Moseley graduated in physics at Oxford in 1910. Determined on a scientific career, he considered the relative merits of research and a fellowship at Oxford versus a recently advertised demonstratorship in Rutherford's department at Manchester. As Heilbron points out, Moseley's deliberations on his immediate future did not include the Cavendish Laboratory. Perhaps, as Heilbron surmises, this was because of Townsend's antagonism towards Thomson, 'whom he wrongly accused of building a reputation on the ideas of other men, and most notably his own' (Heilbron 1974a: 42). Heilbron also tellingly cites Rutherford's reputation for being a more active and assiduous research supervisor than Thomson as a factor in Moseley's decision to opt for Manchester.

Moseley's post-Manchester career, too, is a useful guide to the changing contours of physics in 1910s Britain. Rejecting study in Germany and France, he thought of going to Leeds to learn X-ray technique with W. H. Bragg, with a view to returning to Oxford and a fellowship (Heilbron 1974a: 202). In the event, however, he decided to take his new line of research – X-ray spectroscopy – to Townsend's laboratory in Oxford, not least because 'I will have a much better chance of a research fellowship if I am on the spot clamorous' (Heilbron 1974a: 204). From his base in Townsend's laboratory, Moseley retained strong links with Manchester, entrenching the Electrical Laboratory further in the extended network of laboratories working on modern atomic physics. The difficulties of resuming his research in Oxford (where things 'seem to move . . . slowly compared to Manchester' (Heilbron 1974: 217)) were eased, however, by a 1000 franc (about £40) grant from the Solvay Institute, on whose grants committee Rutherford sat. The ambitious Moseley applied for the chair of physics at Birmingham made vacant by the death of Poynting in April 1914. Hoping to establish Manchester-style research-intensive physics in Birmingham, and with testimonials from Rutherford, W. H. Bragg, and Townsend supporting his application, he left with a number of other delegates for the 1914 meeting of the British Association for the Advancement of Science in Australia (Heilbron 1974a: 110–15).

As Moseley's trajectory indicates, by 1914 there was a new diversity in British physics, with substantial experimental research in 'modern' physics at Manchester, Leeds, Edinburgh, Bristol, the Electrical Laboratory at Oxford, and King's College,

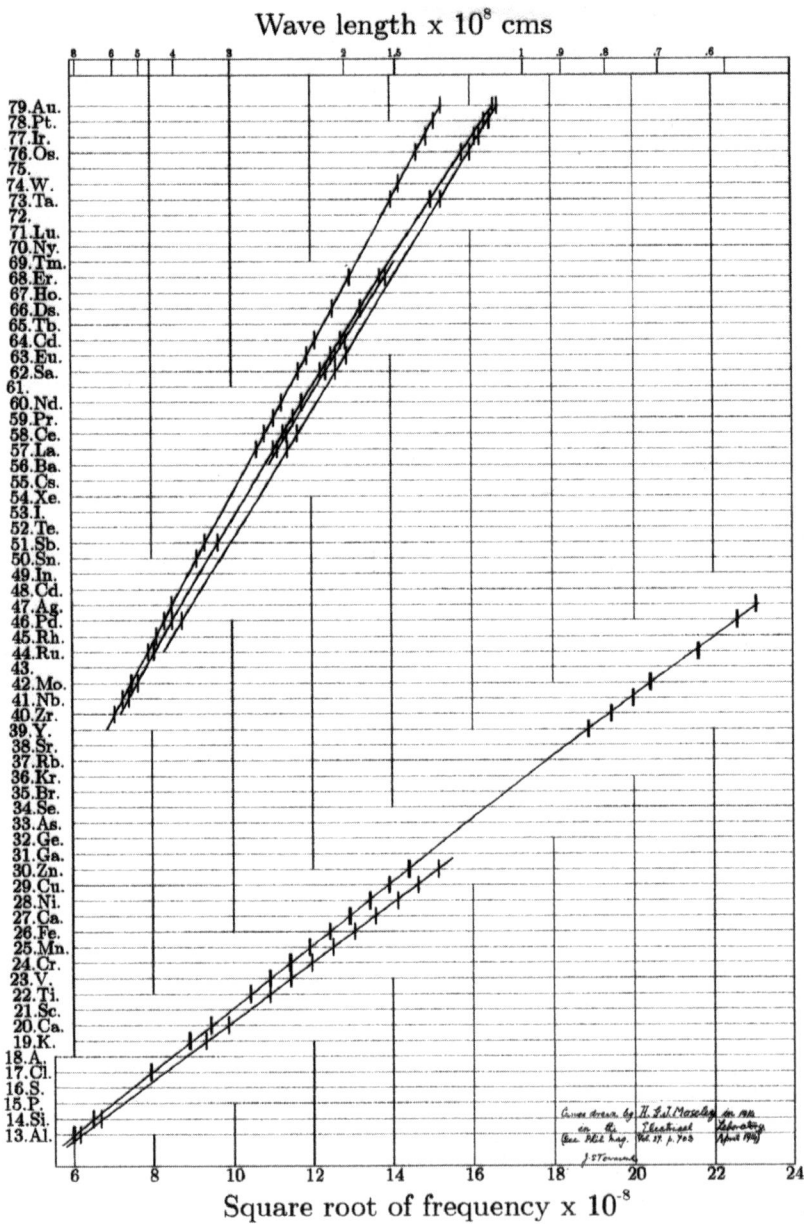

Fig. 8.1. Moseley drew this famous graph early in 1914 while working in Townsend's Electrical Laboratory. The graph, of which the original is now in the Moseley Room in the Clarendon Laboratory, mapped the characteristic X-ray emission spectra (K and L lines) of the available chemical elements. By plotting the square root of the spectra frequency against the atomic number Z, Moseley obtained a straight-line correlation that supported Bohr's recent theory of atomic structure. In addition to yielding a valuable means of identifying substances from their unique spectral emissions, the graph revealed three gaps for values of Z that were later identified with previously unknown elements: rhenium (discovered in 1925), technetium (1937), and promethium (1945). The labelling of the axes, which was added later in place of Moseley's original labelling, introduced errors in the list of elements. For example, 27Ca should be 27Co, and 66Ds should be 66Dy.

London. Neither the Clarendon Laboratory *nor* the Cavendish Laboratory was in the vanguard of this physics. There was also significant research in what the newer centres were coming to regard as more 'traditional' or 'classical' physics (experimental or mathematical) and in teaching instrumentation at Cardiff, Liverpool, Birmingham, East London College, Imperial College, Nottingham, Sheffield, and elsewhere. While active promoters of research in the newer physics (like Rutherford and Moseley) sought to introduce it as widely as possible, it is important to note that the increasing diversity of 'modern' physics in Britain meant that there was no longer a unified programme of reductionist research, as there had been at the Cavendish around 1900. Clearly, much further work is needed in order to explore this diversity as well as the tensions between 'traditional' and 'modern' physics, the conflicting imperatives between promoters of research and advocates of teaching and public lecturing, and the effects of contingent local institutional politics on appointments and the development of physics in Edwardian Britain.

With the outbreak of war in August 1914, the many physicists who had been produced by the growing institutional base for the subject in Britain were among those who rushed to join up. Although it is a historical commonplace that the Second World War was the physicists' war and the First World War the chemists' war, physicists clearly played a crucial role in the services from 1914, as many of them

Fig. 8.2. H. G. J. Moseley in military uniform, a photograph taken shortly before his death in action at Gallipoli in August 1915. For an earlier photograph see Figure 6.2.

were assigned to X-ray or wireless work or to various projects that emerged during the conflict, such as sound-ranging. Building on the precedent that the creation of the Aeronautical Research Council in 1909 had already established (Nayler 1966), the creation of the War Office X-Ray Committee (headed by Hugh Callendar), the Admiralty's Board of Invention and Research (BIR) in July 1915, the Munitions Inventions Department (MID) in August 1915, and the Air Inventions Committee (AIC) in May 1916 brought a new institutional framework and new roles for the country's senior physicists (MacLeod and Andrews 1971; Moseley 1977; Pattison 1983; Hartcup 1988). Among the BIR's projects was submarine detection work directed by Rutherford and Bragg at Manchester University and the Admiralty's experimental stations at Parkeston Quay, Shandon, and Hawkcraig, which brought together physicists, engineers, industrialists, and military men in goal-directed projects (Wilson 1983: 339–85; Hackmann 1984: 11–89). At the Royal Aircraft Factory, Farnborough, a group of physicists including Francis Aston and George Thomson of the Cavendish, Frederick Lindemann, and others gathered to work on problems of aerodynamics and the properties of materials for aviation purposes, while Richard Whiddington and others worked on wireless telegraphy and the problems of aviation signalling. Whiddington later joined a group including the valve manufacturer S. R. Mullard in work on high-vacuum receiving valves at Imperial College – an example that suggests the need for a substantial revision of our understanding of military-industrial-academic cooperation in physics in the Great War (Hartcup 1988).

After decontrol and demobilisation, the new importance of military research continued into the peace, with civilian scientists having found 'a permanent niche in the military technical establishment' (Pattison 1983: 557). Many of the military establishments that had recruited scientists during the war provided new sources of employment for physicists, chemists, and engineers. The Royal Aircraft Establishment at Farnborough (which succeeded the Royal Aircraft Factory in 1918), the Admiralty's Department of Scientific Research and Experiment and Admiralty Research Laboratory, and the Army's Ordnance Research Department and Signals Experimental Establishment were all significant sites in the post-war physics establishment. Similarly, new laboratories were established in industry in an attempt to capitalize on the scientific work of the previous decade. The General Electric Company established new research laboratories at Wembley to work on electronic valves and thermionics, for example, and electrical engineering concerns like Metropolitan-Vickers and wireless companies like Marconi consolidated existing arrangements to form new research centres (Dummelow 1949; Clayton and Algar 1989). Among other governmental initiatives and institutional changes, the creation of the Department for Scientific and Industrial Research (DSIR) in 1916 expanded the institutional infrastructure for science in general, but particularly the physical sciences, not least through the provision of studentships for postgraduate research that created a career structure for scientists within universities and other research establishments (Varcoe 1970; Hull 1999).

DIVERSITY AND MOBILITY: BRITISH PHYSICS RECONSTITUTED, 1919–1930

In order to understand the broader context for the development of Oxford physics in the 1920s, let us now consider the new post-war institutional and intellectual geography of physics at some length in an attempt to evaluate what was 'typical' in interwar British physics. Among the immediate post-war university appointments, pre-war Cavendish men dominated. Invariably, local academic politics was important. At King's College, Royal Holloway College, and University College in London, Owen Richardson (appointed 1914), Frank Horton (1914), and William Bragg (1915), respectively, took up their full duties. Elsewhere there were new appointments from outside. At Birmingham, Samuel Walter Johnson Smith was lured from Imperial College to replace Poynting, while at Leeds Richard Whiddington was recruited fresh from wireless work at Farnborough to head the physics department. At Aberystwyth and Armstrong College, Newcastle, Gwilym Owen and George William Todd, both Cavendish-trained experimentalists, were appointed to chairs. Other than at Oxford, only at Bristol (1919) and Sheffield (1921), where Arthur Tyndall and Samuel Milner formally occupied the physics chairs they had kept warm during the war, and Swansea (1920), where former Manchester lecturer Evan Jenkin Evans was appointed professor, did non-Cavendish men occupy chairs in the post-war round of appointments.

At Cambridge, the appointment of J. J. Thomson as Master of Trinity College in 1918 and his subsequent resignation from the professorship of experimental physics created a vacancy at the Cavendish. He was succeeded by Rutherford, who was given the task of helping to make Cambridge the centre of an imperial research network and of re-establishing its pre-eminence in the physical sciences. This he amply did, transplanting his programme of radioactivity research from Manchester and mobilizing an extensive array of contacts within and outside the university, the new Ph.D. degree, and numerous sources of financial support – including 1851 Exhibitions, DSIR awards, and college fellowships – to sustain it. Despite the fact that Rutherford's programme of nuclear research was criticized within and beyond his laboratory through the 1920s and 1930s, sympathetic coverage in both the professional and the non-scientific press meant that it was widely seen as the most exciting branch of physics. The new regime therefore attracted healthy numbers of students, including a high proportion of Australians, New Zealanders, and Canadians (Stuewer 1985; Hughes 1998a, b).

By the mid-1920s, the Cavendish supported between 20 and 25 research students and visitors per year. Again, it is important to realise that not all came to work on the laboratory's flagship nuclear programme; nor did all the researchers take Ph.D.s (or even register for a degree). Almost as significant as the radioactivity group in terms of numbers of students and resources was the programme of ionospheric and radio research led by Edward Appleton and, later, Jack Ratcliffe. A booming sector of the economy in the 1920s, wireless was intimately associated with industry and the military as well as broadcasting (Ratcliffe 1966; Clark 1971; Budden 1988; Bussey 1990). Elsewhere in the Cavendish, Peter Kapitza pursued the magnetic and low-temperature research that blazed a trail in the development of physics using large and expensive equipment, while 'lone' experimentalists and instrument makers such as Francis Aston (Nobel Prizewinner in chemistry 1922 for his research on isotopes), hydrodynamicist Geoffrey Taylor, and Charles Wilson (Nobel Prizewinner in physics 1927 for his cloud chamber) carried on their individual research in splendid isolation.

Clearly, to talk of a unified, homogeneous 'Cavendish physics' in the 1920s would be a mistake. Much of the success of the Cavendish in the inter-war years was due to the sheer *diversity* of work going on there and the ability of its managers to mobilize financial and other resources to push it forward. The rhetoric of 'pure' research that motivated the laboratory's work on the structure of matter was belied by the close links that many of the researchers had with industrial and military concerns. And, again, we should remember that 'modern' physics and a large research school were not the exclusive preserve of Cambridge. Indeed, with Rutherford's nuclear programme beset by controversy for most of the 1920s (Stuewer 1985; Hughes 1998b), much of the best work in 'modern' physics was done elsewhere. At Manchester, for example, Rutherford's successor Lawrence Bragg, facing difficulties with the organisation of teaching and the management of a large department that included electrochemistry, electrotechnics, and an inherited staff of radioactivity researchers, began to surround himself with men whom he knew from pre-war Cambridge or from his wartime sound-ranging team. He introduced research in X-ray crystallography, exploiting the M.Sc. and Ph.D. degrees to promote original work, and succeeding by the mid-1920s in establishing one of the largest physics research groups outside Cambridge (Phillips 1979; Thomas and Phillips 1990; Hunter 2004).

Nor was Manchester the only centre pursuing the 'modern' physics of atoms and quanta. At Edinburgh, Barkla undertook a comprehensive programme of experimental research into the 'J phenomenon' with a succession of research students. Though Barkla's results were sceptically received and later discredited, he continued his work through to the 1930s, generating over twenty papers and fourteen Ph.D. students (Wynne 1976). Gas-discharge physics featured too at Aberdeen, Leeds, and elsewhere. At Aberdeen, the work of J.J.'s son George Thomson on positive rays in hydrogen led to the elaboration of electron diffraction, for which he shared the 1937 Nobel Prize for physics – aptly reminding us that although we tend to associate the canonical achievements of modern British physics with the Cavendish, they often emerged elsewhere (Moon 1977; Goodman 1981). At the same time, Thomson's achievements form an interesting contrast with those of Richard Whiddington, his close contemporary at the pre-war Cavendish, who resumed his research on gas discharges and characteristic X-rays at Leeds in the early 1920s but whose commitment to teaching and administration meant that his research was eclipsed by the work of de Broglie and others (though Whiddington continued to be known for his ultramicrometer) (Feather 1971). It was the appointment of Edmund Clifton Stoner to a lectureship in 1924 and a readership in 1927 that gave Leeds a national research profile in his area of research, magnetism and atomic structure, despite the fact the Stoner himself seems to have had few research students in this period (Bates 1969; Cantor 1994).

Non-reductionist physics of the various kinds that flourished in the universities of the provinces and the Welsh university colleges, also did so in London. There, spectroscopy figured large at King's and Imperial, and Bragg senior's X-ray crystallography dominated at University College and, from 1923, the Royal Institution. From 1924, Edward Appleton at King's pursued the ionospheric research he had begun in Cambridge, now in conjunction with the DSIR Radio Research Board, while Owen Richardson, Royal Society Yarrow research professor from 1923 to 1944, churned out papers and research students in spectroscopy and was a key figure in the administration of British physics as Chairman of the Physics Committee of the Royal Society and Foreign Secretary of the Physical Society (Wilson 1959). The military and industrial links of many of the London physicists in particular are evident. In 1921, for example, Bragg senior approached a number of academic colleagues in an attempt to recruit a researcher to work on the development of the powder X-ray method with the support of the DSIR and the promise of subsequent lucrative employment at the military research laboratories at Woolwich.[1] Similarly, the effort that was devoted to the investigating and testing of the properties of existing materials and the creation of new ones, shaped as it was by developments in radio, electric lighting, and manufacturing, meant that many of the recruits to the new GEC research laboratories at Wembley from 1923 were London physics graduates (Clayton and Algar 1989; Edgerton and Horrocks 1994).

---

[1] W. H. Bragg to Townsend, Milner, Whiddington, Rutherford, Lees, Callendar, Richardson, Tyndall, S. W. J. Smith, Joly, and Wilberforce, 11 March 1921, marked 'Confidential'; W. H. Bragg papers, Royal Institution, London.

More broadly, the electrical, communications, and broadcasting industries provided increasing numbers of posts in the 1920s, with Western Electric, Standard Telephones and Cables, and many other companies employing physicists for testing and research. In all of these cases, the emphasis on *technique* is important for our understanding of recruitment to non-academic laboratories; in 1926, for example, GEC appointed John Randall not least because of his work on X-ray crystallography with Lawrence Bragg and Reginald James at Manchester (Wilkins 1987: 495–6). Similarly, Research Associations and DSIR Research Stations significantly shaped the institutional landscape of interwar physics. Although the fortunes of the Research Associations fluctuated with economic circumstances, the more successful of them made significant contributions to industrial development, as well as to the career prospects for, and the professional mobility of, physicists (Varcoe 1981). The DSIR's own Research Stations for Fuel (Greenwich), Building (Garston), and Low Temperatures (Cambridge) similarly provided scope for physicists. In particular, the amalgamation of the Meteorological Office and NPL radio research groups at Slough to form the DSIR Radio Research Board's Radio Research Station under Robert Watson-Watt in 1927 emphasizes the enormous significance of wireless and the broadcasting and radio industries for physics and engineering in the 1920s (Bussey 1990).

Just as the increasing scale and scope of research within industry in the 1920s brought about significant changes in the relationship between industry and the universities, the military implications of wireless, too, created some new opportunities for physicists, although acoustic and optical work continued to provide the prime focus for military research through the 1920s. The Army's Signals Experimental Establishment (SEE) at Woolwich was responsible for the development of signalling methods, including fixed and mobile wireless and landline devices, acoustics, and metal and mine detection. In 1924, the acoustic section moved with the Army's Searchlight Experimental Establishment (SLEE) to Biggin Hill and became the Air Defence Experimental Establishment (ADEE). Headed by the Royal College of Science (RCS) graduate and former Callendar research student William Sansome Tucker, ADEE undertook research into gun sound-ranging and acoustics, including the development of concrete acoustic mirrors situated on the south coast for the detection of aircraft (Tomlin 1988; Scarth 1999). Under ex-RCS and NPL physicist Frank Smith, the Admiralty's Department of Scientific Research and Experiment, established in 1919, supported a similar range of studies (Goodeve 1972; Hackmann 1984: 97–231). In 1920 a small group was established at the Navy's Signal School, Portsmouth, to conduct work on thermionic valves, and a year later the Admiralty established the Admiralty Research Laboratory (ARL) alongside the NPL in Teddington (Wood 1961). Another group under Edward Harrison (a former 1851 Exhibitioner who had researched at the

Cavendish between 1904 and 1906) worked on mine design and countermeasures at H.M.S. Vernon, Portsmouth.

In many of these cases, Oxford chemistry graduate Henry Tizard played a key role. Having served as a scientific expert in the RAF during the war, Tizard had returned to Oxford to become university reader in thermodynamics in 1920. He gave up this appointment to become assistant secretary of the DSIR, where he had oversight of the Chemistry, Physics, Engineering, and Radio Research Boards, established to co-ordinate military and civilian research of relevance to the services. He used his position to promote research aggressively within the services. As we shall see, Tizard and other Oxford physical scientists played significant roles in the organization of physics in military and industrial contexts, and these must be considered in any overall assessment of Oxonian and British physics (Farren 1961; Clark 1965).

A number of interesting features emerge from this broad overview of British physics as it developed in the 1920s. First, despite the claim of one job-seeker in 1926 that '[s]o many Physics posts are in fact, though not of course in name, reserved for someone or other who has worked with Rutherford; which I have not had the good fortune to have done',[2] there is little evidence that physicists from either Rutherford's Cavendish or from his pre-war period at Manchester dominated appointments in Britain in the 1920s. The Cavendish physicists who were appointed to senior positions in this period were mostly Thomson students from the years just before the Great War, members of the disappointed generation who by and large did not establish research schools, whether for lack of time, lack of resources, or lack of inclination. Further study of this tail-end of the Thomson tradition is needed, particularly in view of its institutional importance in shaping physics in the 1920s. But it is clear from what we already know that Thomson's students fostered a strong teaching ethic. So too did the growing number of Rutherford's students who were appointed to junior university positions during the later 1920s, few of whom continued the research they had begun at the Cavendish, probably for lack of the necessary radioactive resources.

Secondly, the colleges of the University of London, particularly King's College and Imperial College, played a very significant role in the overall economy of British physics. In the 1920s the London colleges were recognized for their teaching and research in areas relevant to industrial and governmental research, and a large number of physics appointments in this period went to London-educated physicists; it was only later, as we shall see, that Rutherford students came to dominate the upper reaches of British academic physics. The enormous output of spectroscopists from Imperial College and King's College requires further historical investigation, as

[2] L. F. Richardson to O.W. Richardson, 12 August 1926, Richardson papers, Archive for History of Quantum Physics, Niels Bohr Archive, Copenhagen.

does the extent and *modus operandi* of the engagement between the London colleges and military and industrial establishments. Although we have as yet no clear data on the numbers of physicists employed in the military establishments, for example, it seems that few of the military physicists below executive level were from Oxbridge and, as at the NPL, a large proportion were from the London colleges. If correct, this observation ties in nicely with the reputation of the London colleges for 'classical' physics and the predominance of acoustics and optics in the research work of the military establishments, at least up to the mid-1930s.

Thirdly, by considering the industrial, medical, civil, and military, as well as the academic sectors of British physics, we open up a wider consideration of the values and purposes of British physics than is common in the traditional historiography. The importance of spectroscopy, gas-discharge physics, X-ray crystallography, magnetism, and instrumentation in the broader economy of physics both belies the self-evident significance of Cavendish-style reductionist physics and suggests the industrial and military relevance of much of what was going on in British physics laboratories. For example, the creation of a department of textile physics at the University of Leeds in the late 1920s, with William Astbury lured from Bragg's Davy-Faraday Laboratory to develop his X-ray work on the structure of wool molecules, is in some ways typical of the matrix of university, research association, and industry links that had developed by about 1930 (Shimmin 1954; Bernal 1963). Similarly, the widespread development of radio in the 1920s sustained significant links between universities and both industrial and military establishments in ways we have only begun to explore.

HOME OF LOST CAUSES? OXFORD PHYSICS RECONSIDERED

Within this context of British physics as a whole, let us now return to developments at Oxford, where Frederick Lindemann was elected to the professorship of experimental philosophy at the Clarendon Laboratory in April 1919. Lindemann had been a student of Nernst in Berlin before the war and was socially and scientifically well connected, having been part of the secretariat of the first Solvay Congresses. His credentials as a 'modern' physicist were impeccable. Indeed, supported by references from J. J. Thomson, Rutherford, and the Americans Michelson and Millikan, Lindemann explicitly presented himself to Oxford as a broad-based and well-established authority on modern physics. He noted that '[m]y experimental work has dealt mainly with the properties of materials at low temperatures, X-ray spectra and photoelectric photometry . . . I have also published papers on the kinetic theory of solids, photoelectricity, the structure of the atom, the life of radioactive substances, ~~the thermodynamics of isotopes~~ as well as a number of astrophysical problems . . .'.[3] At Farnborough during the war he had worked on the thermodynamics of the (then still speculative) isotopes mooted by his Chudleigh mess-mate Francis Aston based on an earlier suggestion by Frederick Soddy. In keeping with the broad definition of 'modern' physics developed above, Lindemann was indeed a 'modern' physicist, and a wide-ranging one at that (Fort 2004).

Just as with Rutherford at Cambridge, there was an expectation from some of his new colleagues that Lindemann would create a flourishing centre of research at Oxford. What it would take to create such a centre was never specified – not surprisingly, perhaps, given the elaborate structure of the university, as demonstrated by the other chapters in this volume. Nevertheless, Lindemann's early plans certainly went in this direction: his financial estimates for 1920–1 indicate that he intended to provide for between 50 and 60 undergraduates and 30 research students in the Clarendon.[4] Like Bragg at Manchester, Lindemann drew on his wartime contacts to staff his laboratory, with T. C. Keeley and Gordon Dobson among the first to be engaged. With demonstrator Idwal Griffith, who returned from his wartime service at Martlesham Heath, these men used their connections to obtain war surplus equipment for the Clarendon. Lindemann also offered facilities to Thomas Merton and Alfred Egerton in the hope of quickly building up the research profile of the Clarendon.

As Morrell has well described in this volume and elsewhere, however, Lindemann's difficulties in securing funding and students in the early 1920s meant that he was unable to establish low-temperature research at the Clarendon in the manner he would have liked (Morrell 1996: 381–401). Although the temptation to compare the Clarendon with the Cavendish is great, the comparison is perhaps not appropriate, or at least requires significant qualification. Fifteen years younger than Rutherford, Lindemann had not already developed a research and training programme elsewhere that could simply be transplanted and reconstituted after the war. Nor did the Clarendon have the extensive imperial connections that Rutherford inherited from his predecessor at the Cavendish, while Lindemann's scientific connections with Germany could hardly be brought extensively into play in the years after the war to support his work at Oxford, either with students or with other resources. Even so, a steady flow of three or four research students a year into the Clarendon began to produce research in atmospheric electricity under Dobson and spectroscopy under Merton (at least until they both moved out to work in their respective country houses in 1924), in thermodynamics under Egerton (who

---

[3] Lindemann, undated draft MS letter of application for the Oxford chair, LP, A20 f2. This file also contains his testimonials. The deletion of the reference to isotopes perhaps indicates that Lindemann regarded this recent development as too speculative for inclusion in his application.   [4] LP, B8 f1.

attracted research students from outside Oxford), and on a variety of topics under Lindemann.

The arrival of Derek Jackson in 1927 re-established spectroscopy at the Clarendon, demonstrating nicely that not every research student aspired to work at the Cavendish. Having graduated in the Natural Sciences Tripos at Trinity College, Cambridge in 1927, Jackson turned down the opportunity to undertake research in radioactivity with Rutherford, and moved instead at Lindemann's invitation to Oxford, where his work on hyperfine structure in spectroscopy clearly owed much to his undergraduate tutor, Herbert Skinner, as well as to his moneyed background, which allowed him to provide his own apparatus (Kuhn and Hartley 1983). Of course, Jackson may also have found the Clarendon socially and politically a more congenial environment than the Cavendish. Lindemann's well-known involvement in élite social, industrial, and political circles – he was instrumental in the production of Churchill's anti-General Strike propaganda sheet *The British Gazette* in 1926, and later became the politician's scientific adviser – was also reflected in his wider set of values about science and its role in national life (Fort 2004: 108; Wilson 1995). His work with Bolton King on photoelectric cells, for example, led to the establishment of a profitable business run from premises in Keble Road, near the Clarendon Laboratory. This was more than a sideline, for the firm had significant links with the electrical industry (particularly the cinematographic and sound-recording industry) and with the military (especially the Admiralty, where photocells could be used to control torpedoes and other devices). Usually dismissed as peripheral, and perhaps even obstructive, to the history of Oxford physics, Lindemann's business interests and connections deserve to be taken more seriously. His concern for the detection of aircraft using long-wave radiation went back to his Farnborough days, and he maintained the interest through the 1920s, not least via his involvement with military research and industrial policy advisory circles.

In considering the Clarendon, of course, we must also take into account the unusually complicated Oxford scientific landscape, as other authors in this volume have insisted. The Clarendon's supposed 'failure' to develop research in 'modern' atomic physics must be considered against the presence in Oxford of Frederick Soddy, former collaborator of Rutherford and promoter of radioactivity, who was appointed professor of inorganic chemistry in 1919. Widely expected to develop a school of radioactivity research at Oxford, and bolstered by the award of the 1921 Nobel Prize for chemistry for his work on isotopes (he had been proposed by Rutherford, whose reductionist programme his work helped support), Soddy was stymied by a lack both of radioactive material and of support from the university and by political battles with his colleagues (Cruickshank 1986; Morrell 1996: 321–4, 353–7; Merricks 1996). Aptly characterized by Morrell as an 'irritatingly fertile' department-builder, rather than a discipline-builder, Soddy put much of his energy into the social and economic relations of science and the problem of monetary reform. Later he would rail against the Cavendish-dominated history of British physics, which consigned him to the margins of history – a situation not helped by his disastrous 1935 campaign to reform the Royal Society (Merricks 1996: 102–3, 178–84).

Historiographically, Soddy has much in common with John Townsend, who is also often portrayed as having gone into embittered terminal decline in the 1920s. Like Clifton and Soddy, however, Townsend has had an unfairly bad press. Publications continued to flow from Townsend's laboratory in this period, and Townsend's own study of *Motion of Electrons in Gases* (1925) was based on a prestigious lecture at the Franklin Institute, Philadelphia. If Townsend took no further part in international conferences after the 1924 Solvay Congress, he nevertheless continued to engage in scientific collaboration (in 1924–5 Townsend and E. W. B. Gill participated in Appleton and Barnett's radio experiments) and in exchanges in print with other workers over the interpretation of the scattering of slow electrons in gases (Hon 1989; Im 1995). Townsend held what were presumably quite lucrative patents and international licences on wavemeters, again illustrating the business nous of Oxford physicists. And if he was 'antagonistic to new theories' in physics, he was far from alone in being so. In many ways, in fact, he was typical of those who had been trained in physics in the 1890s. Rutherford and Lindemann, like many others in the 1920s, notoriously deprecated the abstractions of wave mechanics (Badash 1987). Hence we should be careful not to conflate the Townsend of the 1920s with the caricature of a drooling septuagenarian that had become current by the late 1930s.

In his development of Thomson-style gas-discharge physics, Townsend maintained a larger establishment of demonstrators than Lindemann, and saw a 'steady (though small) influx of research students into his department' (Morrell 1996: 386; Jones 1957: 109). Like the Clarendon, the Electrical Laboratory was host to a succession of Rhodes Scholars. Roughly evenly divided between these two laboratories, they were never as numerous or as academically successful as the 1851 Exhibitioners at Cambridge, and fewer of them went on to scientific careers, preferring more remunerative employment in business or administration (MacLeod and Andrews 1968; Evans 1996). One Rhodes Scholar who did stay in science was Leonard Huxley, who undertook his D.Phil. research under Townsend on electrical breakdown in gases. He was recruited to the Australian Radio Research Board in 1930 before returning to a lectureship in physics at Nottingham and, soon thereafter, the chair of physics at Leicester (Crompton 1995). Another was Robert Jemison van de Graaff, successively a Rhodes Scholar and an International Education Board Fellow at the Electrical Laboratory before becoming a US National Research Fellow at Princeton and then an assistant professor at MIT,

Fig. 8.3. Robert Jemison van de Graaff, who worked in the Electrical Laboratory as a Rhodes Scholar at The Queen's College from 1925 to 1928. After taking his D.Phil., van de Graaff returned to the USA, where he developed his electrostatic generator and, from 1934, taught at MIT.

where he built his electrostatic particle accelerators and his career (Heilbron and Seidel 1989: 60–6).

If Lindemann was unable to secure the role and status he would have liked for Oxford physics in the 1920s, he nevertheless substantially raised the Clarendon from the low base from which he started in 1919. Yet he also played on a larger stage, being closely involved for example with the promotion of solid-state physics at Bristol in the late 1920s. Arthur Tyndall had been appointed to the physics chair there in 1919, and through a friendship with a member of a local family of tobacco magnates and university benefactors, had secured a gift of £100 000 for a new laboratory. Opened by Rutherford in 1927, the Henry Herbert Wills Physics Laboratory was palatial and probably the best equipped in the country. Following the appointment of John Edward Lennard-Jones as reader in mathematical physics in 1925 and professor of theoretical physics in 1927, support for Lennard-Jones's work came from the DSIR via a route that sheds interesting light on the national politics of physics and particularly the role of Lindemann. In December 1929, Lindemann, a member of both the Scientific Grants Committee and the Industrial Grants Committee of the Advisory Council for Scientific and Industrial Research, suggested to the latter that it should support theoretical research on the relationship between the properties of metals and alloys and their atomic constitution as an alternative to experimental metallurgical work at the NPL. When Lennard-Jones moved to a chair in theoretical chemistry at Cambridge in 1932, his replacement, Nevill Mott, took up the problem of electrons in metals with alacrity, laying the foundations for the future development of solid-state research at Bristol (Keith and Hoch 1986; Pippard 1998; Davis 1998). By the late 1930s, Bristol had established an excellent national and international reputation for its work on the electron theory of metals and alloys and research on electronic processes in ionic crystals, work closely connected with industry.

Keith and Hoch (1986) have correctly given Lindemann credit for his contributions to the promotion of research in the national interest, although they also note that Lennard-Jones's work supported Lindemann's controversial electron lattice model of conductivity in solids. In this respect, we should also remember Lindemann's active involvement in meetings of the Faraday Society, which sponsored discussions on diverse topics in physical chemistry and chemical physics, many of industrial relevance (Sutton and Davies 1996). Either way, the contrast between Lindemann's advocacy of DSIR support for solid-state research at Bristol and Rutherford's marshalling of the Royal Society Mond bequest for the new 'superlab' for Kapitza at Cambridge in 1930 is most instructive (Badash 1985: 15–16). Given our earlier observations about the tension between pedagogy and research, Keith also tellingly points out that Bristol's success in research was to some extent achieved at the expense of its undergraduate teaching. Even so, this seems to have been a price Tyndall thought worth paying (Keith 1984).

Keith notes too the diversity of research carried out at Bristol. Indeed, when Tyndall and Lennard-Jones assiduously sought funding from the Rockefeller Foundation's International Education Board and the Wills family, they stressed their policy of not concentrating research in one area but of collecting a dynamic group of researchers in diverse areas and promoting postgraduate research in the hope of establishing research groups 'each large enough to make an effective attack on its own field' (Keith 1984: 338 n.83), a strategy not dissimilar to Lindemann's at Oxford. After much negotiation, they were successful in 1930 in securing a large award of £50 000 from the Rockefeller Foundation, whose officials supported plans for development at Bristol – 'perhaps the finest physics laboratory in England' – at least partly as a 'counter-irritant' to Oxford and Cambridge (Keith 1984: 352 n.90; Kohler 1991: 254–5), plus £25 000 from the Wills family, bringing financial stability and much scope for fellowships, conferences, and visitors at Bristol. The effect on Bristol, and on British, physics was remarkable, and there began a migration of doctoral and postdoctoral researchers from Cambridge and elsewhere to Bristol, where the number of research staff rose from 13 in 1928 to 22 in 1930 and over 30 by 1939. Swelling the numbers of those working there from 1933 were a number of refugees from Nazi Germany.

REDUCTIONISM, REFUGEES, AND THE COMING OF LARGE-SCALE PHYSICS, 1930–1940

As we have seen, British physics was changing in significant ways, not least in the early 1930s in its response to the Depression. Certainly, demand for research physicists far exceeded supply until the onset of economic crisis in 1931 (Thomson 1931: 26). In much the same way as the DSIR sought to promote microbiological research, the relevance of physical research to industry increasingly became a pressing question and the basis for public and professional judgements as to the significance of a laboratory's work (Vernon 1994). While the Cavendish remained the largest and, in lay eyes, the most successful laboratory, it was also widely attacked for doing irrelevant research and for training men ill-equipped to take up employment in industrial and governmental laboratories. Despite Rutherford's personal power within the upper echelons of British science – he was President of the Royal Society from 1925 to 1930 and was ennobled in 1931 (Cameron 1983; Wilson 1983: 453–95) – it is illuminating that in 1932 Oxford metallurgist William Hume-Rothery and Manchester X-ray powder crystallography maestro Albert Bradley were elected Warren research fellows of the Royal Society from a list of applicants that included Cambridge's finest: Bernal, Blackett, Chadwick, Cockcroft, and Ellis. This no doubt reflected in part the same industrial concerns that had prompted Lindemann to support the development of solid-state physics at Bristol (Morrell 1996: 360).

In this context, the influx of refugee Jewish scientists from Germany from 1933 gives an interesting perspective on the dynamics of British physics as a whole, and of the role of Oxford within that larger picture. While it is difficult to establish exact comparative data because of the transience of many of the refugees, it is clear that Bristol, Manchester, and Oxford were the leading reception centres (Hoch 1983). Many refugees went to Bristol on temporary or longer-term appointments (among them Bethe, Fuchs, Frank, and Fröhlich), not least because several already had connections there, often through Max Born's group at Göttingen or through earlier visiting fellowships sponsored under Tyndall's programme of expansion (Keith and Hoch 1986). With the appointment of Nevill Mott as professor of theoretical physics in 1933, an expanding cast of talented researchers and visitors and several large conferences in the mid- and late-1930s, Bristol consolidated its position as an international centre for both metal physics and cosmic-ray work. Refugee physicists (Italian, Spanish, Polish, and later French, as well as the first wave of Germans) also found other temporary homes in ways that have not yet been systematically explored. Several passed through Manchester en route to other posts, and many individuals found themselves at the smaller universities such as

Southampton, Newcastle, Sheffield, and St Andrews, where they were often able to contribute effectively to departmental activity. The apparent relative lack of response by the London colleges is counterbalanced to some extent by the role of industry and government establishments in accommodating the refugees: the BBC, General Electric, EMI, Kodak, Kew Observatory, the Shirley Institute in Manchester, and other institutions all took some of them in, as did several schools and technical colleges. Much work remains to be done on the differential reception of the new arrivals in the various centres.

At Oxford, as Morrell has shown, the reception of the Jewish refugees is particularly important because Lindemann's success in securing Simon, Kurti, and Mendelssohn and his promotion of low-temperature physics have been widely used to 'rehabilitate' him and the Clarendon. Yet we should be careful not to fall into the same trap as those who retrospectively celebrate the work of the Cavendish. First, Lindemann's efforts after 1930 to introduce machine-style low-temperature physics to the Clarendon with the support of the Rockefeller Foundation must surely be understood in the wider context of developments at Bristol and elsewhere, where Rockefeller largesse was successfully exploited. Similarly, Lindemann's promotion of ICI support for refugee physicists must be seen as a part of his normal (and influential) activities as a member of ICI's Research Council, and it is arguably ICI that should be credited as much as Lindemann himself with building up low-temperature research at Oxford (Rider 1984: 146–50). Furthermore, it may be that in puffing the Clarendon's trumping of Kapitza's Mond Laboratory in the first liquefaction of helium in England, Lindemann started playing the same media game of self-promotion as the Cavendish (Hughes 2000). The Clarendon was not substantially more welcoming to refugee physicists in 1933 than, say, Manchester or Bristol, and it cannot have been clear in the mid-1930s that the Breslau group would remain permanently in Oxford and develop a significant research school there, rather than moving on to the United States or another British university. Simon, for example, applied for the chair of physics at Birmingham in 1936. And, as Morrell points out, the detention of Peter Kapitza in the USSR in 1934 essentially left the field of low-temperature physics to Oxford (Morrell 1996: 404–6).

While Cambridge appears to have done least to support the refugees, developments there were changing the public and institutional face of physics in this period. Experimental nuclear research at the Cavendish, long in the doldrums, revived in 1932 after Chadwick's discovery of the neutron and the 'splitting of the atom' by John Cockcroft and Ernest Walton using a particle accelerator of their own design. Much of the publicity for Cavendish nuclear physics was orchestrated by James Crowther, science correspondent of the *Manchester Guardian*, who told his editor in May 1932 that he was effectively becoming the 'press agent' for the

Cavendish. As nuclear physics underwent rapid development, the scale and cost of accelerators increased rapidly. The Cavendish Laboratory Appeal of 1935 retrospectively constructed 1932 as the '*annus mirabilis* of nuclear physics' in its quest for funds to allow Cambridge to keep up with developments elsewhere, and prompted the Austin bequest of £250 000 in 1936 that at least temporarily sustained Cambridge's place in the international network of nuclear physics (Hughes 2000).

In the wake of the publicity for the achievements of the Cavendish in 1932 and after, nuclear physics, particularly work with neutrons, was taken up at Imperial College, Oxford, Bristol, and elsewhere in the mid-1930s, facilitated in part by the wide distribution of the electronic technique necessary for work on particle detectors (Hughes 1998*b*). This was also the context in which, from the mid-1930s, Rutherford's men increasingly dominated senior university appointments in physics. Following the appointment of George Thomson to the chair of physics at Imperial College in 1930, Aberdeen chose as his successor Arthur Eddington's student John Carroll rather than Patrick Blackett (to Rutherford's fury). In 1933, however, Blackett was appointed at Birkbeck, followed by the appointments of Chadwick at Liverpool in 1935, C. D. Ellis at King's London and Oliphant at Birmingham in 1936, and Blackett again at Manchester in 1937. Members of the new Cavendish-educated network of the 1930s took with them not just the technologies and techniques of nuclear physics but managerial know-how. They would develop in their own laboratories the machines and research programmes that Rutherford only half-heartedly supported at the Cavendish.

In 1935, Tyndall at Bristol could note 'the doubt whether in the future any Laboratory would acquire full international prominence unless some branch of nuclear physics was a subject of experimental investigation within it' (Frank and Perkins 1971: 545). At Bristol, therefore, nuclear physics was developed by Cecil Powell and others, both in the form of a Cockcroft–Walton accelerator and through photographic plate studies of cosmic rays (Frank and Perkins 1971; Galison 1997: 160–78). At Imperial College, Thomson began to promote the development of neutron physics with the appointment of Philip Moon and electronics wizard Eryl Wynn-Williams from the Cavendish Laboratory (the latter quite a catch, as several electrical firms would have given much to have him on their research staff) (Moon 1977; Burcham and Isaak 1996; Ward 1980). Cyclotron-building began at Liverpool and Birmingham, the latter aided by a £60 000 gift in 1938 from Lord Nuffield, Austin's motor-manufacturer rival, and extensive assistance from Ernest Lawrence at Berkeley (Cockburn and Ellyard 1981: 74). The enormous sums of money required for accelerators were often secured by stressing the medical applications of this new technology, continuing a link that had existed from the earliest days of radioactivity. At Oxford, Lindemann accommodated the Hungarian émigré Leo Szilard and his research on neutrons, fresh from

St Bartholomew's Hospital (Lanouette 1992: 149ff). Indeed, he framed his plans for a new Clarendon Laboratory around a Cockcroft–Walton accelerator and a cyclotron, and, like Cockcroft in Cambridge and Oliphant in Birmingham, learned from developments at European high-voltage laboratories, particularly that of Philips at Eindhoven. By 1939, the laboratory was completed, though little money had yet been found to equip it (Morrell 1996: 428–32).

Despite the large sums of money involved, one should be careful not to overstress the shift to accelerator physics and atom-smashing, as contemporaries quickly came to call it (Cathcart 2004); in many places – the Welsh, Scottish, and most of the provincial English universities for example – this style of physics made little or no impact. Other new specialisms such as X-ray crystallography became important, as the co-option of the subject by the Cavendish in the early 1930s indicates; here again divergent disciplinary demarcations are crucial, and we should remember the significant work in this field by Dorothy Hodgkin in Oxford's *chemistry* department in the 1930s (Ferry 1998). And there were counter-examples to the dominance of nuclear physics in major new appointments; for example, Leslie Bates, appointed to the chair at Nottingham in 1936 with Rutherford's support, was a specialist in magnetism (Kurti 1983). Appleton returned to Cambridge in the same year to succeed C. T. R. Wilson as Jacksonian professor, and there he continued his radio research alongside Ratcliffe. Appleton was in turn replaced at King's College by Charles Ellis of the Cavendish, who seems to have had little opportunity to start a new line of research in nuclear physics or anything else (Hutchinson, Gray, and Massey 1981: 214). By and large, however, X-ray studies and spectroscopy remained the staple lines of research in 'modern' physics at many university departments. In several places, and not only those associated with cyclotrons, medical links were important through X-ray or other high-energy machines assigned for medical purposes. Elsewhere, smaller-scale work also drew on medical funding. A good example is Francis Edgar Jones's work at King's College, London, on the absorption of centimetric electromagnetic waves in polar liquids, supported by a grant from the British Empire Cancer Campaign (Macfarlane and Hilsum 1990).

While details of many of the links between physics and medicine in inter-war Britain are far from clear, it is evident that the range of physics research pursued in British universities in the 1930s was enormous. In the shadow of the Depression, the Institute of Physics naturally continued to stress the importance of physics to industry. A well-publicized conference on 'Industrial Physics' at Manchester University in March 1935 brought together nearly 600 academics and industrialists for discussions centred mostly on laboratory vacuum devices (Lang 1935). A striking example of the use of physics in industry was the development of television by EMI at their research laboratory in Hayes, where research managers Isaac Shoenberg and

George Condliffe recruited a team (the buzz-word of much mid-1930s research) of researchers from Bristol, Oxford, and Cambridge (Burns 1986: 87, 291–5; Alexander 1999: 125 ff.; Morgan 1988). But if Cavendish-trained Ph.D.s were disproportionately represented in the team, it is worth noting that they came largely from Ratcliffe's radio side of the laboratory or electron-physics research, not the nuclear physics side – again emphasising the importance of *technique* in industrial research (Lovell 1964; Budden 1988). In 1938 Lindemann told his colleagues that Oxford graduates were of 'more use to industry than Cambridge because more attention was paid here to old-fashioned physics', and that the 'practical problems of industry needed more knowledge of that than modern nuclear physics'.[5] Lest we dismiss this as rhetoric or university politics, there is other evidence to suggest that Lindemann's views were more widely held. In 1937 an influential review of employment trends in science pointed to 'an unsatisfied demand for what may be termed the old-fashioned physicist, the type of worker who can be relied upon to devise satisfactory methods of measuring such physical properties of different materials as viscosity, . . . permeability to water, wearing capacity, and so on'. The reason for this shortage was 'the outstanding success and wide influence of the Cambridge school of atomic physics'. Yet, the author added, 'the "modern" physicist can easily divert his course from the atom to wireless and television, to problems of the textile trade or of the photographic industry, and because of that he is also in growing demand' (Haslett 1937: 355).

Even as these words were written, however, the situation was changing. By the mid-1930s, Cambridge might still have been the nuclear nirvana for some but it was no longer the 'obvious' place for an ambitious physics postgraduate *even in reductionist physics*, as it had been in 1900. Bristol-educated Bernard Lovell coveted an appointment in Patrick Blackett's department at Birkbeck but had to accept a post in Bragg's department at Manchester; he was delighted to find that Blackett was elected to succeed Bragg after the latter's departure to the NPL in 1937. Blackett imported his cosmic ray work (and much of his research group) from Birkbeck, and encouraged others like Lovell and George Rochester to join it. He also supported the development by Douglas Hartree (who moved from the Beyer chair of applied mathematics to a chair of theoretical physics in 1937) of the differential analyser that built on the department's links with Metropolitan-Vickers and tied it into an extensive network of industrial and military users (Saward 1984: 20–41; Lovell 1987: 156–8; Darwin 1958). At Birmingham, Oliphant attracted John Randall from GEC with a Royal Society Warren research fellowship, while at Liverpool Chadwick recruited machine-builders such as Bernard Kinsey and Gerald Pickavance to push forward the cyclotron project. The institutional and technical diversification of

---

[5] 'Report of the Advisory Committee' [on the new Clarendon Laboratory], typescript dated 19 May 1938, LP, B30.

nuclear and cosmic ray physics made the subject significantly more accessible to more researchers than it had been when concentrated at Cambridge, and it vastly increased the opportunities for research in the field.

Rutherford's death in 1937 sent shock-waves through the scientific world, although the Cavendish nuclear physicists recovered quickly enough to clamour for a favourable succession at Cambridge. From India C.V. Raman privately expressed an interest in the position to Owen Richardson, one of the electors, but Chadwick was the obvious candidate in the eyes of the nuclear physicists. The appointment of Lawrence Bragg in 1938 was therefore highly significant. *Nature* noted that the Cavendish was now 'so large that no one man can control it all closely', and added that 'Bragg's tact and gift of leadership form the best possible assurance of the happy co-operation of its many groups of research workers' (Anon. 1938). With accelerator physics, cosmic ray and gas discharge research, crystallography, ionospheric work under Ratcliffe and Appleton, and the Mond Laboratory, now under John Allen, all going strong, in addition to the various other individual researchers such as Aston, Taylor, and Wilson, one can see *Nature*'s point. In this sense the election of Bragg was not necessarily a result of the difficulty in adjudicating between the claims of the nuclear physicists, as some have suggested. Rather, it reflected a broader consensus as to the relative position of nuclear physics in the context of the discipline as a whole and of the wider social and economic imperatives of the late 1930s.

By this time, anyway, the threat of war was changing the relations between universities, industry, and the military establishments, where work on the problems of air defence continued. At ADEE Biggin Hill, Tucker continued to oversee work on sound detection techniques, while the SEE at Woolwich developed a wireless communication system for tanks (Hartcup 1970: 159–87; Burns 1988; Scarth 1999). In the Admiralty's research departments, extensive work was undertaken on magnetic mines and countermeasures against them, and on the infra-red detection of ships and aircraft. Growing recognition of the air defence issue led to the establishment of the Air Ministry's Committee for the Scientific Survey of Air Defence (under Tizard) and the Committee for Imperial Defence's Air Defence Research Committee (under Swinton) in 1935–6, with Blackett and Lindemann as key players. As is well known, Lindemann was close to Churchill on these questions and he promoted the infra-red detection of aircraft and the use of aerial mines, involving Reginald Jones and George Pickard in infra-red work at the Clarendon and later at the Admiralty Research Laboratory (Jones 1961; Jones 1979: 38–76). Following the demonstration of the feasibility of radiolocation of aircraft (Radio Direction Finding, or RDF) by Watson-Watt and Radio Research Station staff early in 1935, the Clarendon's long-standing military links were further strengthened in 1936 when Gerald Touch was among the first to be recruited by Watson-Watt for the

Bawdsey Research Establishment, where construction of the Chain Home RDF system was masterminded (Burns 1988; Latham and Stobbs 1996, 1999; Zimmerman 2001; Bragg, M. 2002).

As radar took shape in the spring and summer of 1939, physicists from Cambridge, Oxford, Manchester, and Birmingham universities and from numerous industrial research establishments were mobilized and inducted into the secrets of Chain Home. Although many of those who led the way to the coastal CH stations were Cavendish research students, the reader will by now not be surprised to learn that they were those whose primary interests lay in amplifier technique and electronics. Early in 1939, the Clarendon and the physics laboratories at Liverpool and Bristol were better placed than the Cavendish to explore the phenomenon of nuclear fission, and when Tube Alloys (atomic bomb) work was shared out among the universities, it was Simon and the Oxford low-temperature group who made significant contributions to isotope separation, while the depleted Cavendish housed only a small group working on the characteristics of chain reactions. When Appleton resigned the Jacksonian professorship in 1939 to become Secretary of the DSIR in succession to Heath, John Cockcroft was appointed to the Jacksonian chair. This must have been scant comfort to the nuclear physicists, though. For on 1 September 1939 Cockcroft was at the Rye Chain Home station '100 feet up with nice views and oast houses lending variety' when he heard that Germany had invaded Poland (Hartcup and Allibone 1984: 89).

CONCLUSION

Asking the rhetorical question 'What is a physicist?' in 1941, *Nature* observed that 'recently the literature on popular science has helped to spread the idea that a physicist is essentially a splitter of atoms', so that 'fears are expressed that the services of physicists are liable to be neglected because service and business officers are still under the impression that physicists "being only interested in atoms" cannot be expected to be of much assistance' (Anon. 1941a: 617). As I have demonstrated, this emphasis on reductionist physics was a contingent phenomenon of the later 1930s. Combined with a post-war historiography that sought to naturalize and legitimize nuclear weapons in the workings of nature, such an emphasis has significantly shaped the historiography of British physics. Any revisionist understanding of the history of the Clarendon Laboratory and Oxford physics therefore requires a revisionist understanding of the history of British physics more generally. This chapter has attempted a very preliminary broad-brush re-evaluation of the wider history of the subject with a view to repositioning Oxford physics in the wider disciplinary context. I have tried to indicate some of the many limitations and gaps in our current understanding and have raised several questions that I have had to leave unanswered. Nevertheless, I think several significant points emerge from the analysis presented here.

It is clear that early-twentieth-century British physics was large and diverse – institutionally, intellectually, and materially. I hope I have shown by the range of examples that the bulk of physics was not reductionist atomic and nuclear physics, nor in most places did it even aspire to be: rather, the traditional fields of optics, heat, sound, properties of materials, measurement, and instrumentation continued to play an important role throughout the period under consideration, particularly when laboratories in the industrial, military, and civil sectors are taken into account. The importance of teaching and the difficulties of prosecuting research in teaching-heavy departments (especially where large classes of medical students or other service teaching had to be catered for) are also important: research was not a universal imperative, and research in 'modern' physics still less so. Through their bread-and-butter teaching, many more departments sought to produce physicists capable of using their knowledge and skills in the world of work and to develop links with local industrial concerns than sought to develop 'pure' reductionist physics. Here, the importance of the London colleges and several of the provincial universities in fostering non-reductionist physics through to the 1930s offers at once a significant corrective to Cambridge-centred views of the subject and significant scope for expanding the context within which Oxford physics should be understood.

I have also shown that even if we focus on the development of research in 'modern' physics in Britain after 1900, the Cavendish did not necessarily dominate, and that the term itself was given a much wider meaning by contemporaries than a later, nuclear-dominated historiography would suggest. Gas-discharge physics, thermionics and electron physics, X-rays, and even spectroscopy were all important elements of the new physics, and all were widely distributed. Much physics research was concerned with industrial problems or instrumentation and existed on the borders of chemistry, engineering, and what would later become metallurgy. While this question of disciplinary boundaries demands further work, it is also clear in my period that much of what would later come to be recognized as 'physics' research was being done in other departments. When these factors are taken into account, the presence of Townsend and his students at the Electrical Laboratory, the location of Soddy and William Hume-Rothery in chemistry, and Lindemann's promotion of diverse research interests (not least in connection with his own political and commercial concerns) at the Clarendon in the 1920s help us to locate Oxford as much more in the vanguard of 'modern' physics than is usually appreciated.

We have seen that the inter-war Cavendish (and perhaps even the Cavendish pre-1914) was far from the homogeneous institution it is usually assumed to have been; wireless research (and later, too, crystallography) was at least as important as radioactivity and nuclear research there in the 1920s and 1930s. In terms of financial outlay, Kapitza's low-temperature research was more expensive than the nuclear programme at the Cavendish in the 1920s. In this sense, Lindemann's promotion of low-temperature physics at Oxford through the German refugees propelled Oxford to the forefront of a research field that encapsulated the ways in which British physics as a whole was fundamentally shaped by institutional imperatives coming from government and industry. In the inter-war years, leading Oxford scientists were active in shaping and directing those imperatives to the benefit not just of Oxford but of the discipline more broadly. In this wider context, then, Oxford physics begins to look more closely attuned to national priorities than the reductionist programme at the Cavendish. Indeed, in this sense, one wonders not why Lindemann failed to develop a large research school in nuclear physics, but why the Clarendon had so little involvement in *wireless* research, as the science that bridged academic, military, and governmental concerns. Further research is needed here, but it could be argued that the Cavendish Laboratory dominated British physics in the 1930s not because of its emphasis on nuclear physics, but *despite it*. Perhaps, in fact, the real reasons for the continuing pre-eminence of the Cavendish lay in the breadth and diversity of research carried out there, its managers' control over sources of funding, its efficiency in attracting and turning out research students within a self-perpetuating imperial network, and its pro-active publicity machine and media allies.

If we are properly to understand the Clarendon Laboratory, especially in the face of a continuing heroic historiography of the Cavendish (Cathcart 2004), much work remains to be done on the history of the whole of early-twentieth-century British physics, at both local and national levels. We need careful, contextual studies of individual appointments, career choices, the mobility of physicists between university and industrial, governmental, military, and medical laboratories, the tensions between teaching, administration, and research in different institutions, and the porosity and fluidity of boundaries between physics, chemistry, and electrical engineering, and between experimental and theoretical traditions (which I have barely touched in this chapter). While the chapters in this volume contribute substantially to a sensitive local contextual analysis of the Clarendon Laboratory and mark a rejection of the apologetic accounts that have sometimes characterized histories of non-Cantabrigian physics, the task now is to set it in its proper wider institutional and disciplinary context. This chapter is just a beginning.

9

Epilogue

Robert Fox and Graeme Gooday

This volume's emphasis on the dispersed physical locations and diverse disciplinary settings for physics has allowed its authors to present the history of the discipline in Oxford in a distinctive new light. Such a perspective takes us beyond the work of the professors of the subject to embrace that of college fellows (including some who regarded themselves primarily as chemists or mathematicians), independent men of means, and laboratory demonstrators and assistants; in doing so, it opens the door on areas of activity in physics that have hitherto passed largely unnoticed. When this broader view is taken, it becomes apparent that physics in Oxford in our period could boast substantial achievements and that it had, at different times, real sources of vigour. Certainly, by the later 1930s the field was progressing briskly and adjusting, as best its practitioners could manage, to the ways in which the physical sciences everywhere had come to be pursued since the First World War. In Oxford, as in other universities, new departures such as low-temperature physics and atomic physics were calling for more costly apparatus and larger buildings, and links with industry were transforming the nature and pace of work at the interface between academic science and the industrial economy.

Oxford's openness to these trends was demonstrated in 1937 by Lindemann's success in persuading the University of the need for a new Clarendon Laboratory that would replace the by now ill-adapted building of 1870 and provide facilities on a scale well beyond the means of individual colleges. By 1937, only three of the college laboratories still survived, all of them devoted primarily to chemistry and each closed within a decade (Christ Church and Balliol–Trinity in 1941, Jesus in 1947). At a time of financial stringency and competing demands from other disciplines, Lindemann's case had to be carefully argued, the more so as undergraduate numbers in physics, almost static since the mid-1920s, would have undermined any justification founded on the needs of teaching and aroused the suspicions of the more conservative college-based dons. Lindemann, however, was a forceful,

We are grateful to Dr Dennis Shaw for his careful reading and criticism of this chapter.

Fig. 9.1. The south front of the Clarendon Laboratory towards the end of its life, from a rare early 35-mm colour slide by T. C. Keeley. The first meeting of the University's Physical Society to be held in the new Clarendon Laboratory, on 5 December 1939, heard a talk by Keeley on 'Colour photography'. It is possible that Keeley took the photograph for that occasion.

well-connected negotiator, and by the autumn of 1939 he had seen his new laboratory through to completion, at a cost of £77 000. It was a mark of the status that physics had come to occupy in the Oxford intellectual hierarchy that, in an act of conspicuous and significant generosity, the funds were provided entirely by the University, following an intricate loan transaction with the Ministry of Agriculture and Fisheries (Morrell 1992: 303–7; Morrell 1997: 428–32).

To outward appearances, therefore, physics in Oxford entered the years of war in a healthy state, and Lindemann displayed a justified pride in the progress that had been made in his two decades in the chair. Despite his own detachment from active research since 1924, he could draw a persuasive contrast between the laboratory he led and the lacklustre institution he had inherited in 1919. Although this contrast has done something of a disservice to historical understanding by contributing to the unflattering judgements of Clifton's occupancy of the chair that have informed (unduly, in our view) the secondary literature, it served its immediate polemical purpose in helping to secure the University's support for the new building. Yet Lindemann's public satisfaction did not blind him to the limitations of what he had achieved. Most frustrating of all were the chronic monetary constraints that led the University systematically to favour capital expenditure over recurrent costs (Morrell 1992: 306–7; Morrell 1994: 431–2). Hence, while Lindemann now occupied a building that bore comparison with modern laboratories anywhere, he lacked the costly equipment – in particular, a cyclotron and a large magnet on the lines of the great Bellevue magnet in Paris – that he saw as indispensable if Oxford was to compete with the world's leading centres for physics, not least with Cambridge. No less importantly, he commanded too few of the permanent university demonstratorships (just two, compared with Townsend's four) on which effective long-term laboratory planning depended. The *émigré* physicists who had settled in Oxford had certainly helped to raise the research profile of the Clarendon; in doing so, they had also contributed to the eclipse of the Electrical Laboratory, where by the 1930s Townsend, more entrenched than ever in his opposition to much of modern physics, had set his face against expansion. Yet theirs was an unstable presence, and the precariousness of their short-term contracts (like those of most of the 23 researchers in the Clarendon on the eve of the war) remained, for Lindemann, a nagging source of anxiety.

The role of chance at crucial points in the history of Oxford physics has been alluded to in several chapters. It is almost inconceivable, for example, that Clifton would have secured the funds for his laboratory in the late 1860s had it not been for the protracted nature of the discussions between the Clarendon Trustees and the University; this meant that, for all their efforts to find a worthy objective for their accumulated funds, the Trustees still had the right amount of unallocated money to disburse just when Clifton needed it. Almost three-quarters of a century later, chance played its part again in helping to fashion an unanticipated role for the new Clarendon at a time when the calls of military service risked draining the premises of staff and students. The fact that the new laboratory became ready for occupation within weeks of the beginning of hostilities proved to be an extraordinarily favourable coincidence. Once talk of requisitioning the building as a military hospital had passed, the way was open for the empty space to be allocated for war-work. Already in the summer of 1939 ten or so of those working in the laboratory who had experience of electronics had been initiated in the early development of radar at Pevensey on the Sussex coast, and by January 1940 they were engaged back in the Clarendon on research on microwave components intended for use in operational radar. Reflex klystrons for wavelengths as little as 1.25 cm were developed, as were a range of improvements in receiver sensitivity and, most notable of all, A. H. Cooke's transmit-receive tube, which became a standard piece of radar equipment (Keeley 1945–6; Collie 1946–7; Bleaney 1988: 285; Sanders 1997: 13–14). The other focus of war-related research was the study of the gaseous diffusion method of separating the isotopes of uranium, ^{235}U and ^{238}U. Through this work, led from

Fig. 9.2. Sir Francis Simon in his Oxford days, c.1950. By this time, Simon was a prominent figure in Oxford science, as reader in thermodynamics since 1937 and a Student of Christ Church since 1945. He occupied the Dr Lee's chair of experimental philosophy for less than a month before his death in October 1956.

late 1940 by Francis Simon and his fellow-refugees in the low-temperature group, the Clarendon made a significant contribution to the Tube Alloys project, the cover under which Oxford (even after the main centre of activity moved to the USA and the nascent Manhattan project in 1943) and other British universities played a central role in the early development of the atomic bomb (Simon 1945–6; Gowing 1964: esp. Chapters 2–5 and 8–9; Attwood 2004).

War inevitably put strains on life in the Clarendon and Electrical Laboratories. One all too visible source of turbulence was Townsend's refusal to participate in the scientific instruction of Royal Air Force cadets, which resulted, in 1941, in his enforced and, as Lindemann and most other observers believed, overdue resignation from the Wykeham chair (a position to which Townsend had been appointed with no compulsory retirement age). The usual rhythms of undergraduate instruction too were changed by departures on military service (including Jackson's for characteristically colourful activity as a night-pilot) and the introduction of a densely packed unclassified Honour School, completed in two rather than the customary three years. But the resolve to maintain the Clarendon as the setting for a relatively normal research-oriented academic life combined with the continued presence of a strong core of the leading pre-war researchers to preserve the gathering momentum of the 1930s. The debt that Oxford owed to T. C. Keeley for maintaining this level of continuity was immense and, in its long-term repercussions, crucial for the laboratory's post-war future. In Lindemann's absence as advisor to

Fig. 9.3. Thomas Clews Keeley. A graduate of St John's College, Cambridge, Keeley became a fellow of Wadham in 1924, where he continued to live until his death in 1988. He was always close to Lindemann, whom he had met at the Royal Aircraft Factory at Farnborough during the First World War. From the mid-1920s until ten years after his official retirement (in 1961), he was a dedicated and skilful administrator of the Clarendon Laboratory.

Churchill, Keeley displayed (in Collie's words) 'voracious industry' in running the laboratory, organizing undergraduate teaching, and acting as liaison officer with the Admiralty, services that were duly recognized by the award of the CBE in 1944 (Collie 1946–7: 56).

The contributions that the Clarendon made to the war-effort played to strengths that had developed within the Oxford tradition of what Jeff Hughes calls, in his chapter, 'non-reductionist physics'. While Cambridge had its greatest triumphs in the discovery and study of new particles and the quest for the 'ultimate' constituents of matter, Oxford's distinctiveness lay above all in invention, often with a powerful streak of mechanical ingenuity that both the radar and atomic bomb projects called for. When the war ended, the experimental skills that this tradition fostered continued to bear fruit as peace-time research resumed, notably in the application of microwave devices, developed for radar, to a new field of spectroscopy. The University's physicists could boast not only a proven record of utility over the preceding six years but also a certain distance from the more esoteric areas of the discipline in which the Cavendish had excelled. Lindemann's strong personality and continuing visibility as a public figure, within and beyond Oxford, also played its part, not least in helping to undermine lingering arts-based reticence about the right of his discipline to be counted among Oxford's leading intellectual endeavours. Anti-science prejudice could still take confrontational form, as when the Principal of Brasenose College, William Stallybrass, is said to have asserted in Council that the expansion of the sciences in the University 'had got to stop' (Roche 1994: 284). But by the late 1940s such sentiments were few and, for the most part, muted.

Although the war had helped to create a solid platform for the future development of physics in Oxford, careful management remained necessary if the potential was to be realized. One piece of administrative streamlining was achieved uncontroversially by ending the all too destructive rivalry between the Clarendon and Electrical Laboratories, which in 1945 were brought together in a single Department of Physics under Lindemann's leadership. In a climate that had been transformed by public recognition of the contributions of scientific research to the victory, other impediments that had frustrated Lindemann in the 1930s began to be addressed. The low-temperature group, far better-housed than had been possible in the old Clarendon, was one beneficiary of the post-war optimism. As its work resumed, Simon was promoted from his pre-war position of reader to a personal chair (in thermodynamics) in 1945, and permanent posts were found for Kurti and Mendelssohn; benefiting from this new stability, the Clarendon moved quickly to the cutting edge of research into high-field superconducting magnets and key areas of solid-state physics. In 1945 Dobson too received a personal chair, and

Fig. 9.4. Lindemann in later life with one of Sir Francis Simon's daughters, Dorothee.

atmospheric physics expanded as a Clarendon speciality; here, the appointment three years later of Alan Brewer, Dobson's wartime collaborator in work on the stratosphere, to a lectureship marked a significant step on the way to the establishment of Oxford as the headquarters of the International Ozone Commission (in 1948) and the creation of the semi-autonomous department of atmospheric, oceanic, and planetary physics (Dobson 1968). Continuity with pre-war strengths was similarly achieved by the return of Jackson and Kuhn to high-resolution spectroscopy, work that led on, via James Griffiths's discovery of ferromagnetic resonance (1945), to a wide range of studies of nuclear magnetic resonance and pioneering research on lasers.

In addition to consolidating existing traditions, Lindemann rectified what many had seen as an imbalance in the profile of Oxford physics by promoting an important new departure in theoretical physics. As Sedleian professors of natural philosophy, Bartholomew Price (1853–98) and A. E. H. Love (1898–1940) had worked in areas of applied mathematics that impinged on physics, as had E. A. Milne, the first holder of the Rouse Ball chair of mathematics (1928–50). Love's research on the elasticity of solids, especially as applied to the earth's crust, illustrates the point, as does the collaboration in astrophysics that Milne undertook during the 1930s with

Fig. 9.5. Ida Busbridge (*left*) and Madge Gertrude Adam. Madge Adam was the first woman to take a first in physics at Oxford (at St Hugh's in 1934). She went on to postgraduate work in astronomy, taking a D.Phil. in 1940, and worked for many years in the University Observatory. As a fellow of St Hugh's from 1941 and physics tutor there from 1941 until 1957 (when she resigned her tutorship to concentrate on research), she made an important contribution to the teaching of physics, notably during the Second World War. Ida Busbridge, a London graduate who took a D.Phil. at St Hugh's in 1938, had an equally significant role in mathematics. Like Adam, she became a fellow of St Hugh's, the first woman mathematics fellow of an Oxford college.

the Savilian professor of astronomy, Harry Plaskett (Adam 1996: 161). But, in a manner typical of the fragmented structures of Oxford physics before the war, the involvement of the holders of the Sedleian and Rouse Ball chairs in the experimental research of the Clarendon had never been close. Now, despite Lindemann's undiminished partiality for Oxford's experimental tradition and his coolness with regard to much mathematical physics (Einstein's work being, for him, a conspicuous exception), theoreticians were drawn formally within the realm of physics by the reallocation of the Wykeham chair, vacant since 1941, to theoretical physics. With Maurice Pryce as Wykeham professor from 1946 to 1955, electron paramagnetic resonance emerged as an ideal field in which experimentalists and theoreticians could and did work together and in which Oxford for some years led the world.

Nuclear physics, always in the shadow of the Cavendish, was another field ripe for post-war expansion. The work that C. H. Collie had initiated in the 1930s continued, aided by the new facility of a 'high-tension' room with a 1-million-volt Cockcroft–Walton set. For a decade from 1946 Hans von Halban, an Austrian physicist who came to Oxford with an established reputation in the field, was a leading figure, with the title of professor of nuclear physics. But significant further growth had to await the review of resources that followed Lindemann's retirement in 1956. The personal chair that Simon liberated on his appointment as Lindemann's successor in the Dr Lee's chair (a chair that he held for only a month until his sudden death in October 1956) was now assigned to nuclear physics, with Denys Wilkinson, whom Collie and Douglas Roaf had enticed from Cambridge, as its first occupant from 1957 to 1976. Using his forceful advocacy, Wilkinson soon established an ambitious nuclear research laboratory on the Keble Road site, where he installed a 28-MV tandem van de Graaff machine and a large team of doctoral students and other researchers, most of them working in close collaboration with CERN in Geneva and the Rutherford Laboratory at Harwell (Roche 1994: 278–9).

The appointments of the immediate post-war years were just one facet of far broader changes that affected all the sciences in Oxford in the new age of what Derek de Solla Price characterized long ago as Big Science, bred of the Second World War (Price 1963: esp. 1–32). With the centralization of resources in expanded laboratories, the old individualism could not survive; if any doubted this, the Tube Alloys project, which engaged about thirty full-time researchers, as well as thirty or so support-staff, offered compelling evidence that the future lay with team-work. Inexorably, the power of the leading professors grew, though not always to everyone's satisfaction, as the bitter debates of the early 1960s concerning the proposals of the recently arrived Linacre professor of zoology, John Pringle, for an intrusive multistorey zoology building on the edge of the Parks show (Roche 1994: 278–9; *The Letters of Mercurius* 1970: 49–55). After the near-stagnation of the interwar years, there was growth, too, in the number of undergraduates in the sciences, fostered in part by the introduction of new degree-schemes (biochemistry from 1947 and metallurgy from 1956, for example) and the expansion of others (rather slow in geology and botany but dramatic in engineering science, despite a scare in the 1950s that the subject might be closed down on the grounds that its small size made it unviable). In this expansion, physics stood among the pacemakers. An average of roughly twenty candidates a year who took Honours in physics in the Natural Science School in the 1930s rose to over sixty in the early 1950s and more than 130 a decade later (Roche 1994: 254–6). While engineering science experienced the same pace of change – a six-fold increase between the 1930s and the early 1960s – the pre-war base for that school was much smaller (fewer than ten Honour candidates a year between 1935 and 1940). With regard to total numbers, physics in the same period never quite overtook chemistry, which throughout the inter-war years had been the biggest player among the physical sciences at undergraduate level, with an average of over forty candidates a year taking Honours. But by 1960 physics had very nearly closed the gap, following a decade in which undergraduate admissions in the subject increased twice as rapidly as those in chemistry.

Within fifteen years of the war, therefore, physics and chemistry were level-pegging and both could claim to have the largest undergraduate schools in their disciplines in the country. Expansion on this scale raised new challenges. It also resurrected an old one that had faced the discipline for decades: the relative isolation of physicists from the life of the colleges. Additional demonstratorships and fixed-term research appointments met the enhanced demand for tutorial and practical instruction, but they did not necessarily bear with them a college attachment and hence an *entrée* to the traditional heart of Oxford life. Here, the still high degree of collegiate autonomy made for an uphill struggle, and it was some time before the relentless efforts of Lindemann and Simon bore fruit. Early successes in the 1940s saw three former graduate students of the Clarendon appointed to tutorial fellowships: R. A. Hull (at Brasenose), A. H. Cooke (at New College), and Brebis Bleaney (at St John's). But still in 1951 only ten of the University's 28 undergraduate colleges had a physicist among their fellows. It took the undergraduate explosion of the 1950s to change things. By the early 1960s, only Trinity among the men's colleges and Somerville and St Anne's among the women's colleges were without at least one fellow in the subject.

On a number of fronts, therefore, physics during the two decades after the war might be seen to have finally shed the constraints that for so long had crippled its progress and impeded its adjustment to the norms of 'big' physics in the twentieth century. Although there is a core of truth in such a perception, *Physics in Oxford* has sought to avoid sweeping retrospective assessments of this kind by analysing the history of the discipline in the University in ways that discount judgements founded on how physics has come to be done in our own day. The book has

Fig. 9.6. Fellows of Brasenose College, including three physicists: W. N. Stocker (front row, extreme left), N. Kurti (back row, extreme left), and R. A. Hull (middle row, fifth from the left). The Principal of Brasenose, William Stallybrass, a distinguished academic lawyer who voiced reservations about the continued expansion of the sciences, is in the middle of the front row. The photograph dates from the late 1940s, but before October 1948 (when Stallybrass died). Photograph by Gilman & Soame, Limited, Oxford.

certainly not attempted a rehabilitation of Oxford physics, nor even (despite some low points) suggested that rehabilitation is needed. Instead it has sought to map out the perspectives that its authors see as most appropriate to the diverse contexts in which physics in a university with a very particular structure and often unhelpful entrenched traditions was pursued. In the hundred years we treat, many options were open with regard to the definition and goals of the discipline and the ways in which it should be pursued, and decisions about which of these could and should be adopted were neither unequivocal nor easy. There was, in short, no obvious royal road to success. The authors of *Physics in Oxford* have reflected this in their various chapters, all of which seek to convey, in addition to the broader national and international picture of disciplinary change, the local and frequently contingent circumstances that coloured rhetoric and decision-making within the University. The result is a multifaceted analysis of a complexity that has made the book rewarding, if often exceedingly difficult, to write.

Appendix I

The Classification of the Oxford B.A.

The main changes in the system of classification of candidates for the degree of Bachelor of Arts can be summarized as follows:

- 1801   Candidates were required to offer themselves either for the ordinary examination or for the 'more strict examination', held once a year, at which Honours were awarded. A maximum of twelve candidates were to be distinguished as 'Candidati qui se Examinatoribus Publicis MAXIME commendaverunt'. Other candidates of merit could be place in the group of 'Candidati qui se Examinatoribus Publicis EGREGIE commendaverunt'

- 1807   Candidates worthy of distinction were now placed in one of two classes in either or both of the two schools: Literae Humaniores or Disciplinae Mathematicae et Physicae. Candidates 'not deserving any honour' but worthy of the degree were placed in the third class

- 1809   The names in the second class in each school were now divided into two groups, according to attainment

- 1825   The lower part of the second class became the third class, with candidates considered 'worthy of their degree, but not of distinction' placed in a new fourth class

- 1830   A reordering created four classes of honours, with successful candidates not deserving honours being now placed in a new fifth class. The fourth class included candidates for the Pass degree who had performed above the level normally required of them

- 1860   From Easter Term 1861, the fifth class was abolished, and candidates who previously had been placed in it were henceforth simply stated to have passed

Appendix II

The Syllabus for the Oxford B.A., 1831–1872

A. 1831–1850

Candidates for the B.A. had to undergo two examinations:

(a) *Responsions*, taken between the sixth and ninth terms after matriculation. The examination, conducted three times a year by the appointed 'Masters of the Schools', covered the classical languages (with a special emphasis on grammar) and either the rudiments of logic or Euclid's 'Elements of geometry'

(b) *The Public Examination*, held twice a year (in the Easter and Michaelmas Terms) and taken no earlier than the beginning of the fourth year after matriculation (except for the sons of certain categories of noblemen and knights, who could be examined at the beginning of their third year). Candidates were to be examined in:

  (i) *The Rudiments of Religion*, including a knowledge of the gospels in Greek, the evidences of religion (natural and revealed), the Thirty-Nine Articles of the Church of England, and the history of the Old and New Testament

  (ii) *Literae Humaniores*, including the Latin and Greek languages and ancient history, rhetoric and moral and political science (drawn from Greek and Roman authors), logic, and Latin composition

  (iii) *Disciplinae Mathematicae et Physicae*. The elements of the mathematical sciences and physics.

Whereas all candidates for the B.A. had to be examined and to pass in the Rudiments of Religion, the examiners had discretion with regard to the other parts of the syllabus. The minimum requirement was that candidates be examined in at least three Greek and Latin authors, translation from English into Latin, and either logic or the first four books of Euclid. The effect of this flexibility was that mathematical or scientific study could be, and almost invariably was, completely avoided, with even elementary Euclidean geometry being no more than an option.

B. From 1850

Candidates for the B.A. had to undergo three examinations:

(a) *Responsions* Responsions, taken between the third and seventh terms after matriculation, the examination being conducted three times a year by the 'Masters of the Schools', as before. The requirements remained predominantly classical, with the emphasis on a command of the languages, tested through one Greek and one Latin author and an exercise in Latin prose composition. Arithmetic, however, was required of all candidates, together with either two books of Euclid or elementary algebra.

(b) *The First Public Examination* Candidates were to be examined by Moderators between their eighth and twelfth term after matriculation. The examination took place twice a year.

   (i) The subjects listed as 'minimum' requirements, for a Pass, were: the Gospels in Greek, one Greek and one Latin author (one a poet, the other an orator), and Latin prose composition. There was also to be a paper of grammatical questions and a paper on either mathematics (testing algebra and three books of Euclid) or logic.

   (ii) Honours could be awarded in either Classics or Mathematics, the latter defined as 'pure' mathematics. In Classics, the chosen authors would normally be poets and orators (Homer, Virgil, Demosthenes, and Cicero being especially recommended), and candidates had the opportunity of answering philological and critical questions as well as offering Greek and Latin translations in prose and verse. Candidates for the highest honours were required to be examined in logic. Those not aspiring to the highest honours could opt instead to be examined in Euclid and algebra. For Honours in Mathematics, the mathematical exercises carried greater weight.

(c) *The Public Examination* (known, from 1862, as *Second Public Examination*) The examination was held twice a year, with candidates being required to achieve at least a Pass (not necessarily in the same term) in two of the four schools, in any of which Honours could be awarded and one of which had to be Literae Humaniores. The requirements were as follows:

   (i) *Literae Humaniores* For a Pass, candidates would be tested on the Four Gospels and Acts of the Apostles in Greek, sacred history, a knowledge of the books of the Old and New Testaments, the evidences of Christianity and the Thirty-Nine Articles, the work of a philosopher and an historian (either Greek or Latin), and translation from Greek or Latin into English. For Honours, a more extensive command of theology, the ancient languages and history, and moral and political philosophy was required, with logic being essential for candidates for either first or second class Honours.

   (ii) *Mathematics* For a Pass, the first six books of Euclid or the 'first part' of algebra were required; for Honours, candidates were examined in 'mixed' as well as 'pure' mathematics.

   (iii) *Natural Science* Candidates for a Pass had to show an acquaintance with two of the three branches: Mechanical Philosophy, Chemistry, or Physiology. In addition, they were to be examined in one of the branches of Mechanical Philosophy. For Honours, candidates were examined in all three of Mechanical Philosophy, Chemistry, and Physiology, and in any one of the sciences covered by the NSS. The latter could be either an area of one of the 'primary' sciences (including mechanics, pneumatics, sound, light, heat, and electricity, for example) or one of the 'special' sciences, such as geology, physical geography, botany, zoology, and mineralogy, that did not form part of the main syllabus.

(iv) *Law and Modern History* For a Pass, the normal requirements were a period of English history and, in law, Blackstone. For Honours, a more extensive knowledge of both history (up to 1789) and law was required.

C. The amendments of 1860, 1870, and 1872

The main changes affecting candidates in the School of Natural Science were those of 1860, which allowed candidates who gained at least third-class honours in any school to take their B.A. without passing in a second school, and of 1870, which introduced the Preliminary Honour examination as an additional hurdle to be passed by candidates aspiring to Honours. The Preliminary examination, which candidates could sit at any time after the first public examination, was in chemistry and the more elementary parts of mechanics and physics. In the final examination, any one or more of (in the terminology now used) physics, chemistry, or biology could be offered. A modification in 1870 also exempted candidates who achieved fourth-class honours in any school from the requirement to pass in a second school.

Statutes of February 1872 transformed the requirements for the Pass degree across all the schools; see Fox, Chapter 2, 70.

It was not until 1886 that the class lists specified the particular science in which candidates had specialized.

Appendix III

Letter from Robert Clifton to Sir William Thomson

Reproduced by permission of Glasgow University Library, Department of Special Collections (MS Kelvin C32)

Oxford
Sepr 16th/67

Dear Sir William,

I hope you will again forgive my applying to you for assistance and advice and give me your opinion on the following matter.

I found on my appointment at Oxford that the department of Physics was provided only with a lecture room, a small laboratory attached, which is completely absorbed for lecture purposes [p. 2] and a small office. No arrangements had been made for students to work practically themselves to gain a thorough knowledge of Physics. I borrowed a small room from one of my colleagues who does not at present reside in Oxford and have endeavoured to let a few work practically at Physics but could not receive all who applied, nor could those whom I received get on well as in one small room (not capable [p. 3] of being darkened) many exp^{ts} were altogether impossible and those which were possible in many cases so interfered with one another that accuracy was in general out of the question.

In consequence of my complaints the University appointed a Committee to consider my wants and a sum of money was found available for the purposes of building an addition to our Museum where as you know the scientific [p. 4] Professors have their lecture rooms &c. I was requested by this committee to provide plans showing the number and arrangement of the rooms required for a moderately complete course of instruction in Physics, including lecture theatre, my present lecture room being given up to some other Profr. I have accordingly devised the accompanying plans & I shall be exceedingly thankful for any advice on [p. 5] the matter. If you agree to the arrangement and would kindly write me a few lines expressing such agreement and an opinion as to the necessity for students to receive practical instruction which I know you entertain and will allow me to use your letter to support my application it will be of the greatest possible use to me. I will add a few words of explanation of the plans.

[p. 6] As the men here will not in general be able to devote much time to the subject it will I think be necessary to have several laboratories so that several men taking different branches (they are allowed a choice for the degree) may work simultaneously without interfering with one another.

The building proposed is to consist of two stories; in the middle, what is called the "court" is the apparatus [p. 7] cabinet. Going from the floor to the roof with a gallery, it will hold a large quantity of apparatus. The lecture theatre also goes to the roof. There are twelve students laboratories, five facing south (for sunlight expts) and seven facing north. Each will accommodate from two to four students according to the nature of their work. There are two private Laboratories one to be light while the other is [p. 8] dark if necessary. A store room, an examn room to be used also for students to reduce their observations & so leave the laboratories free for others using the same instruments.

A lecture room where lectures on Mathematical Physics (which I think are much wanted) can be given to a small class without interfering with the arrangements for the experial lectures in the theatre.

[p. 9] In the roof of the front part is a long optical gallery intended to go the whole length of the roof, not as in the plan only to the side roof; a portion of the roof on the south side is flat so that it can be used for operations in the open air when needed. There are also some small Workshops in a sort of yard where apps may be repaired or models constructed.

The battery rooms and store [p. 10] rooms for rough things are to be in the cellars.

The demand for scientific instruction appears to be rapidly increasing here & I think the accompanying plans represent no more than will be needed in a few years tho perhaps more than will be required immediately. The Chemical & Physiological departments are well provided on about the same scale as is here proposed [provided *deleted*] for Physics.

[p.11] As soon as proper arrangements are made I believe a class of Senior Students (young fellows of Colleges) will arise. Some individuals of the class already exist who are anxious to devote themselves to Physical research and with twelve laboratories it will I hope be possible to accommodate some of them with rooms where they may carry on their work without interruption.

I very much regret that [p. 12] I did not feel well enough to go to Dundee, where I should have had an opportunity of consulting you personally but I hope you will excuse this long letter & I need not add that any suggestions for alterations in the plans will be thankfully received by

        Yours most sincerely
        [signed] R B Clifton

Sir William Thomson FRS

BIBLIOGRAPHY

[Anon.] (1855). Report of meeting of 14 May 1855. *Proceedings of the Ashmolean Society*, 3, no. 32: 72.

—— (1861). *Catalogue of the Philosophical Apparatus, Minerals, Geological Specimens, &c. in the Possession of Dr. Daubeny*. Oxford: printed by James Wright.

—— (1876). 'The University of Oxford Commission'. *Nature*, 18: 23–4.

—— (1877). *Catalogue of the Special Loan Collection of Scientific Apparatus at the South Kensington Museum. MDCCCLXXVI* (3rd edition). London: HMSO.

—— (1879). Obituary of William Froude. *Proceedings of the Royal Society of London*, 29: 'Obituary notices of fellows deceased', ii–vi.

—— (1880). 'Mr. William Froude, LL.D., F.R.S.'. *Minutes of Proceedings of the Institution of Civil Engineers*, 60: 395–404.

—— (1887). 'The Clarendon Laboratory grant'. *Oxford Magazine*, 5: 249 (1 June 1887).

—— (1892a). 'High speed photography'. *Scientific American*, 67: 216.

—— (1892b). 'Inductoscripts'. *The Electrical Review*, 31: 693.

—— (1894). *Biographical Sketches and Recollections (with early letters) of Henry John Stephen Smith, M.A., F.R.S. late Savilian Professor of Geometry in the University of Oxford*. Oxford: printed for private circulation.

—— (1896). 'The storage and explosion of compressed gaseous mixtures'. *Journal of Gas Lighting . . .* , 10 March: 507–8.

—— (1897–8). 'The Jervis-Smith chronograph'. *The Electrician*, 40: 136.

—— (1898). 'Smith's tramway chronograph constructed by Messrs. Elliott Brothers, London'. *Engineering*, 65: 692–3.

—— (1904). 'New buildings of the University of Liverpool: the George Holt Physics Laboratory'. *Nature*, 71: 63–5.

—— (1906). *The Physical Laboratories of the University of Manchester. A Record of 25 Years' Work. Prepared in Commemoration of the 25th Anniversary of the Election of Dr. Arthur Schuster, F.R.S., to a Professorship in the Owens College, by his Old Students and Assistants*. Manchester: Manchester University Press.

—— (1907). 'University and educational intelligence'. *Nature*, 75: 453–4.

—— (1910). *A History of the Cavendish Laboratory 1871–1910*. London: Longmans, Green.

—— (1911a). 'The Rev. F. J. Jervis-Smith F.R.S.'. *Nature*, 87: 318 (7 September 1911); correction and additions by Jervis-Smith's son, E. J. Jervis-Smith, 390 (21 September 1911).

—— (1911b). Announcement of the death of the Revd F. J. Jervis-Smith. *Science*, new ser., 34: 343.

—— (1911c). Obituary of F. J. Jervis-Smith. *Manchester Guardian*, 1 September 1911.

—— (1913–14). 'George James Burch'. *The Oxford Magazine*, 32: 301.

—— (1914). 'Death of Dr. G. J. Burch. A distinguished citizen and scientist'. *The Oxford Chronicle, Berks and Bucks Gazette*, 20 March 1914, 12.

[Anon.] (1921a). 'Obituary. Prof. R. B. Clifton, F.R.S.'. *Nature*, 107: 18–19.
—— (1921b). 'Obituary. Prof. A. W. Reinold'. *Nature*, 107: 276.
—— (1934) *The Balliol College Register . . . 1833–1933*. Oxford.
—— (1938). 'News and views. Prof. W. L. Bragg, O.B.E., F.R.S.'. *Nature*, 141: 403.
—— (1941a). 'What is a physicist?'. *Nature*, 147: 617–18.
—— (1941b). 'Obituaries. Mr. I. O. Griffith'. *Nature*, 148: 589.
—— (1946). 'John Job Manley'. *The Proceedings of the Physical Society*, 58: 332–3.
—— (1994). 'Lord Berkeley, scientist and mature student before his time', in Lorna Fullard (ed.), *The Open University Oxford Research Unit* (4th edition). Milton Keynes: Open University Press, 34–5.
Abraham, H. A., and Langevin, P. (eds.) (1905). *Les Quantités élémentaires d'électricité. Ions, électrons, corpuscules*. Paris: Gauthier-Villars.
Achard, F. (1998). 'La publication du *Treatise on Electricity and Magnetism* de James Clerk Maxwell'. *Revue de synthèse*, 4th ser., 119: 511–44.
Acland, H. W. (1890). *Oxford and Modern Medicine. A Letter to Dr. James Andrew, M.D. Oxon., F.R.C.P. Lond.* Oxford and London: Frowde.
Adam, M. G. (1996). 'The changing face of astronomy in Oxford (1920–60)'. *Quarterly Journal of the Royal Astronomical Society*, 37: 153–79.
Adams, P. (1996). *Somerville for Women. An Oxford College, 1879–1993*. Oxford: Oxford University Press.
Alexander, R. C. (1999). *The Inventor of Stereo. The Life and Works of Alan Dower Blumlein*. Oxford: Focal Press.
Allen, H. S. (1946). 'Obituaries. Prof. William Peddie'. *Nature*, 158: 50–1.
—— (1945–8). 'Charles Glover Barkla 1877–1944'. *Obituary Notices of Fellows of the Royal Society* 5: 341–66.
Allibone, T. E., and Clarke, J. R. (1959). 'Samuel Roslington Milner 1875–1958'. *Biographical Memoirs of Fellows of the Royal Society*, 5: 129–147.
Andrade, E. N. da C. (1942–4). 'William Henry Bragg 1862–1942'. *Obituary Notices of Fellows of the Royal Society*, 4: 277–300 [bibliography prepared by K. Lonsdale].
Andrews, P. W. S., and Brunner, E. (1955). *The Life of Lord Nuffield. A Study in Enterprise and Benevolence*. Oxford: Blackwell.
Applebey, M. P. (1948). 'Ernald George Justinian Hartley 1875–1947'. *Journal of the Chemical Society*, 1948: 899–901.
Application of Robert Bellamy Clifton, M.A. for the Professorship of Experimental Philosophy, in the University of Oxford, with Testimonials [1865]. The only known copy is in the Clarendon Laboratory Archive.
Arms, N. (1966). *A Prophet in Two Countries. The Life of F. E. Simon*. Oxford and New York: Pergamon Press.
Attwood, T. V. (2004). 'Uranium isotope separation in the U.K. during World War II'. Ph.D. Thesis, University of Liverpool.
Avent, C., and Pipe, H. (eds.) (1991). *Lady Margaret Hall Register 1879–1990*. Oxford: Lady Margaret Hall.
Ayrton, W. E., and Perry, J. (1878a). 'The contact theory of voltaic action'. *The Telegraphic Journal*, 6: 99–100.

—— (1878b). 'Contact theory of voltaic action. Parts I and II'. *Proceedings of the Royal Society of London*, 27: 196–238.

Badash, L. (1985). *Kapitza, Rutherford, and the Kremlin*. New Haven: Yale University Press.
—— (1987). 'Ernest Rutherford and theoretical physics', in R. H. Kargon and P. Achinstein (eds.), *Kelvin's Baltimore Lectures and Modern Theoretical Physics. Historical and Philosophical Perspectives*. Cambridge, Mass., and London: MIT Press, 349–73.
Bagguley, D. M. S., and Bleaney, B. (1990). 'Ferromagnetic resonance at the Clarendon Laboratory, Oxford: a tribute to J. H. E. Griffiths (1908–1981)'. *Contemporary Physics*, 31: 35–42.
Baker, H. B. (1910). 'Ionisation of gases and chemical change'. *Nature*, 84: 388–9.
B[aker], H. B. (1921–2). 'Obituary. Robert Edward Baynes'. *The Oxford Magazine*, 40: 48.
——, and B[one], W. A. (1931–2). 'Harold Baily Dixon – 1852–1930'. *Proceedings of the Royal Society of London*, series A, 134: 'Obituary notices', i–xvii.
Ball, R. S. (1871). *Experimental Mechanics. A Course of Lectures . . . with Illustrations*. London and New York: Macmillan.
Bannon, J., and Brose, H. L. (1928). 'The motions of electrons in ethylene'. *Philosophical Magazine*, 7th ser., 6: 817–24.
Barrow, R. F., and Danby, C. J. (1991). *The Physical Chemistry Laboratory. The First Fifty Years*. Oxford: Physical Chemistry Laboratory, University of Oxford.
Bates, L. F. (1969). 'Edmund Clifton Stoner 1899–1968'. *Biographical Memoirs of Fellows of the Royal Society*, 15: 201–37.
Baynes, R. E. (1878a). *Lessons on Thermodynamics*. Oxford: Clarendon Press.
—— (1878b). *The Book of Heat*. London: Stewart.
Bellamy, F. A. (1908). *A Historical Account of the Ashmolean Natural History Society of Oxfordshire, 1880–1905*. Oxford: published by the author.
Bensaude-Vincent, B. (1987). *Langevin 1872–1946. Science et vigilance*, Paris: Belin.
Bentwich, N. de M. (1953). *The Rescue and Achievement of Refugee Scholars. The Story of Displaced Scholars and Scientists 1933–1952*. The Hague: M. Nijhoff.
Berkeley, R., Eighth Earl of (1895–7). 'On an accurate method of determining the densities of solids'. *The Mineralogical Magazine*, 11: 64–8.
Berman, R. (1987). 'Lindemann in physics'. *Notes and Records of the Royal Society of London*, 41: 181–9.
Bernal, J. D. (1963). 'William Thomas Astbury 1898–1961'. *Biographical Memoirs of Fellows of the Royal Society*, 9: 1–35.
Beveridge, W. H. (1959). *A Defence of Free Learning*. London: Oxford University Press.
Beyerchen, A. D. (1977). *Scientists under Hitler. Politics and the Physics Community in the Third Reich*. New Haven and London: Yale University Press.
Birkenhead, Second Earl of [F. W. F. Smith] (1961). *The Prof in Two Worlds. The Official Life of Professor F. A. Lindemann, Viscount Cherwell*. London: Collins.
Bleaney, B. (1988). 'Physics at the University of Oxford'. *European Journal of Physics*, 9: 283–8.
—— (1994). 'The physical sciences in Oxford, 1918–1939 and earlier'. *Notes and Records of the Royal Society of London*, 48: 247–61.
—— (1996). 'Heinrich Gerhard Kuhn, 10 March 1904–26 August 1994'. *Biographical Memoirs of Fellows of the Royal Society*, 42: 219–32.

Bleaney, B., Cooke, A. H., Kurti, N., and Stevens, K. W. H. (1986). 'F. A. Lindemann, Viscount Cherwell (1886–1957): head of the Clarendon Laboratory, University of Oxford (1919–56)'. *Physics Bulletin*, 37: 261–3.

Bloch, E. (1904). 'L'ionisation par le phosphore et par les actions chimiques', *Le Radium*, 2: 33–9.

Bloch, E. (1912). 'Sur l'emploi des cellules photoélectriques comme photophones'. *Comptes rendus hebdomadaires des séances de l'Académie des Sciences*, 154: 427–8.

Bloch, L. and E. (1914). 'Sur les spectres d'étincelle du nickel et du cobalt dans l'ultraviolet extrême'. *Comptes rendus hebdomadaires des séances de l'Académie des Sciences*, 158: 784–7.

Board of Education (1900). *Reports from University Colleges. 1900* [Cd 331]. London: HMSO.

Boase, F. (1892–1921). *Modern English Biography* (6 volumes). Truro: Netherton and Worth, for the author.

Bosanquet, R. H. M. (1873). 'On just intonation in music: with a description of a new instrument for the easy control of all systems of tuning other than the ordinary equal temperament'. *Proceedings of the Royal Society of London*, 21: 131–2.

—— (1876a). *An Elementary Treatise on Musical Intervals and Temperament, with an Account of an Enharmonic Harmonium exhibited in the Loan Collection of Scientific Instruments, South Kensington, 1876: also of an Enharmonic Organ exhibited to the Musical Association*. London: Macmillan; reprint edited by Rudolf Rasch, Utrecht: Diapason Press (1987).

—— (1876b). 'On a new form of polariscope, and its application to the observation of the sky'. *Philosophical Magazine*, 5th ser., 2: 20–8.

—— (1877a). 'On instruments of just intonation', in *South Kensington Museum. Conferences held in connection with the Special Loan Collection of Scientific Apparatus. 1876. Physics and Mechanics*. London: Chapman and Hall, 55–8.

—— (1877b). 'Notes on the theory of sound'. *Philosophical Magazine*, 5th ser., 3: 271–8, 343–9, 418–24.

—— (1877c). 'Notes on the theory of sound'. *Philosophical Magazine*, 5th ser., 4: 25–39, 125–36, 216–22.

—— (1879). 'On the present state of experimental acoustics, with suggestions for the arrangement of an acoustic laboratory, and a sketch of research'. *Philosophical Magazine*, 5th ser., 8: 290–305.

—— (1880). 'Note on the laboratory at St. John's College, Oxford'. *Philosophical Magazine*, 5th ser., 10: 217–26.

—— (1881a). 'On the beats of consonances of the form $h : 1$'. *Philosophical Magazine*, 5th ser., 11: 420–36.

—— (1881b). 'Note on the laboratory at St. John's College, Oxford [continued]'. *Philosophical Magazine*, 5th ser., 12: 178–84.

—— (1883). 'Preliminary paper on a uniform rotation machine; and on the theory of electromagnetic tuning-forks'. *Proceedings of the Royal Society of London*, 34: 445–7.

—— (1884). 'On electromagnets – No. I'. *Philosophical Magazine*, 5th ser., 17: 531–6.

—— (1885a). 'Electromagnets. – II'. *Philosophical Magazine*, 5th ser., 19: 73–94.

—— (1885b). 'Electromagnets. – III'. *Philosophical Magazine*, 5th ser., 19: 333–40.

—— (1885c). 'Electromagnets. – IV'. *Philosophical Magazine*, 5th ser., 20: 318–23.

—— (1886a). 'On electromagnets – V'. *Philosophical Magazine*, 5th ser., 22: 298–309.

—— (1886b). 'On electromagnets – VI'. *Philosophical Magazine*, 5th ser., 22: 535–9.

—— (1887). 'On electromagnets – VII'. *Philosophical Magazine*, 5th ser., 23: 338–50.
Bosanquet, R. H. M., and Sayce, A. H. (1878–9). 'Preliminary paper on the Babylonian astronomy'. *Monthly Notices of the Royal Astronomical Society*, 39: 454–61.
—— (1879–80a). 'The Babylonian astronomy. No. 2'. *Monthly Notices of the Royal Astronomical Society*, 40: 105–123.
—— (1879–80b). 'The Babylonian astronomy. No. 3: The Venus tablet'. *Monthly Notices of the Royal Astronomical Society*, 40: 565–78.
Boullin, D. J. (1989). 'The tram chronograph'. *Bulletin of the Scientific Instrument Society*, no. 23: 16–18.
Bowen, E. J. (1958). 'David Leonard Chapman 1869–1958'. *Biographical Memoirs of Fellows of the Royal Society*, 4: 35–44.
—— (1969). 'Chemistry at Oxford (with special reference to physical chemical studies)'. Oxford: unpublished typescript (copies in Museum of the History of Science and Radcliffe Science Library, Oxford).
—— (1970). 'The Balliol–Trinity laboratories, Oxford 1853–1940'. *Notes and Records of the Royal Society of London*, 25: 227–36.
—— (1973). ' Sir Harold Hartley, G.C.V.O., C.H., C.B.E., M.C., F.R.S.'. *Balliol College Annual Record 1973*: 8–12.
Boys, C. V. (1890). 'On the Cavendish experiment'. *Proceedings of the Royal Society of London*, 46: 253–68.
—— (1895). 'On the Newtonian constant of gravitation'. *Philosophical Transactions of the Royal Society of London*, series A, 186: 1–72.
B[oys], C. V. (1913). 'Frederick John Jervis-Smith, 1848–1911'. *Proceedings of the Royal Society of London*, series A, 88: 'Obituary notices of fellows deceased', iv–vi; reprinted in Jervis-Smith 1915.
—— (1932–5). 'Henry Reginald Arnulph Mallock 1851–1933'. *Obituary Notices of Fellows of the Royal Society*, 1: 95–100.
Bragg, M. (2002). *RDF 1. The Location of Aircraft by Radio Methods, 1935–1945*. Paisley: Hawkhead.
Bragg, W. L. (1965). 'Reginald William James 1891–1964'. *Biographical Memoirs of Fellows of the Royal Society*, 11: 115–25.
Brewer, F. M. [misprinted as J. M.] (1961). 'Oxford: a home of chemistry and industry'. *Chemistry and Industry*, 24 June 1961: 845–53.
British Association for the Advancement of Science (1926). *Daily Time-Table: Supplementary Issue . . .* Oxford: British Association for the Advancement of Science.
British Medical Association (1936). *The Book of Oxford. Printed for the 104th annual meeting of the British Medical Association*. Oxford: British Medical Association.
Brittain, V. (1960). *The Women at Oxford. A Fragment of History*. London: Harrap.
Broadbent, T. E. (1998). *Electrical Engineering at Manchester University. 125 Years of Achievement*. Manchester: Manchester University School of Engineering.
Brock, M. G. (1997). 'The Oxford of Peel and Gladstone 1800–1833', in Brock and Curthoys (1997), 7–71.
——, and Curthoys, M. C. (eds.) (1997). *The History of the University of Oxford. Volume VI. Nineteenth-Century Oxford, Part 1*. Oxford: Clarendon Press.

Brock, M. G., and Curthoys, M. C. (eds.) (2000). *The History of the University of Oxford. Volume VII. Nineteenth-Century Oxford, Part 2*. Oxford: Clarendon Press.

Brock, W. H. (ed.) (1967). *The Atomic Debates. Brodie and the Rejection of the Atomic Theory. Three Studies*. Leicester: Leicester University Press.

—— (2002). 'Exploring the Hyperarctic: James Dewar at the Royal Institution', in James (2002), 169–90.

de Broglie, M. (1909). 'Etude sur les suspensions gazeuses'. *Le Radium*, 6: 203.

—— (1913a). 'Sur les images multiples que présentent les rayons de Röntgen après avoir traversé des cristaux'. *Comptes rendus hebdomadaires des séances de l'Académie des Sciences*, 166: 1011–12.

—— (1913b). 'Sur la réflexion des rayons de Röntgen', *Comptes rendus hebdomadaires des séances de l'Académie des Sciences*, 156: 1153–5.

——, and Brizard, L. (1910). 'Contribution à l'étude de l'ionisation des gaz en présence de réactions chimiques'. *Le Radium*, 7: 164.

Brooke, C. (1867). *The Elements of Natural Philosophy; or an Introduction to the Study of the Physical Sciences . . . based on the Treatise by the late Golding Bird, M.A., M.D., F.R.S., F.L.S.* (6th edition). London: John Churchill and Sons.

Brose, H. L. (1919). *The Theory of Relativity. An Introductory Sketch based on Einstein's Original Writings, including a Biographical Note*. Oxford: Basic Blackwell.

—— (1925). 'The motions of electrons in oxygen'. *Philosophical Magazine*, 6th ser., 50: 536–46.

Brown, J. A. (1906). 'An investigation of the potential required to maintain a current between parallel plates in a gas at low pressures'. *Philosophical Magazine*, 6th ser., 12: 210–32.

Brown, J. D., and Stratton, S. S. (1897). *British Musical Biography*. Birmingham: S. S. Stratton.

Browne, J. (2002). *Charles Darwin. The Power of Place*. London: Jonathan Cape.

Buchanan, R. A. (1985). 'The rise of scientific engineering in Britain'. *The British Journal for the History of Science*, 18: 218–33.

Buchwald, J. Z., and Warwick, A. (eds.) (2001). *Histories of the Electron. The Birth of Microphysics*. Cambridge, Mass., and London: MIT Press.

Budden, K. G. (1988). 'John Ashworth Ratcliffe, 12 December 1902–25 October 1987'. *Biographical Memoirs of Fellows of the Royal Society*, 34: 671–711.

Burcham, W. E., and Isaak, G. R. (1996). 'Philip Burton Moon, 17 May 1907–9 October 1994'. *Biographical Memoirs of Fellows of the Royal Society*, 42: 249–64.

Burns, R. W. (1986). *British Television. The Formative Years*. London: Peter Peregrinus.

—— (ed.) (1988), *Radar Development to 1945*. London: Peter Peregrinus.

Bussey, G. (1990). *Wireless, the Crucial Decade. History of the British Wireless Industry 1924–34*. London: Peter Peregrinus.

Cahan, D. (1985). 'The institutional revolution in German physics, 1865–1914'. *Historical Studies in the Physical Sciences*, 15, part 2: 1–65.

Calvert, H. R. (1957). 'Lord Cherwell', letter in *Nature*, 180: 1146.

Cameron, J. G. P. (1960). *A Short History of the Royal Engineering College Coopers* [sic] *Hill*. Cooper's Hill Society, for private circulation.

Cameron, N. (1983). 'The politics of British Science in the Munich era', in W. R. Shea (ed.), *Otto Hahn and the Rise of Nuclear Physics*. Dordrecht: D. Reidel, 181–99.

Campbell, N. P. (1912–13). 'On the application of Manley's differential densimeter to the study of sea-waters on board ship'. *Proceedings of the Royal Society of Edinburgh*, 33: 124–36.

Cantor, G. N. (1994). 'The making of a British theoretical physicist – E. C. Stoner's early career'. *The British Journal for the History of Science*, 27: 277–90.

Caroe, G. M. (1978). *William Henry Bragg 1862–1942. Man and Scientist*. Cambridge: Cambridge University Press.

Cathcart, B. (2004). *The Fly in the Cathedral. How a Small Group of Cambridge Scientists Won the Race to Split the Atom*. London: Viking.

Cattermole, M. J. G., and Wolfe, A. F. (1987). *Horace Darwin's Shop. A History of the Cambridge Scientific Instrument Company 1878 to 1968*. Bristol and Boston: Adam Hilger.

Channell, D. F. (1982). 'The harmony of theory and practice: the engineering science of W. J. M. Rankine'. *Technology and Culture*, 23: 39–52.

Chapman, D. L. (1899). 'The rate of explosion in gases'. *Philosophical Magazine*, 5th ser., 47: 90–104.

—— (1913). 'A contribution to the theory of electrocapillarity'. *Philosophical Magazine*, 6th ser., 25: 475–81.

Charlton, P. (1984). *John Stainer and the Musical Life of Victorian Britain*. Newton Abbot: David & Charles.

Chayut, M. (1991). 'J. J. Thomson: the discovery of the electron and the chemists'. *Annals of Science*, 48: 527–44.

Church, A. H., and Hayes, E. H. (1901–4). *On the Relation of Phyllotaxis to Mechanical Laws* (3 parts). London and Oxford: Williams and Norgate.

Clark, R. W. (1965). *Tizard*. London: Methuen.

—— (1971). *Sir Edward Appleton*. Oxford and New York: Pergamon.

Clayton, R., and Algar, J. (1989). *The GEC Research Laboratories 1919–1984*. London: Peter Peregrinus.

Clifton, R. B. (1860). 'On the conical refraction of a straight line'. *Quarterly Journal of Mathematics*, 3: 360–3.

—— (1866a). 'An attempt to refer some phenomena attending the emission of light to mechanical principles'. *Proceedings of the Literary and Philosophical Society of Manchester*, 5: 24–8.

—— (1866b). 'Note on Professor De Morgan's paper entitled "On the early history of the signs + and −" '[dated 26 January 1865]. *Transactions of the Cambridge Philosophical Society*, 11, part 1: 213–18.

—— (1877). 'On the difference of potential produced by the contact of different substances'. *Proceedings of the Royal Society of London*, 26: 299–314.

—— (1901–2). 'Report of the professor of experimental philosophy, 1901'. *OUG*, 32: 647–8 (17 June 1902).

—— (1902–3). 'Report of the professor of experimental philosophy, 1902'. *OUG*, 33: 549 (19 May 1903).

—— (1903–4). 'Report of the professor of experimental philosophy, 1903'. *OUG*, 34: 560–1 (3 May 1904).

—— (1904–5). 'Report of the professor of experimental philosophy, 1904'. *OUG*, 35: 572–3 (16 May 1905).

Clifton, R. B. (1905–6a). 'Report of the professor of experimental philosophy, 1905'. OUG, 36: 612 (15 May 1905).

C[lifton], R. B. (1905–6b). 'Rev. Bartholomew Price. 1818–1898', *Proceedings of the Royal Society of London*, 75: 'Obituary notices of fellows deceased', 30–4.

Clifton, R. B. (1909–10). 'Report of the professor of experimental philosophy, 1909'. OUG, 40: 746 (31 May 1910).

—— (1910–11). 'Report of the professor of experimental philosophy, 1910'. OUG, 41: 958–9 (14 June 1911).

—— (1912–13). 'Report of the professor of experimental philosophy, 1912'. OUG, 43: 967 (4 June 1913).

—— (1913–14). 'Report of the professor of experimental philosophy, 1913'. OUG, 44: 863 (10 June 1914).

——, and Roscoe, H. E. (1860–2). 'On the effect of increased temperature upon the nature of the light emitted by the vapour of certain metals or metallic compounds'. *Proceedings of the Literary and Philosophical Society of Manchester*, 2: 227–30; also in *Chemical News*, 5 (1862), 233–4.

Cockburn, S., and Ellyard, D. (1981). *Oliphant. The Life and Times of Sir Mark Oliphant.* Adelaide: Axiom Books.

Cohen, M. N. (1995). *Lewis Carroll. A Biography.* London: Macmillan.

Collie, C. H. (1946–7). 'Oxford physics and the war'. *Oxford*, 9, no. 2: 54–8.

——, and Griffiths, J. H. E. (1936). 'The passage of neutrons through matter'. *Proceedings of the Royal Society of London*, series A, 155: 434–46.

——, and Roaf, D. (1940). 'On the mode of action of the Geiger–Müller counter'. *Proceedings of the Physical Society*, 52: 186–190.

——, Hasted, J. B., and Ritson, D. M. (1948). 'The dielectric properties of water and heavy water'. *Proceedings of the Physical Society*, 60: 145–160.

Conroy, J. (1878–9). 'Some experiments on metallic reflexion'. *Proceedings of the Royal Society of London*, 28: 242–50.

—— (1880–1). 'Some experiments on metallic reflexion. No. II'. *Proceedings of the Royal Society of London*, 31: 486–500.

—— (1883a). 'Some experiments on metallic reflexion. No. III. On the amount of light reflected by metallic surfaces' [including a note by G. G. Stokes, 39–41]. *Proceedings of the Royal Society of London*, 35: 26–41.

—— (1883b). 'A new photometer'. *Philosophical Magazine*, 5th ser., 15: 423–6.

—— (1883–4). 'Some experiments on metallic reflection. No. IV: On the amount of light reflected by metallic surfaces'. *Proceedings of the Royal Society of London*, 36: 187–98.

—— (1884). 'Some experiments on metallic reflection. No. V: On the amount of light reflected by metallic surfaces'. *Proceedings of the Royal Society of London*, 37: 36–42.

—— (1889). 'Some observations on the amount of light reflected and transmitted by certain kinds of glass'. *Philosophical Transactions of the Royal Society of London*, series A, 180: 245–89.

Cook, A. (1999). 'Reginold Victor Jones, 29 September 1911–17 December 1997'. *Biographical Memoirs of Fellows of the Royal Society*, 45: 239–54.

Corsi, P. (1988). *Science and Religion. Baden Powell and the Anglican Debate 1800–1860.* Cambridge: Cambridge University Press.

Croft, A. J. (1986). 'Oxford's Clarendon Laboratory'. Typescript, with an almost complete set of illustrations, in Clarendon Laboratory Archive. Copy, without illustrations, in the History of Science and Technology Seminar Room, Modern History Faculty, Oxford. Imperfect copy in the Radcliffe Science Library.

Crompton, R. W. (1995). 'Leonard George Holden Huxley 1902–1988', http://www.asap.unimelb.edu.au/bsparcs/aasmemoirs/huxley.htm.

Crosland, M. P. (2003). 'Difficult beginnings in experimental science at Oxford: the gothic Chemistry Laboratory'. *Annals of Science*, 60: 399–421.

Crowther, J. G. (1970). *Fifty Years with Science*. London: Barrie and Jenkins.

—— (1974). *The Cavendish Laboratory, 1874–1974*. London: Macmillan.

Cruickshank, A. D. (1979). 'Soddy at Oxford'. *The British Journal for the History of Science*, 12: 277–88.

—— (1986). 'Soddy at Oxford', in G. B. Kauffman (ed.), *Frederick Soddy (1877–1956). Early Pioneer in Radiochemistry*. Dordrecht, Boston, and Lancaster: D. Reidel, 157–70.

Cryer, T., and Jordan, H. G. (1888). *Text-Book of Applied Mechanics*. Manchester and London: John Heywood.

Curthoys, M. C. (1997). 'The examination system', in Brock and Curthoys (1997), 339–74.

Darwin, C. G. (1958). 'Douglas Rayner Hartree, 1897–1958'. *Biographical Memoirs of Fellows of the Royal Society*, 4: 103–16.

Daubeny, C. G. B. (1855). *A Dream of the New Museum*. Oxford: J. Vincent.

Daunt, J. G., and Mendelssohn, K. (1938a). 'Transfer of helium II on glass'. *Nature*, 141: 911–12.

——, and Mendelssohn, K. (1938b). 'Transfer effect in liquid helium II'. *Nature*, 142: 475.

Davis, E. A. (ed.) (1998). *Nevill Mott. Reminiscences and Appreciations*. London: Taylor & Francis.

Dean, K. (2003). 'Inscribing settler science: Ernest Rutherford, Thomas Laby and the making of careers in physics'. *History of Science*, 41: 217–240.

Dixon, H. B. (1920a). 'Obituary. D. H. Nagel'. *Nature*, 106: 186.

D[ixon], H. B. (1920b). 'A. G. Vernon Harcourt, 1834–1919'. *Proceedings of the Royal Society of London*, series A, 97: 'Obituary notices of fellows deceased', vii–xi.

Dixon, H. B., and Lowe, H. F. (1885). 'The decomposition of carbonic acid gas by the electric spark'. *Journal of the Chemical Society. Transactions*, 47: 571–6.

Dobson, G. M. B., Griffith, I. O., and Harrison, D. N. (1926). *Photographic Photometry. A Study of Methods of Measuring Radiation by Photographic Means*. Oxford: Clarendon Press.

Dobson, G. M. B. (1926). *The Uppermost Regions of the Earth's Atmosphere, being the Halley Lecture delivered on 5 May 1926*. Oxford: Clarendon Press.

—— (1968). 'Forty years' research on atmospheric ozone at Oxford: a history'. *Applied Optics*, 7: 387–405.

Donkin, W. F. (1870). *Acoustics. Theoretical. Part I*. Oxford: Clarendon Press.

Donnan, F. G. (1936). 'Obituary. Prof. James Rice'. *Nature*, 137: 807–8.

Dummelow, J. (1949). *1899–1949*. Manchester: Metropolitan-Vickers Electrical Company Limited.

Dunbabin, J. P. D. (1997). 'Finance and property', in Brock and Curthoys (1997), 375–437.

Dunbar, R. T. (1942). 'Obituaries. Prof. A.L. Selby'. *Nature*, 150: 285.

Earwaker, J. P. (1870–1). 'Natural science at Oxford'. *Nature*, 3: 170–1.

Edgerton, D. E. H. (1991). *England and the Aeroplane. An Essay on a Militant and Technological Nation*. Basingstoke: Macmillan.

Edgerton, D. E. H. (1996). 'British scientific intellectuals and the relations of science, technology and war', in P. Forman and J. M. Sánchez-Ron (eds.), *National Military Establishments and the Advancement of Science and Technology. Studies in 20th Century History*. Dordrecht, Boston, and London: Kluwer, 1–35.

——, and Horrocks, S. M. (1994). 'British industrial research and development before 1945'. *Economic History Review*, 47: 213–38.

Edwards, C. A. (1939–1941). 'Henry Cort Harold Carpenter 1875–1940'. *Obituary Notices of Fellows of the Royal Society*, 3: 611–25.

Egerton, R. (1963). *Sir Alfred Egerton F.R.S. 1886–1959. A Memoir with Papers*. Privately printed.

Einstein, A. (1931). *Rhodes Lecture 1931. Theory of Relativity. Its Formal Content and Present Problems*. Oxford: Oxford University Press.

—— (1933). *On the Method of Theoretical Physics*. Oxford: Clarendon Press.

E[lliott], E. B. (1916–17). 'William Esson (1839–1916)'. *Proceedings of the Royal Society of London*, series A, 93. 'Obituary notices of fellows deceased', liv–lvii.

Engel, A. J. (1983). *From Clergyman to Don. The Rise of the Academic Profession in Nineteenth-Century Oxford*. Oxford: Clarendon Press.

Evans, R. (ed.) (1996). *Register of Rhodes Scholars, 1903–1995*. Oxford: Rhodes Trust.

Falconer, I. (1985). 'Theory and experiment in J. J. Thomson's work on gaseous discharge'. Ph.D. thesis, University of Bath.

—— (1987). 'Corpuscles, electrons and cathode rays: J. J. Thomson and the "Discovery of the electron" '. *The British Journal for the History of Science*, 20: 241–76.

—— (1988). 'J. J. Thomson's work on positive rays, 1906–1914'. *Historical Studies in the Physical Sciences*, 18, part 2: 265–310.

—— (1989). 'J. J. Thomson and "Cavendish Physics" ', in James (1989), 104–17.

Farren, W. S. (1961). 'Henry Thomas Tizard 1885–1959'. *Biographical Memoirs of the Fellows of the Royal Society*, 7: 313–48.

Fauvel, J., Flood, R., and Wilson, R. (eds.) (2000). *Oxford Figures. 800 Years of the Mathematical Sciences*. Oxford: Oxford University Press.

Feather, N. (1971). 'Richard Whiddington 1885–1970'. *Biographical Memoirs of Fellows of the Royal Society*, 17: 741–56.

Ferry, G. (1998). *Dorothy Hodgkin. A Life*. London: Granta.

Final Report of Her Majesty's Commissioners appointed to inquire into Accidents in Mines and the Possible Means of preventing their Occurrence or limiting their Disastrous Consequences, together with Evidence and Appendices [Cd. 4699] (London: HMSO, 1886).

Fleming, J. A. (1934). *Memories of a Scientific Life*. London: Marshall, Morgan and Scott.

Fletcher, L. (1892). *The Optical Indicatrix and the Transmission of Light in Crystals*. London: Henry Frowde.

Forgan, S., and Gooday, G. J. N. (1994). ' "A fungoid assemblage of buildings": diversity and adversity in the development of college architecture and scientific education in nineteenth-century South Kensington'. *History of Universities*, 13: 153–92.

Forman, P., Heilbron, J. L., and Weart, S. (1975). 'Physics circa 1900: personnel, funding, and productivity of the academic establishments'. *Historical Studies in the Physical Sciences*, 5: 1–185.

Fort, A. (2003). *Prof. The Life of Frederick Lindemann*. London: Jonathan Cape.

Foster, J. (1885). *Men-at-the-Bar. A Biographical Handlist of the members of the Various Inns of Court, including Her Majesty's Judges, etc.* London: Reeves and Turner.

—— (1893). *Oxford Men & Their Colleges*. Oxford: James Parker.

Fox, R. (1997). 'The University Museum and Oxford science, 1850–1880', in Brock and Curthoys (1997), 641–91.

——, and Guagnini, A. (1999). *Laboratories, Workshops, and Sites. Concepts and Practices of Research in Industrial Europe, 1800–1914*. Berkeley: Office for the History of Science and Technology, University of California; also in *Historical Studies in the Physical and Biological Sciences*, 29, part 1 (1988), 53–139, and part 2 (1999), 191–294.

Frank, F. C., and Perkins, D. H. (1971). 'Cecil Frank Powell 1903–1969'. *Biographical Memoirs of Fellows of the Royal Society*, 17: 541–63.

Freundlich, E. (1920). *The Foundations of Einstein's Theory of Gravitation* (trans. by H. L. Brose). Cambridge: Cambridge University Press.

Froude, W. (1877). 'On a new dynamometer for measuring the power delivered to the screws of large ships'. *Proceedings of the Institution of Mechanical Engineers*, 28: 237–60.

Fudano, J. (1990). 'Early X-ray research at physical laboratories in the United States of America, circa 1900: a reappraisal of American physics'. Ph.D. thesis, University of Oklahoma.

Galison, P. (1997). *Image and Logic. A Material Culture of Microphysics*. Chicago and London: University of Chicago Press.

——, and Assmus, A. (1989). 'Artificial clouds, real particles', in Gooding *et al.* (1989).

Ganot, A. (1866). *Elementary Treatise on Physics Experimental and Applied for the Use of Colleges and Schools* (trans. by E. Atkinson, 2nd edition). London: H. Baillière.

Gardiner, R. B. (1895). *The Registers of Wadham College, Oxford. (Part II.) From 1719 to 1871*. London: George Bell and Sons.

Gardner, J. A. (1934). 'Victor Herbert Veley. 1856–1933'. *Journal of the Chemical Society*, 1934, Part I: 570–3.

Gavroglu, K. (1995). *Fritz London. A Scientific Biography*. Cambridge: Cambridge University Press.

Gill, E. W. B. (1934). *War, Wireless and Wangles*. Oxford: Basil Blackwell.

——, and Pidduck, F. B. (1908). 'The genesis of ions by collision of positive and negative ions in a gas: experiments on argon and helium'. *Philosophical Magazine*, 6th ser., 16: 280–90.

Gingras, Y. (1987). 'La réception des rayons X au Québec: radiographie des pratiques scientifiques', in M. Fournier, Y. Gingras, and O. Keel (eds.), *Sciences et médecine au Québec. Perspectives sociohistoriques*. Montreal: Institut québecois de Recherches sur la Culture, 69–86.

—— (1991). *Physics and the Rise of Scientific Research in Canada* (trans. by Peter Keatring). Montreal: McGill-Queen's University Press.

Girtin, T. (1964). *The Triple Crowns. A Narrative History of the Drapers' Company, 1364–1964*. London: Hutchinson.

Glazebrook, R. T. (1910). 'The Rayleigh Period', in [Anon.] (1910), 40–74.

G[lazebrook], R. T. (1921). 'Robert Bellamy Clifton, 1836–1921'. *Proceedings of the Royal Society*, series A, 99: 'Obituary notices of fellows deceased', vi-viii.

Goldman, L. N. (1995). *Dons and Workers. Oxford and Adult Education since 1850*. Oxford: Clarendon Press.

Goldsmith, M. (1980). *Sage. A Life of J. D. Bernal*. London: Hutchinson.

Gooday, G. J. N. (1989). 'Precision measurement and the genesis of physics teaching laboratories in Victorian Britain'. Ph.D. thesis, University of Kent at Canterbury.

—— (1990). 'Precision measurement and the genesis of physics teaching laboratories in Victorian Britain'. *The British Journal for the History of Science*, 23: 25–51.

—— (1991). 'Teaching telegraphy and electrotechnics in the physics laboratory: William Ayrton and the creation of an academic space for electrical engineering in Britain 1873–84'. *History of Technology*, 13: 73–111.

—— (1997). 'Instrumentation and interpretation: managing and representing the working environments of Victorian experimental science', in Lightman (1997), 409–37.

—— (2001). 'The questionable matter of electricity: the reception of J. J. Thomson's "corpuscle" among electrical theorists and technologists', in Buchwald and Warwick (2001), 101–34.

—— (2004). *The Morals of Measurement. Accuracy, Irony, and Trust in late Victorian Electrical Practice*. Cambridge: Cambridge University Press.

—— (2005). 'Fear, shunning, and valuelessness: controversy over the use of "Cambridge" mathematics in late Victorian electro-technology', in D. Kaiser (ed.), *Pedagogy and the Practice of Science*. London and Cambridge, MA: MIT Press, 111–49.

——, and Low, M. F. (1998–9). 'Technology transfer and cultural exchange: Western scientists and engineers encounter late Tokugawa and Meiji Japan'. *Osiris*, 2nd ser., 13: 99–128.

Goodeve, C. F. (1972). 'Frank Edward Smith 1876–1970'. *Biographical Memoirs of Fellows of the Royal Society*, 18: 525–48.

Gooding, D. C., Pinch, T. J., and Schaffer, S. J. (eds.) (1989). *The Uses of Experiment. Studies in the Natural Sciences*. Cambridge: Cambridge University Press.

Goodman, P. (ed.) (1981). *Fifty Years of Electron Diffraction. In Recognition of Fifty Years of Achievement by the Crystallographers and Gas Diffractionists in the Field of Electron Diffraction*. Dordrecht, Boston, and London: D. Reidel.

Gould, P. (1997). 'Women and the culture of university physics in late nineteenth-century Cambridge'. *The British Journal for the History of Science*, 30: 127–49.

Gowing, M. (1964). *Britain and Atomic Energy 1939–1945*. London: Macmillan.

Gray, A. (1896–7). 'Lord Kelvin's laboratory in the University of Glasgow'. *Nature*, 55: 486–92.

Green, R. L. (1953–4). *The Diaries of Lewis Carroll* (2 volumes, continuously paginated). London: Cassell.

Griffin, J. J., & Sons (1910). *The Harcourt Chloroform Inhaler*. London: John J. Griffin & Sons.

Griffiths, E. (1941). 'George William Clarkson Kaye 1880–1941'. *Obituary Notices of Fellows of the Royal Society*, 3: 881–95.

Gunther, A. E. (1967). *Robert T. Gunther. A Pioneer in the History of Science 1869–1940*. Oxford: Printed at the University Press, Oxford, for the subscribers.

Gunther, R. T. (1924). *The Daubeny Laboratory Register 1849–1923*. Oxford. [3 volumes in one, consisting of *A History of The Daubeny Laboratory, Magdalen College, Oxford* (London, 1904), *The Daubeny Laboratory Register 1904–1915* (Oxford, 1916), and *The Daubeny Laboratory Register 1916–1923* (Oxford, 1924)]. Oxford: Oxford University Press.

—— (1933). *Lewis Evans Collection Annual Report*.

—— (1937). *Early Science in Oxford* (14 volumes, 1923–45). Oxford: Printed at the University Press, Oxford, for the author. Volume 11. *Oxford Colleges and their Men of Science*.

—— (1939). 'A historic X-ray apparatus'. *The Times*, 10 November 1939.

Hackmann, W. D. (1984). *Seek & Strike. Sonar, Anti-Submarine Warfare and the Royal Navy, 1914–54*. London: Her Majesty's Stationery Office.

—— (1988). 'Sonar, wireless telegraphy and the Royal Navy: scientific development in a military context, 1890–1939', in Nicolaas A. Rupke (ed.), *Science, Politics and the Public Good. Essays in Honour of Margaret Gowing*. London: Macmillan, 90–118.

Hannabuss, K. C. (2000a). 'Mathematics', in Brock and Curthoys (2000), 443–55.

—— (2000b). 'The mid-nineteenth century', in Fauvel *et al.* (2000), 186–201.

—— (2000c). 'Henry Smith', in Fauvel *et al.* (2000), 202–17.

Harcourt, A. G. V.: *see* Vernon Harcourt, A. G.

Harley, H. C. (1997). 'The Radcliffe Science Library', in Brock and Curthoys (1997), 692–3.

Harman, P. M. (ed.) (1995). *The Scientific Letters and Papers of James Clerk Maxwell. Volume 2: 1862–1873*. Cambridge: Cambridge University Press.

—— (ed.) (2002). *The Scientific Letters and Papers of James Clerk Maxwell. Volume 3: 1874–89*. Cambridge: Cambridge University Press.

Harrison, B. H. (ed.) (1994). *The History of the University of Oxford. Volume VIII. The Twentieth Century*. Oxford: Clarendon Press.

Harrod, R. F. (1959). *The Prof. A Personal Memoir of Lord Cherwell*. London: Macmillan.

Hartcup, G. (1970). *The Challenge of War. Scientific and Engineering Contributions to World War Two*. Newton Abbot: David and Charles.

—— (1988). *The War of Invention. Scientific Developments, 1914–18*. London: Brassey's.

——, and Allibone, T. E. (1984). *Cockcroft and the Atom*. Bristol: Adam Hilger.

Hartley, H. (1942–4). 'Randal Thomas Mowbray Rawdon Berkeley, Earl of Berkeley 1865–1942'. *Obituary Notices of Fellows of the Royal Society*, 4: 167–82.

—— (1971a). *Studies in the History of Chemistry*. Oxford: Clarendon Press.

—— (1971b). 'The contribution of the college laboratories to the Oxford School of Chemistry', in Hartley (1971a), 223–32.

——, and Gabor, D. (1970). 'Thomas Ralph Merton 1888–1969'. *Biographical Memoirs of Fellows of the Royal Society*, 16: 421–40.

Haselfoot, C. E., and Kirkby, P. J. (1904). 'The electrical effects produced by the explosion of hydrogen and oxygen'. *Philosophical Magazine*, 6th ser., 8: 471–81.

Haslett, A. W. (1937). 'The unending quest: scientific careers in the next thirty years. II – Employment trends'. *Journal of Careers*, 16: 355–9.

Hawkins, C. C. (1922–5). *The Dynamo* (6th edition, 3 volumes). London: Sir Isaac Pitman & Sons.

——, and Wallis, F. (1893). *The Dynamo*. London: Whittaker & Co.

Heilbron, J. L. (1968). 'The scattering of α and β particles and Rutherford's atom', *Archive for History of Exact Sciences*, 4: 247–307.

—— (1974a). *H. G. J. Moseley. The Life and Letters of an English Physicist, 1887–1915*. Berkeley, Los Angeles, and London: University of California Press.

—— (1974b). 'Moseley, Henry Gwyn Jeffreys (1887–1915)', in *DSB*, vol. 9, 542–5.

—— (1979). 'Physics at McGill in Rutherford's time', in M. Bunge and W. R. Shea (eds.), *Rutherford and Physics at the Turn of the Century*. New York: Dawson and Science History Publications, 42–73.

——, and Kuhn, T. S. (1969). 'The genesis of the Bohr atom'. *Historical Studies in the Physical Sciences*, 1: 211–90.

——, and Seidel, R. W. (1989). *Lawrence and his Laboratory. A History of the Lawrence Berkeley Laboratory. Volume 1*. Berkeley, CA: University of California Press.

Hevesy, G. (1948). 'Francis William Aston 1877–1945'. *Obituary Notices of Fellows of the Royal Society*, 5: 635–50.

Hibbert, C. (ed.) (1988). *The Encyclopaedia of Oxford*. London: Macmillan.

Hilliard, E. (1914). *The Balliol College Register 1832–1914*. Oxford: printed for private circulation by Horace Hart at the University Press.

Hoch, P. K. (1983). 'The reception of Central European refugee physicists of the 1930s: USSR, UK, USA'. *Annals of Science*, 40: 206–46.

—— (1991). 'Some contributions to physics by German-Jewish emigrés in Britain and elsewhere', in W. E. Mosse (ed.), *Second Chance. Two Centuries of German-speaking Jews in the United Kingdom*. Tübingen: Mohr, 229–41.

——, and Yoxen, E. J. (1987). 'Schrödinger at Oxford: a hypothetical national cultural synthesis which failed'. *Annals of Science*, 44: 593–616.

Hoddeson, L., Braun, E., Teichmann, J., and Weart, S. R. (eds.) (1992). *Out of the Crystal Maze. Chapters from the History of Solid-State Physics*. New York and Oxford: Oxford University Press.

Home, R. W. (1990). *Physics in Australia to 1945. Bibliography and Biographical Register*. Melbourne: Department of History and Philosophy of Science, University of Melbourne, and National Centre for Research and Development in Australian Studies, Monash University.

Hon, G. (1989). 'Franck and Hertz versus Townsend: a study of two types of experimental error'. *Historical Studies in the Physical Sciences*, 20, part 1: 79–106.

Hong, S. (1994). 'Controversy over voltaic contact phenomena, 1862–1900'. *Archive for History of Exact Sciences*, 47: 233–89.

Hopkins, T. H. T. (1878). 'Specimens of ornamental surface turning'. *The Forge and Lathe*, new ser., no. 17: 278 and 2 plates.

Houghton, J. T. (1988). 'Address [on Claude Hurst 1907–88] in Jesus College chapel'. *Jesus College Record 1988*: 7–9.

——, and Walshaw, C. D. (1977). 'Gordon Miller Bourne Dobson 25 February 1889–11 March 1976'. *Biographical Memoirs of Fellows of the Royal Society*, 23: 41–57.

Howarth, J. (1987). 'Science education in late-Victorian Oxford: a curious case of failure?'. *English Historical Review*, 102: 334–71.

—— (2000). ' "Oxford for Arts": the Natural Sciences, 1880–1914', in Brock and Curthoys (2000), 457–97.

Howorth, M. I. (1953). *Atomic Transmutation. The Greatest Discovery Ever Made from Memoirs of Professor Frederick Soddy, M.A., LL.D., F.R.S. Nobel Laureate 1921.* London: New World Publications.

Hughes, J. (1998a). ' "Modernists with a vengeance": changing cultures of theory in nuclear science, 1920–1930'. *Studies in History and Philosophy of Modern Physics*, 29B: 339–67.

—— (1998b). 'Plasticine and valves: industry, instrumentation and the emergence of nuclear physics', in J.-P. Gaudillère and I. Löwy (eds.), *The Invisible Industrialist. Manufacturers and the Production of Scientific Knowledge.* London: Macmillan, 58–101.

—— (1998c). 'Rutherford, the Cavendish Laboratory and the Solvay Councils', in P. Marage and G. Wallenborn (eds.), *The Solvay Councils and the Birth of Modern Physics.* Basel: Birkhäuser, 24–34.

—— (2000). '1932: The *annus mirabilis* of nuclear physics?'. *Physics World*, 13, number 7: 43–8.

—— (2002). 'Craftsmanship and social service: W. H. Bragg and the modern Royal Institution', in James (2002), 225–47.

—— (2003). 'Radioactivity and nuclear physics', in M. J. Nye (ed.), *The Cambridge History of Science. Volume 5. The Modern Physical and Mathematical Sciences.* Cambridge: Cambridge University Press, 350–74.

Hull, A. (1999). 'War of words: the public science of the British scientific community and the origins of the Department of Scientific and Industrial Research, 1914–16'. *The British Journal for the History of Science*, 32: 461–81.

Humphries, A. E. (1970). 'A history of the college laboratories in Oxford'. Honour School of Natural Science, Chemistry Part II thesis, University of Oxford.

Hunter, G. K. (2004). *Light is a Messenger. The Life and Science of William Lawrence Bragg.* (Oxford and New York: Oxford University Press.

Hurst, H. E. (1906). 'Genesis of ions by collision and sparking-potentials in carbon dioxyde and nitrogen'. *Philosophical Magazine*, 6th ser., 11: 535–52.

—— (1969–70). 'Recollections of the study of physics in Oxford at the beginning of the twentieth century'. *The Oxford Magazine*, 14 November 1969: 59–60.

Hurst, H. E., and Lattey, R. T. (1910). *A Text-Book of Physics.* London: Constable.

—— (1912). *A Text-Book of Physics* (2nd edition, 3 volumes). London: Constable.

Hutchins, R. D. (1990). 'Magdalen's astronomy observatory'. *Magdalen College Record 1990*: 44–51.

—— (1992). 'Professor John Phillips of Oxford, 1853–74. Catalyst for the University Observatory'. Thesis, Honour School of Modern History, University of Oxford.

—— (1994). 'John Phillips, "geologist-astronomer", and The origins of the Oxford University Laboratory, 1853–1873'. *History of Universities*, 13: 193–249.

Hutchinson, K., Gray, J. A., and Massey, H. (1981). 'Charles Drummond Ellis, 11 August 1895–10 January 1980'. *Biographical Memoirs of Fellows of the Royal Society*, 27: 199–233.

Im, G. S. (1995), 'The formation and development of the Ramsauer effect'. *Historical Studies in the Physical Sciences*, 25, part 2: 269–300.

Ing, H. R. (1957). 'Allan Frederick Walden 1871–1956'. *Proceedings of the Chemical Society*, August 1957: 237.

James, F. A. J. L. (ed.) (1989). *The Development of the Laboratory. Essays on the Place of Experiment in Industrial Civilization*. Basingstoke: Macmillan.

—— (2002). *'The Common Purposes of Life'. Science and Society at the Royal Institution of Great Britain*. Aldershot: Ashgate.

Jenkin, C. F. (1908). *Engineering Science. An Inaugural Lecture on the Training for the Engineering Profession delivered before the University, Oct. 16, 1908*. Oxford: Clarendon Press.

—— (1911–12). 'Report of the professor of engineering science, 1911'. OUG, 42: 988 (19 June 1912).

Jervis-Smith, F. J. (1879). 'A liquid rheostat'. *Nature*, 20: 552.

—— (1883). 'On a new form of ergometer'. *Philosophical Magazine*, 5th ser., 15: 87–90.

—— (1884). *On Some New Forms of Work-Measuring Machines as applied to Dynamos and Electro-Motors*. London: E. & F. N. Spon.

—— (1885). 'On an ergometer for small electromotors' [abstract], *Proceedings of the Bristol Naturalists' Society*, new ser., 4: 143–4.

—— (1888). 'The application of hydraulic power to mercurial pumps'. *Philosophical Magazine*, 5th ser., 25: 313–14.

—— (1889a). 'A continuous heat and electrical-current measuring-instrument'. *Philosophical Magazine*, 5th ser., 27: 28–9.

—— (1889b). 'A water-jacketted flexible tube'. *Chemical News*, 60: 187.

—— (1889c). 'An experimental investigation of the circumstances under which a change of the velocity in the propagation of the ignition of an explosive gaseous mixture takes place in closed and open vessels. Part I. Chronographic measurements' [abstract]. *Proceedings of the Royal Society of London*, 45: 451–2.

—— (1890a). 'A mercury-still for the rapid distillation of mercury in a vacuum', *Philosophical Magazine*, 5th ser., 29: 501–3.

—— (1890b). 'A new form of electric chronograph'. *Philosophical Magazine*, 5th ser., 29: 377–83.

—— (1890–2). 'Inductoscript'. *Proceedings of the Physical Society*, 11: 353–6.

—— (1891a). 'On some new methods of investigating the points of recalescence in steel and iron'. *Philosophical Magazine*, 5th ser., 31: 433–6.

—— (1891b). *Notes of a Lecture on some Methods of Recording some of the Phenomena of Recalescence, and also Electrical and Magnetic Phenomena*. Oxford, privately printed.

—— (1892). 'An acoustic method whereby the depth of water in a river may be measured at a distance'. *Nature*, 46: 246.

—— (1893a). 'Inducto-script'. *Report of the Sixty-second Meeting of the British Association for the Advancement of Science held at Edinburgh in August 1892*. London: John Murray, 644–5.

—— (1893b). 'Inducto-script'. *The Camera*, 6: 139–40.

—— (1893c). 'A water-cooled brake ergometer'. *Engineering*, 56: 280.

—— (1894a). *A Torsion Ergometer (Dynamometer) . . . and Mechanical Integrator*. Oxford.

—— (1894b). 'The penetrative power of bullets'. *Nature*, 50: 124 (7 June 1894); also 174 (21 June 1894), in reply to E. Ball's letter on 'Bullet-proof shields', 148 (14 June 2004).
—— (1896a). 'Photography and chronographic measurements'. *Nature*, 53: 206.
—— (1896b). 'A note on the Tesla spark and X-ray photography'. *Nature*, 54: 594–5.
—— (1896c). 'An optical rotostat'. *Engineering*, 61: 590.
—— (1897–8). 'A carbon-detector or receiver for Hertz waves', *The Electrician*, 40: 84–5.
—— (1898a). 'A new method of measuring the torsional angle of a rotating shaft or spiral spring', *Philosophical Magazine*, 5th ser., 45: 183–5.
—— (1898b). 'Electric light wires as telephonic circuits'. *Nature*, 58: 51.
—— (1901a). 'Copies by phosphorescence'. *The British Journal of Photography*, 48: 146 (8 March 1901).
—— (1901b). 'Phosphorescence as a source of illumination in photography', *Nature*, 63: 421.
—— (1901c). 'The rolling angle of a ship found by photography'. *Nature*, 64: 576.
—— (1902). 'A high pressure spark-gap used in connexion with the Tesla coil'. *Philosophical Magazine*, 6th ser., 4: 224–6.
—— (1903). *The Tram Chronograph*. Oxford: Horace Hart, printer to the University.
—— (1908a). 'On the generation of a luminous glow in an exhausted receiver moving near an electrostatic field, and the action of a magnetic field on the glow so produced'. *Proceedings of the Royal Society of London*, series A, 80: 212–17.
—— (1908b). 'Further note on a luminous glow generated by electrostatic induction in an exhausted vessel made of silica'. *Proceedings of the Royal Society of London*, series A, 81: 214–16.
—— (1908c). *Evangelista Torricelli written on the Occasion of the Tercentenary Commemoration of the Italian Philosopher, at Faenza, October 15 and 16, 1908*. Oxford: Oxford University Press.
—— (1910a). 'Liquid microphones'. *The Electrician*, 65: 616.
—— (1910b). 'Chronograph', in *Encyclopaedia Britannica* (11th edition), vol. 6. Cambridge: Cambridge University Press, 301–5.
—— (1915). *Dynamometers. Edited and amplified by Charles Vernon Boys, F.R.S.* London: Constable.
—— (1926). 'Chronograph' in *Encyclopaedia Britannica* (13th edition), vol. 6. London and New York: Encyclopaedia Britannica Company: 301–5.
Johnson, A. H. (1914–1922). *The History of the Worshipful Company of the Drapers of London* (5 volumes). Oxford: Clarendon Press.
Jones, F. L.: *see* Llewellyn Jones, F.
Jones, K. W. (1915). *Life of John Viriamu Jones*. London: Smith, Elder.
Jones, M. J. (1997). 'The agricultural depression, collegiate finances, and provision for education at Oxford, 1871–1913'. *Economic History Review*, 5: 57–81.
Jones, R. V. (1957). 'Obituaries. The Right Hon. Viscount Cherwell, P.C., C.H., F.R.S.'. *Nature*, 180: 579–81.
—— (1961). 'Infra-red detection in British air defence, 1935–38'. *Infrared Physics*, 1: 153–62.
—— (1966). 'Winston Leonard Spencer Churchill 1874–1965'. *Biographical Memoirs of Fellows of the Royal Society*, 12: 35–105.
—— (1978). *Most Secret War*. London: Hamish Hamilton.
—— (1979). *Most Secret War*. London: Coronet.

—— (1986–7). 'Lindemann beyond the Laboratory'. *Notes and Records of the Royal Society of London*, 41: 191–210.

—— (1987). 'Oxford physics in transition: 1929–39', in Williamson (1987), 113–26.

F.S.K. (1926). 'Edwin Henry Barton, 1858–1925'. *Proceedings of the Royal Society of London*, series A, 111: 'Obituary notices of fellows deceased', xl–xliii.

Kargon, R. H. (1977). *Science in Victorian Manchester. Enterprise and Expertise*. Manchester: Manchester University Press.

K[eeley], T. C. (1945–6). 'Physics in Oxford during the war (the Clarendon Laboratory). I. Radar'. *The Oxford Magazine*, 64: 289–91.

Keeley, T. C. (1958). 'Lord Cherwell'. *Year Book of the Physical Society 1958*: 79–82.

Keesom, W. H. (1942). *Helium*. Amsterdam and London: Elsevier.

Keith, S. T., and Hoch, P. K. (1986). 'Formation of a research school: theoretical solid state physics at Bristol 1930–54'. *The British Journal for the History of Science*, 19: 19–44.

Keith, S. T. (1984). 'Scientists as entrepreneurs: Arthur Tyndall and the rise of Bristol physics'. *Annals of Science*, 41: 335–57.

Kennedy, A. B. W. (1886). 'The use and equipment of engineering laboratories'. *Minutes of Proceedings of the Institution of Civil Engineers*, 88: 1–80.

—— (1894). 'The critical side of mechanical training'. *Report of the Sixty-Fourth Meeting of the British Association for the Advancement of Science held at Oxford in August 1894*. London: John Murray, 1894, 739–47; also in *The Electrician*, 33 (1894), 450–3.

Kent, P. W. (2001). *Some Scientists in the Life of Christ Church, Oxford*. Oxford: Oxford University Press.

Kerst, D. W. (1946). 'Historical development of the betatron'. *Nature*, 157: 90–5.

Kim, Dong-Won (1995). 'J. J. Thomson and the emergence of the Cavendish School, 1885–1900'. *The British Journal for the History of Science*, 28: 191–226.

—— (2002). *Leadership and Creativity. A History of the Cavendish Laboratory, 1871–1919*. Dordrecht and London: Kluwer Academic.

King, M. C. (1984). 'The course of chemical change: the life and times of Augustus G. Vernon Harcourt (1834–1919)'. *Ambix*, 31: 16–31.

Kirkby, P. J. (1902). 'On the electrical conductivities produced in air by the motion of negative ions'. *Philosophical Magazine*, 6th ser., 3: 212–25.

—— (1904). 'The effect of the passage of electricity through a mixture of oxygen and hydrogen at low pressures'. *Philosophical Magazine*, 6th ser., 7: 223–32.

—— (1905). 'The union of hydrogen and oxygen at low pressures through the passage of electricity'. *Philosophical Magazine*, 6th ser., 9: 171–85.

——, and Marsh, J. E. (1913). 'Some electrical and chemical effects of the explosion of azoimide'. *Proceedings of the Royal Society of London*, series A, 88: 90–9.

Koenigsberger, L. (1906). *Hermann von Helmholtz* (trans. by F. A. Welby). Oxford: Clarendon Press.

Kohler, R. E. (1991). *Partners in Science. Foundations and Natural Scientists 1900–1945*. Chicago and London: University of Chicago Press.

Kuhn, H. G. (1934). *Atomspektren*. Leipzig: Akademischer Verlag Gesellschaft.

Kuhn, H. G., and Hartley, C. (1983). 'Derek Ainslie Jackson, 23 June 1906–20 February 1982'. *Biographical Memoirs of Fellows of the Royal Society*, 29: 269–96.

Kuhn, T. S. (1962). *The Structure of Scientific Revolutions*. Chicago: University of Chicago Press.

Kurti, N. (1958). 'Franz Eugen Simon 1893–1956'. *Biographical Memoirs of Fellows of the Royal Society*, 4: 225–56.

—— (1983). 'Leslie Fleetwood Bates, 7 March 1897–20 January 1978'. *Biographical Memoirs of Fellows of the Royal Society*, 29: 1–25.

—— (1984). 'Opportunity lost in 1865?'. *Nature*, 308: 313–14.

—— (1985). 'Helmholtz's choice'. *Nature*, 314: 499.

—— (1987). 'Undergraduate in Paris 1926–8; graduate student and Dr. Phil. in Berlin, 1928–31', in Williamson (1987), 86–90.

—— (1990). 'Reflections of an amateur "historian of science" ', in J. J. Roche (ed.), *Physicists Look Back. Studies in the History of Physics*. Bristol and New York: Adam Hilger, 78–87.

—— (1991). *Oxford Reminiscences. The after-dinner Speech at the Workshop Banquet by Nicholas Kurti, F.R.S. With an introduction by Norman Booth*. Privately printed. Copy in CLA.

Laidler, K. J. (1988). 'Chemical kinetics and the Oxford college laboratories'. *Archive for History of Exact Sciences*, 38: 197–283.

Lang, H. R. (1935). 'Conference on Industrial Physics'. *Nature*, 135: 555–6.

Langevin, P. (1900a). 'Sur l'ionisation des gaz'. *Bulletin des séances de la Société française de physique*: $1^{\text{ère}}$ partie, 39.

—— (1900b). 'Les ions dans les gaz'. *Bulletin de la Société internationale des électriciens*, 17: 203–22.

—— (1902). 'Recherches sur les gaz ionisés'. *Bulletin des séances de la Société française de physique*: 2e partie, 45–7.

—— (1903). 'Recombinaison et mobilité des ions dans les gaz'. *Annales de chimie et de physique*, 7th ser., 28: 433–530.

—— (1906). 'Recherches récentes sur le mécanisme de la décharge disruptive'. *Le Radium*, 3: 107–15.

—— (1909). *Notice sur les travaux scientifiques de M. Paul Langevin*. Paris: Société générale d'imprimerie et d'édition.

—— (1931). 'La physique au Collège de France', in *Le Collège de France (1530–1930). Livre jubilaire composé à l'occasion de son quatrième centenaire*. Paris: Presses universitaires de France, 61–79.

Lanouette, W. [with B. Silard] (1992). *Genius in the Shadows. A Biography of Leo Szilard. The Man behind the Bomb*. New York: Charles Scribner's Sons.

Latham, C., and Stobbs, A. (1996). *Radar. A Wartime Miracle*. Stroud: Alan Sutton.

—— (1999). *Pioneers of Radar*. Stroud: Alan Sutton.

Law, B. R. (1998). *Building Oxford's Heritage. Symm & Company from 1815*. Oxford: Prelude Promotion, for Symm and Company.

Lelong, B. (2001). 'Paul Villard, J.-J. Thomson, and the composition of cathode rays', in Buchwald and Warwick (2001), 135–67.

The Letters of Mercurius (1970). London: John Murray.

Lightman, B. (ed.) (1997). *Victorian Science in Context*. Chicago and London: University of Chicago Press.

Lindemann, F. A. (1932). *The Physical Significance of the Quantum Theory*. Oxford: Clarendon Press.

—— (1933). 'The place of mathematics in the interpretation of the universe'. *Philosophy*, 8: 14–29.

—— (1938). 'Designing a new physics laboratory'. *Oxford*, 5: 53–62.

L[indemann], F. A. (1938–9). 'Scientific research in national life'. *Oxford Magazine*, 57: 496–7.

Lindemann, F. A., and Keeley, T. C. (1933). 'Helium liquefaction plant at the Clarendon Laboratory, Oxford'. *Nature*, 131: 191–2.

——, and Simon, F. E. (1942). 'Walther Nernst 1864–1941'. *Obituary Notices of Fellows of the Royal Society*, 4: 101–12.

Little, R. (1995). 'Prof Ozone and his garden-shed'. *Oxford Times*, 19 May 1995.

Llewellyn Jones, F. (1957). 'John Sealy Edward Townsend'. *Year Book of the Physical Society 1957*: 106–10.

Lodge, O. (1914). 'The discovery of radioactivity, and its influence on the course of physical science [Becquerel Memorial Lecture]'. *Memorial Lectures delivered before the Chemical Society, 1901–1913. Volume II*. London: Gurney & Jackson, 217–54.

London, F., and London, H. (1935). 'The electromagnetic equations of the supraconductor'. *Proceedings of the Royal Society of London*, series A, 149: 71–88.

Long, D. A. (1989). 'The Sir Leoline Jenkins laboratories'. *Jesus College Record 1989*: 17–20.

—— (1995–6). 'Sir Leoline Jenkins laboratories 1907–47'. *Jesus College Record 1995–6*: 46–57.

Lovell, A. C. B. (1964). 'Joseph Lade Pawsey 1908–1962'. *Biographical Memoirs of Fellows of the Royal Society*, 10: 229–43.

—— (1987). 'Bristol and Manchester – the years 1931–9', in Williamson (1987), 148–60.

Lukes, S. (1987). 'Interview with Heini Kuhn'. *Balliol College Record 1987*: 42–53.

Macfarlane, G. G., and Hilsum, C. (1990). 'Francis Edgar Jones, 16 January 1914–10 April 1988'. *Biographical Memoirs of Fellows of the Royal Society*, 35: 181–99.

McGraw-Hill Modern Scientists and Engineers. Vol 2. H-Q. New York: McGraw-Hill.

MacLeod, R. M. (1972). 'Resources of science in Victorian England: the Endowment of Science movement, 1868–1900', in P. Mathias (ed.), *Science and Society 1600–1900*. Cambridge: Cambridge University Press, 111–66.

MacLeod, R. M., and Andrews, E. K. (1968). 'Scientific careers of 1851 Exhibition scholars'. *Nature*, 218: 1011–16.

—— (1971). 'Scientific advice in the war at sea, 1915–1917: the Board of Invention and Research'. *Journal of Contemporary History*, 6, part 2: 3–40.

MacLeod, R. M., and Moseley, R. (1980). 'The "Naturals" and Victorian Cambridge: reflections on the anatomy of an elite, 1851–1914'. *Oxford Review of Education*, 6: 177–95.

—— (1982). 'Breaking the circle of the sciences: the Natural Sciences Tripos and the "examination revolution" ', in R. M. MacLeod (ed.), *Days of Judgment. Science, Examinations and the Organization of Knowledge in late Victorian England*. Driffield: Nafferton, 189–92.

Magnello, E. (2000). *A Century of Measurement. An Illustrated History of the National Physical Laboratory*. Bath: Canopus.

Makower, W., and Geiger, H. (1912). *Practical Measurements in Radio-activity*. London: Longmans, Green.

Mallock, A. (1917–19). 'Growth of trees, with a note on interference bands formed by rays at small angles'. *Proceedings of the Royal Society of London*, series B, 90: 186–99.

Manley, J. J. (1906–7). 'On the application of a differential densimeter to the study of some Mediterranean waters'. *Proceedings of the Royal Society of Edinburgh*, 27: 210–32.

Marage, P., and Wallenborn, G. (eds.) (1995). *Les Conseils Solvay et les débuts de la physique moderne*. Brussels: Université libre de Bruxelles.

Marsden, B. (1992). 'Engineering science in Glasgow: economy, efficiency and measurement as prime movers in the differentiation of an academic discipline'. *The British Journal for the History of Science*, 25: 319–46.

Maxwell, J. C. (1873). *A Treatise on Electricity and Magnetism* (2 volumes). Oxford: Clarendon Press.

—— (ed.) (1879). *The Electrical Researches of the Honourable Henry Cavendish, F.R.S. written between 1771 and 1881, edited from the Original Manuscripts in the Possession of the Duke of Devonshire, K.G.* Cambridge: Cambridge University Press.

McCrea, W. H. (1950–1). 'Edward Arthur Milne 1896–1950'. *Obituary Notices of Fellows of the Royal Society*, 7: 421–43.

McLennan, J. C. (1912). 'On the series lines in the arc spectrum of mercury'. *Proceedings of the Royal Society of London*, series A, 87: 256–68.

——, and Henderson, J. P. (1915). 'Ionisation potentials of mercury, cadmium, and zinc, and the single- and many-lined spectra of these elements'. *Proceedings of the Royal Society of London*, series A, 91: 485–91.

Mehra, J. (1975). *The Solvay Conferences on Physics. Aspects of the Development of Physics since 1911*. Dordrecht and Boston: D. Reidel.

Mendelssohn, K. (1960). 'The coming of the refugee scientists'. *The New Scientist*, 7: 1343–4.

—— (1964). 'Prewar work on superconductivity as seen from Oxford'. *Reviews of Modern Physics*, 36: 7–12.

—— (1966). 'The world of cryogenics. IV: The Clarendon Laboratory Oxford'. *Cryogenics*, 6: 129–40.

—— (1973). *The World of Walther Nernst. The Rise and Fall of German Science*. London: Macmillan.

Merricks, L. (1996). *The World Made New. Frederick Soddy, Science, Politics, and Environment*. Oxford: Oxford University Press.

Meyer, O. E. (1877). *Die kinetische Theorie der Gase. In elementarer Darstellung mit mathematischen Zusätzen*. Breslau: Maruschke and Berendt.

—— (1899). *The Kinetic Theory of Gases. Elementary Treatise with Mathematical Appendices* (trans. by R. E. Baynes from the second revised edition). London: Longmans, Green.

Milne, E. A. (1929). *The Aims of Mathematical Physics. An Inaugural Lecture delivered before the University of Oxford on 19 November 1929*. Oxford: Clarendon Press.

Moon, P. B. (1977). 'George Paget Thomson, 1892–1975'. *Biographical Memoirs of Fellows of the Royal Society*, 23: 529–56.

Moore, J. J. (1878). *The Historical Handbook and Guide to Oxford, embracing a succinct History of the University and the City from the Year 912* (2nd edition). Oxford: Thos. Shrimpton and Son.

Moore, W. J. (1989). *Schrödinger. Life and Thought*. Cambridge: Cambridge University Press.

Morgan, B. L. (1988). 'James Dwyer McGee, 1903–1987'. *Biographical Memoirs of Fellows of the Royal Society*, 34: 513–51.

Morrell, J. B. (1992). 'Research in physics at the Clarendon Laboratory, Oxford, 1919–1939'. *Historical Studies in the Physical and Biological Sciences*, 22: 263–307.

Morrell, J. B. (1993). 'W. H. Perkin, Jr., at Manchester and Oxford: from Irwell to Isis'. *Osiris*, 8: 104–26.

—— (1994). 'The non-medical sciences, 1914–1939', in Harrison (1994), 139–63.

—— (1997). *Science at Oxford 1914–1939. Transforming an Arts University*. Oxford: Clarendon Press.

—— (2005). *John Phillips and the Business of Victorian Science*. Aldershot: Ashgate.

Morton, A. Q. (2003). 'The electron made public: the exhibition of pure science in the British Empire Exhibition, 1924–5', in B. S. Finn, R. Bud, and H. Trischler (eds.), *Exposing Electronics*. London: Science Museum, 25–43.

Morton, V. (1987). *Oxford Rebels. The Life and Friends of Nevil Story Maskelyne 1823–1911, Pioneer Oxford Scientist, Photographer and Politician*. Gloucester: Alan Sutton.

Moseley, R. (1976). 'Science, government and industrial research: the origins and development of the National Physical Laboratory, 1900–1975'. Ph.D. thesis, University of Sussex.

—— (1977). 'Tadpoles and frogs: some aspects of the professionalization of British physics, 1870–1939'. *Social Studies of Science*, 7: 423–46.

Moss, M. S., and Russell, I. (1988). *Range and Vision. The First Hundred Years of Barr & Stroud*. Edinburgh: Mainstream.

M[ott], N. F. (1932). 'Quantum theory'. *Nature*, 130: 330–1.

Nayler, J. L. (1966). 'The Aeronautical Research Council'. *Journal of the Royal Aeronautical Society*, 70: 79–82.

New, R. (1990). 'Lord Berkeley 1865–1942 and the Foxcombe laboratory'. *Boars Hill Association Newsletter*, 87: 4.

Newitt, D. M. (1960). 'Alfred Charles Glyn Egerton 1886–1959'. *Biographical Memoirs of Fellows of the Royal Society*, 6: 39–64.

O'Dea, W. T. (1934). *Handbook of the Collections Illustrating Electrical Engineering. II. Radio Communication. Part I. – History and Development*. London: HMSO.

Ogston, A. G. (1973). 'Harold Brewer Hartley 1878–1972'. *Biographical Memoirs of Fellows of the Royal Society*, 19: 349–73.

Oldroyd, D. R., and Hutchings, D. W. (1979). 'The chemical lectures at Oxford (1822–1854) of Charles Daubeny, M.D., F.R.S'. *Notes and Records of the Royal Society of London*, 33: 217–59.

Pais, A. (1991). *Niels Bohr's Times, in Physics, Philosophy, and Polity*. Oxford: Clarendon Press.

Pattison, M. (1868). *Suggestions on Academical Organisation with especial reference to Oxford*. Edinburgh: Edmonston and Douglas.

Pattison, M. (1983). 'Scientists, inventors and the military in Britain, 1915–19: the Munitions Inventions Department'. *Social Studies of Science*, 13: 521–68.

Perry, J. (1903–4). 'Oxford and science'. *Nature*, 69: 207–14 (with additional comments, 270).

Phillips, D. (1979). 'William Lawrence Bragg, 31 March 1890–1 July 1971'. *Biographical Memoirs of Fellows of the Royal Society*, 25: 75–143.

Phillips, J. (1868). Obituary of Charles Daubeny. *Proceedings of the Ashmolean Society*, 3, new ser., no. 5: 8–22 (meeting of 17 February 1868).

Phillips, V. (comp.) (2001). *Royal Commission for the Exhibition of 1851. Record of Award Holders in Science, Engineering and the Arts, 1891 to 2000*. London: Royal Commission for the Exhibition of 1851.

Picken, D. K. (1948). 'Thomas Howell Laby 1880–1946'. *Obituary Notices of Fellows of the Royal Society*, 5: 733–55.

Pine, L. G. (1972). *The New Extinct Peerage 1884–1971. Containing Extinct, Abeyant, Dormant and Suspended Peerages with Genealogies and Arms*. London: Heraldry Today.

Pippard, B. (1998). 'Sir Nevill Francis Mott, 30 September 1905–8 August 1996'. *Biographical Memoirs of Fellows of the Royal Society*, 44: 315–28.

Plarr, V. G. (ed.) (1899). *Men and Women of the Time* (15th edition). London: George Routledge & Sons.

Poulton, E. B. (1911). *John Viriamu Jones and other Oxford Memories*. London: Longmans, and New York: Green.

Prest, J. M. (ed.) (1993). *The Illustrated History of Oxford University*. Oxford and New York: Oxford University Press.

Price, B. (1848). *A Treatise on the Differential Calculus, and its Application to Geometry, founded chiefly on the Method of Infinitesimals*. London: George Bell; Oxford: J. H. Parker; Cambridge: J. and J. J. Deighton.

—— (1852–60). *A Treatise on Infinitesimal Calculus, containing Differential and Integral Calculus, Calculus of Variations, Applications to Algebra and Geometry, and Analytical Mechanics* (4 volumes). Oxford: Clarendon Press.

Price, D. de Solla (1963). *Little Science, Big Science*. New York and London: Columbia University Press.

Pyatt, E. C. (1983). *The National Physical Laboratory. A History*. Bristol: Adam Hilger.

A.R. (1926). 'Andrew Gray – 1847–1925'. *Proceedings of the Royal Society of London*, series A, 110: 'Obituary notices of fellows deceased', xvi–xix.

Ratcliffe, J. A. (1966). 'Edward Victor Appleton 1892–1965'. *Biographical Memoirs of Fellows of the Royal Society*, 12: 1–21.

—— (1975). 'Robert Alexander Watson-Watt, 13 April 1892–5 December 1973'. *Biographical Memoirs of Fellows of the Royal Society*, 21: 549–68.

Rayleigh, Lord (third Baron), and Sidgwick, Mrs H. (E.) (1884). 'On the electro-chemical equivalent of silver, and on the absolute electromotive force of Clark cells'. *Philosophical Transactions of the Royal Society of London*, 175: 411–60.

Reader, W. J. (1975). *Imperial Chemical Industries. A History. Volume II. The First Quarter-Century 1926–1952*. London: Oxford University Press.

Reiche, F. (1922). *The Quantum Theory* (trans. by H. S. Hatfield and H. L. Brose). London: Methuen.

Rider, R. (1984). 'Alarm and opportunity: emigration of mathematicians and physicists to Britain and the United States, 1933–1945'. *Historical Studies in the Physical Sciences*, 15, part 1: 107–76.

Robb-Smith, A. H. T. (1997). 'Medical education', in Brock and Curthoys (1997), 563–82.
Roberts, R. W. (1944). 'Obituaries. Prof. L. R. Wilberforce'. *Nature*, 153: 517–18.
Roche, J. J. (1994). 'The non-medical sciences, 1939–1970', in Harrison (1994), 251–89.
Rolleston, G. (1870). *Forms of Animal Life, being Outlines of Zoological Classification based upon Anatomical Investigation and Illustrated by Descriptions of Specimens and of Figures.* Oxford: Clarendon Press.
Roscoe, H. E. (1906). *The Life & Experiences of Sir Henry Enfield Roscoe, D.C.L., LL.D., F.R.S. written by himself.* London and New York: Macmillan.
Rücker, A. (1873–4). 'On the adiabatics and isothermals of water'. *Proceedings of the Royal Society*, 22: 451–61.
Rücker, A. W. (1894). 'Address to Section A – Mathematical and Physical Science'. *Report of the Sixty-fourth Meeting of the British Association for the Advancement of Science held at Oxford in August 1894.* London: John Murray, 542–54; also in *The Electrician*, 33 (1894), 417–21.
Ruhemann, M., and Ruhemann, B. (1937). *Low Temperature Physics.* Cambridge: Cambridge University Press.
Rupke, N. A. (1997). 'Oxford's scientific awakening and the role of geology', in Brock and Curthoys (1997), 543–62.
Rutherford, E. (1915–16). 'Henry Gwyn Jeffreys Moseley'. *Nature*, 96: 33–4.
——, and Geiger, H. (1908). 'An electrical method of counting the number of α-particles from radio-active substances'. *Proceedings of the Royal Society*, series A, 81: 141–61.

Sanders, J. H. (1997). 'The Clarendon Laboratory Archive at Oxford'. *Bulletin of the Scientific Instrument Society*, no. 54: 10–14.
—— (2000). 'Nicholas Kurti, 14 May 1908–24 November 1998'. *Biographical Memoirs of Fellows of the Royal Society*, 46: 299–316.
Sanderson, M. (1972). *The Universities and British Industry 1850–1970.* London: Routledge & Kegan Paul.
Sarton, G. (1927). 'Moseley: the numbering of the elements'. *Isis*, 9: 96–111.
Saward, D. (1984). *Bernard Lovell. A Biography.* London: R. Hale.
Scarth, R. N. (1999). *Echoes from the Sky. A Story of Acoustic Defence.* Hythe: Hythe Civic Society.
Schaffer, S. J. (1992). 'Late Victorian metrology and its instrumentation: a manufactory of ohms', in R. Bud and S. Cozzens (eds.), *Invisible Connections. Instruments, Institutions and Science.* Bellingham, Wash: SPIE Optical Engineering Press, 23–56.
—— (1995). 'Accurate measurement is an English science', in M. N. Wise (ed.), *The Values of Precision.* Princeton, NJ: Princeton University Press, 135–72.
Schlick, M. (1920). *Space and Time in Contemporary Physics. An Introduction to the Theory of Relativity and Gravitation* (trans. by H. L. Brose). Oxford: Clarendon Press.
Scholes, P. A. (1938). *The Oxford Companion to Music, Self-indexed and with a Pronouncing Glossary.* London, New York, and Toronto: Oxford University Press.
—— (1945). *The Oxford Companion to Music Self-indexed and with a Pronouncing Glossary* (6th edition). London, New York, and Toronto: Oxford University Press.
—— (1947). *The Mirror of Music 1844–1944. A Century of Musical Life in Britain as reflected in the Pages of the Musical Times* (2 volumes). London: Novello and Oxford University Press.

Scriblerus Redivivus [pseudonym for E. Caswall] (1836). *Pluck Examination Papers for Candidates at Oxford and Cambridge in 1836*. Oxford: Henry Slatter.

Scurlock, R. G. (ed.) (1992). *History and Origins of Cryogenics*. Oxford: Clarendon Press.

Searby, P. (1992). *A History of the University of Cambridge. Volume III, 1750–1870*. Cambridge: Cambridge University Press.

Sherman, A. J. (1973). *Island Refuge. Britain and Refugees from the Third Reich 1933–1939*. London: Elek.

Shimmin, A. N. (1954). *The University of Leeds. The First Half-Century*. Cambridge: Cambridge University Press.

Shinn, T. (1993). 'The Bellevue grand électroaimant, 1900–1940: birth of a research-technology community'. *Historical Studies in the Physical and Biological Sciences*, 24, part 1: 157–87.

Shipley, A. E. (1913). *'J'. A Memoir of John Willis Clark*. London: Smith, Elder.

Shoenberg, D. (1971). 'Heinz London 1907–1970'. *Biographical Memoirs of Fellows of the Royal Society*, 17: 441–61.

—— (1983). 'Kurt Alfred Georg Mendelssohn, 7 January 1906–18 September 1980'. *Biographical Memoirs of Fellows of the Royal Society*, 29: 361–98.

—— (1985). 'Piotr Leonidovich Kapitza, 9 July 1894–8 April 1984'. *Biographical Memoirs of Fellows of the Royal Society*, 31: 325–74.

Shorter, J. (1980). 'A. G. Vernon Harcourt: a founder of chemical kinetics and a friend of "Lewis Carroll" '. *Journal of Chemical Education*, 57: 411–16.

Sibum, H. O. (1995). 'Reworking the mechanical value of heat: instruments of precision and gestures of accuracy in early Victorian England'. *Studies in History and Philosophy of Science*, 26: 73–106.

Simcock, A. V. (1984). *The Ashmolean Museum and Oxford Science, 1683–1983*. Oxford: Museum of the History of Science.

—— (ed.) (1985). *Robert T. Gunther and the Old Ashmolean*. Oxford: Museum of the History of Science.

—— (1993a). 'Robert Walker (1801–1865)', in *DNB Missing Persons*, 695–6.

—— (1993b). 'John Whiteside (1679–1729)', in *DNB Missing Persons*, 712–13.

S[imon], F. E. (1945–6). 'Physics in Oxford during the war (the Clarendon Laboratory). II. Atomic energy'. *The Oxford Magazine*, 64: 352–4.

Simon, F. E. (1956). 'The third law of thermodynamics: an historical survey'. *Year Book of the Physical Society 1956*: 1–22.

Smith, E. C. (1937). *A Short History of Naval and Marine Engineering*. Cambridge: Cambridge University Press.

Smith, E. E. (1975). *Radiation Science at the National Physical Laboratory 1912–1955*. London: HMSO.

Smith, F. G. (1986). 'Martin Ryle, 27 September 1918–14 October 1984'. *Biographical Memoirs of Fellows of the Royal Society*, 32: 497–524.

Smith, F. W. F. (1961): *see* Birkenhead, Second Earl of (1961).

S[mith], S. W. J. (1932). 'Hugh Longbourne Callendar, 1863–1930'. *Proceedings of the Royal Society of London*, series A, 134: 'Obituary notices', xviii-xxvi.

Smith, T. W. M. (1979). 'The Balliol–Trinity Laboratories'. Honour School of Natural Science, Chemistry Part II thesis, University of Oxford.

—— (1982). 'The Balliol–Trinity Laboratories', in J. M. Prest (ed.), *Balliol Studies*. London: Leopard's Head Press, 185–224.

Snow, C. P. (1961). *Science and Government*. London: Oxford University Press.

S[oddy], F. (1933). 'John Watts 1843–1933'. *Journal of the Chemical Society 1933*: 1652–3.

Soddy, F. (1956). *The Cubic Equation with Three Real Roots. Its Geometrical Diagram and a Machine that Solves it* (2nd edition, with addition describing an improved machine). London: New World Publications.

Southwell, R. V. (1939–41). 'Charles Frewen Jenkin 1865–1940'. *Obituary Notices of Fellows of the Royal Society*, 3: 575–85.

Sparrow, J. (1937). *Mark Pattison and the Idea of a University*. Cambridge: Cambridge University Press.

Stanier, R. S. (1958). *Magdalen School. A History of Magdalen College School Oxford*. Oxford: Basil Blackwell.

Steel, J. L. S. (1957). 'Malcolm Percival Applebey 1884–1957'. *Proceedings of the Chemical Society*, July 1957: 214–15.

Stewart, B. (1866). *An Elementary Treatise on Heat*. Oxford: Clarendon Press.

—— (1895). *An Elementary Treatise on Heat* (6th edition). Oxford: Clarendon Press.

S[tocker], W. N. (1921). 'Prof. A. W. Reinold, F.R.S.', *Nature*, 107: 276.

Stogdon, J. H. (ed.) (1937). *The Harrow School Register 1845–1937* (5th edition, 2 volumes). London: Longmans.

Strutt, J. W.: see Rayleigh, Lord (third Baron), and Sidgwick, Mrs. H. (E.) (1884).

Strutt, R. J., Fourth Baron Rayleigh (1924). *John William Strutt. Third Baron Rayleigh, O.M., F.R.S.* London: Edward Arnold.

—— (1942). *Life of Sir J. J. Thomson, O.M., sometime Master of Trinity College, Cambridge*. Cambridge: Cambridge University Press.

Stuewer, R. (1985). 'Artificial disintegration and the Cambridge-Vienna controversy', in P. Achinstein and O. Hannaway (eds.), *Observation, Experiment and Hypothesis in Modern Physical Science*. Cambridge, Mass., and London: MIT Press, 239–307.

Sutcliffe, P. (1978). *The Oxford University Press. An Informal History*. Oxford: Clarendon Press.

Sutherland, L. S., and Mitchell, L. G. (eds.) (1986). *The History of the University of Oxford. Volume V. The Eighteenth Century*. Oxford: Clarendon Press.

Sutton, L. E., and Davies, M. (1996). *The History of the Faraday Society*. Cambridge: Royal Society of Chemistry.

Sviedrys, R. (1976). 'The rise of physics laboratories in Britain'. *Historical Studies in the Physical Sciences*, 7: 405–36.

Szilard, L. (1978). *Leo Szilard. His Version of the Facts. Selected Recollections and Correspondence* (ed. by Spencer R. Weart and Gertrud Weiss Szilard). Cambridge, Mass., and London: MIT Press.

Tansley, A. G. (1936–8). 'Arthur Harry Church 1865–1937'. *Obituary Notices of Fellows of the Royal Society*, 2: 433–43.

Temple, W. (1921). *The Life of Bishop Percival*. London: Macmillan.

ter Haar, D. (1982). 'Obituary. James Howard Eagle Griffiths, 6 December 1908–28 August 1981'. *Magdalen College Record 1982*: 38–43.

Testimonials in favour of George Griffith, M.A., F.C.S., Candidate for the Professorship of Experimental Philosophy of the University of Oxford [1865]. Oxford.

Thomas, J. M., and Phillips, D. (eds.) (1990). *Selections and Reflections. The Legacy of Sir Lawrence Bragg*. London: The Royal Institution of Great Britain.

Thomas, J. S. G. (1949). 'Samuel Walter Johnson Smith 1871–1948'. *Obituary Notices of Fellows of the Royal Society*, 6: 579–98.

Thompson, H. W. (1973). 'Cyril Norman Hinshelwood 1897–1967'. *Biographical Memoirs of Fellows of the Royal Society*, 19: 375–431.

Thompson, S. P. (1884). *Dynamo-Electric Machinery. A Manual*. London: E. & F. N. Spon.

Thomson, G. P. (1958). 'Frederick Alexander Lindemann, Viscount Cherwell 1886–1957'. *Biographical Memoirs of Fellows of the Royal Society*, 4: 45–71.

Thomson, J. J. (1900). 'The genesis of ions in the discharge of electricity through gases'. *Philosophical Magazine*, 5th ser., 50: 278–83.

—— (1903). *Conduction of Electricity through Gases*. Cambridge: Cambridge University Press.

—— (1906). *Conduction of Electricity through Gases* (2nd edition). Cambridge: Cambridge University Press.

—— (1932). 'The growth in opportunities for education and research in physics during the past fifty years'. *British Association for the Advancement of Science. Report of the Centenary Meeting. London–1931. September 23–30*. London: British Association for the Advancement of Science, 19–30.

Thomson, J. J., and Thomson, G. P. (1928–33). *Conduction of Electricity through Gases* (3rd edition, 2 volumes). Cambridge: Cambridge University Press.

Thorpe, J. F. (1932–5). 'Herbert Brereton Baker 1862–1935'. *Obituary Notices of Fellows of the Royal Society*, 1: 523–6

—— (1936–8). 'Sir Robert Mond 1867–1938'. *Obituary Notices of Fellows of the Royal Society*, 2: 627–32.

T[horpe], T. E. (1916). 'Sir Arthur Rücker (1848–1915)', *Proceedings of the Royal Society*, series A, 92: 'Obituary notices of fellows deceased', xxi–xlv.

Tizard, H. T. (1954). 'Nevil Vincent Sidgwick, 1873–1952', *Obituary Notices of Fellows of the Royal Society*, 9: 237–58.

Tomlin, D. H. (1988). 'The RSRE: a brief history, from earliest times to present day'. *IEE Review*, 34: 403–7.

Townsend, J. S. E. (1898a). 'Electrical properties of newly prepared gases'. *Philosophical Magazine*, 5th ser., 45: 125–51.

—— (1898b). 'Applications of diffusion to conducting gases'. *Philosophical Magazine*, 5th ser., 45: 469–80.

—— (1900a). 'The diffusion of ions into gases'. *Philosophical Transactions of the Royal Society*, series A, 193; 129–58.

—— (1900b). 'The conductivity produced in gases by the motion of negatively charged ions'. *Nature*, 62: 340–1.

—— (1901–2). 'Report of the Wykeham professor of physics, 1901'. *OUG*, 32: 648 (17 June 1902).

—— (1902–3). 'Report of the Wykeham professor of physics, 1902'. *OUG*, 33: 549–50 (19 May 1903).

—— (1903–4). 'Report of the Wykeham professor of physics, 1903'. *OUG*, 34: 561–50 (3 May 1904).

Townsend, J. S. E. (1904–5). 'Report of the Wykeham professor of physics, 1904'. *OUG*, 35: 573–50 (16 May 1905).

—— (1905–6). 'Report of the Wykeham professor of physics, 1905'. *OUG*, 36: 612–13 (15 May 1906).

—— (1909–10). 'Report of the Wykeham professor of physics, 1909'. *OUG*, 40: 746–7 (31 May 1910).

—— (1910). *The Theory of Ionization of Gases by Collision*. London: Constable.

—— (1910–11). 'Report of the Wykeham professor of physics, 1910'. *OUG*, 41: 959–60 (14 June 1911).

—— (1911–12). 'Report of the Wykeham professor of physics, 1911'. *OUG*, 42: 988 (19 June 1912).

—— (1912–13). 'Report of the Wykeham professor of physics, 1912'. *OUG*, 43: 967–8 (4 June 1913).

—— (1913–14). 'Report of the Wykeham professor of physics, 1913'. *OUG*, 44: 863–4 (10 June 1914).

—— (1915). *Electricity in Gases*. Oxford: Clarendon Press.

—— (1915–16). 'Report of the Wykeham professor of physics, 1915'. *OUG*, 46: 545–6 (14 June 1916).

—— (1916–17). 'Report of the Wykeham professor of physics, 1916'. *OUG*, 47: 556 (13 June 1917).

—— (1917–18). 'Report of the Wykeham professor of physics, 1917'. *OUG*, 48: 479 (12 June 1918).

—— (1920). 'Oscillations obtained by coupling a secondary circuit with a continuous wave valve oscillator'. *Radio Review*, number 8 (May).

——, and Hurst, H. E. (1904). 'The genesis of ions by the motion of positive ions, and a theory of the sparking potential', *Philosophical Magazine*, 6th ser., 8: 738–49.

——, and Kirkby, P. J. (1901). 'Conductivity produced in hydrogen and carbonic acid gas by the motion of negatively charged ions'. *Philosophical Magazine*, 6th ser., 1: 630–42.

——, and Morrell, J. H. (1921). 'Electric oscillations in straight wires and solenoids'. *Philosophical Magazine*, 6th ser., 42: 265–78.

Travers, M. W. (1956). *A Life of Sir William Ramsay K.C.B., F.R.S.* London: Edward Arnold.

Trenn, T. J. (1976). 'John Sealey Edward Townsend, 1868–1957', in *DSB*, vol. 13, 445–7.

—— (1986). 'The Geiger–Müller counter of 1928'. *Annals of Science*, 43: 111–35.

Tuckwell, W. (1900). *Reminiscences of Oxford*. London: Cassell.

Turner, G. L'E. (1965). 'The discovery of atomic numbers'. *Institute of Physics and the Physical Society. Bulletin*, 16: 54–5.

—— (1986). 'The physical sciences', in Sutherland and Mitchell (1986), 659–81.

Tyerman, C. (2000). *A History of Harrow School 1324–1991*. Oxford: Oxford University Press.

Ubbelohde, A. R. (1960). 'Obituary. Sir Alfred Egerton F.R.S. (1886–1959)'. *Fuel*, 39: 100–3.

Varcoe, I. (1970). 'Scientists, government and organised research in Great Britain 1914–16: the early history of the DSIR'. *Minerva*, 8: 192–216.

—— (1981). 'Co-operative research associations in British industry, 1918–34'. *Minerva*, 19: 433–63.

V[ernon], H. M. (1914). 'Dr. G. J. Burch, F.R.S.'. *Nature*, 93: 114–15.

Vernon, H. M., and Vernon, K. D. (1909). *A History of the Oxford Museum*. Oxford: Clarendon Press.

Vernon, K. (1994). 'Microbes at work: micro-organisms, the D.S.I.R. and industry in Britain, 1900–1936'. *Annals of Science*, 51: 593–613.

Vernon Harcourt, A. G. (c.1880). *The Pentane Standard and Pentane Lamp*. Birmingham: John Wright & Co.

V[ernon] H[arcourt], A. [G.] (1905). 'Sir John Conroy. 1845–1900'. *Proceedings of the Royal Society of London*, 75: 246–52.

Vernon Harcourt, A. G. (1910). 'The Oxford Museum and its founders'. *The Cornhill Magazine*, new ser., 28: 350–63.

von Engel, A. (1957). 'John Sealy Edward Townsend, 1868–1957'. *Biographical Memoirs of Fellows of the Royal Society*, 3: 257–72.

—— (1971). 'Townsend, Sir John Sealey Edward (1868–1957)', in *DNB 1951–1960*: 983–5.

Walden, A. F., and Manley, J. J. (1901). *An Introduction to the Study of Physics*. London: A. & C. Black.

Walker, R. (1848). *A Letter addressed to the Rev. the Vice-Chancellor, on Improvements in the Present Examination Statute, and the Studies of the University*. Oxford: J. Vincent.

—— (1851–2). *Text Book of Mechanical Philosophy. For the Use of Students* (in two parts, continuously paginated). Oxford: John Henry Parker.

—— (1857). *Remarks on Certain Parts of the Proposed Form of the Statute respecting the Examinations for the Degree of B.A.* Oxford: J. Vincent.

—— (1860). *The Physical Constitution of the Sun. A Discourse delivered in the Sheldonian Theatre at Oxford, before the British Association, June 29, 1860*. London: Taylor and Francis.

Ward, F. A. B. (1980). 'Obituary. C. E. Wynn-Williams'. *Nature*, 283: 117–18.

Warwick, A. (1992). 'Cambridge mathematics and Cavendish physics: Cunningham, Campbell and Einstein's relativity 1905–1911. Part I: The uses of theory'. *Studies in History and Philosophy of Science*, 23: 625–56.

—— (1993). 'Cambridge mathematics and Cavendish physics: Cunningham, Campbell and Einstein's relativity 1905–1911. Part II: Comparing traditions in Cambridge physics'. *Studies in History and Philosophy of Science*, 24: 1–25.

—— (1994). 'The worlds of Cambridge physics', in R. Staley (ed.), *The Physics of Empire. Public Lectures*. Cambridge: Whipple Museum of the History of Science, 57–86.

—— (2003). *Masters of Theory. Cambridge and the Rise of Mathematical Physics*. Chicago and London: University of Chicago Press.

Watson, K. D. (1994). *Sources for the History of Science in Oxford*. Oxford: Modern History Faculty, University of Oxford.

—— (2002). ' "Temporary hotel accommodation"? The early history of the Davy-Faraday Research Laboratory, 1894–1923', in James (2002), 191–223.

Watson-Watt, R. A. (1957). *Three Steps to Victory. A Personal Account by Radar's Greatest Pioneer*. London: Odhams.

Webster, C. (1994). 'Medicine', in Harrison (1994): 317–43.

—— (2000). 'The Medical School under Osler', in Brock and Curthoys (2000): 504–7.

Weinberg, F. J. (1990). 'Alfred Rene Jean Paul Ubbelohde, 14 December 1907–7 January 1988'. *Biographical Memoirs of Fellows of the Royal Society*, 35: 383–402.

Wheaton, B. R. (1983). *The Tiger and the Shark. Empirical Roots of Wave-Particle Dualism*. Cambridge: Cambridge University Press.

Whitrow, G. J. (1970). 'Edward Arthur Milne, 1896–1950', in *DSB*, vol. 9, 404–6.

Who's Who in Oxfordshire (1936). London: Who's Who in the Counties.

Wilberforce, L. R. (1910). 'The development of the teaching of physics in Cambridge', in Anon. (1910), 250–80.

Wilkins, M. H. F. (1987). 'John Turton Randall, 25 March 1905–16 June 1984'. *Biographical Memoirs of Fellows of the Royal Society*, 33: 493–535.

Williamson, R. (ed.) (1987). *The Making of Physicists*. Bristol: Adam Hilger.

Willis, R., and Clark, J. W. (1886). *The Architectural History of the University of Cambridge, and of the Colleges of Cambridge and Eton* (4 volumes). Cambridge: Cambridge University Press.

Willsher, A. P. (1961). 'Daubeny and the development of the Chemistry School in Oxford, 1822–1867'. Chemistry Part II thesis, University of Oxford.

Wilson, D. (1983). *Rutherford. Simple Genius*. London: Hodder and Stoughton.

Wilson, D. B. (1981–2). 'Experimentalists among the mathematicians: physics in the Cambridge Natural Sciences Tripos, 1851–1900'. *Historical Studies in the Physical Sciences*, 12: 325–71.

Wilson, T. (1995). *Churchill and the Prof.* London: Cassell.

Wilson, W. (1959). 'Owen Willans Richardson 1879–1959'. *Biographical Memoirs of Fellows of the Royal Society*, 5: 207–15.

Wood, A. B. (1961). 'Obituaries. Dr. C. V. Drysdale, C.B., O.B.E.'. *Nature*, 190: 214–15.

Woods, H. G. (1887). *The Clarendon Laboratory Grant*. Oxford: privately printed. Pamphlet in MHS, MS Gunther 65.

Wormell, R. (1883). *Lectures on Sound*. London: Thomas Murby.

Wynne, B. (1976). 'C. G. Barkla and the J Phenomenon: a case study in the treatment of deviance in physics'. *Social Studies of Science*, 6: 307–47.

Zimmerman, D. (2001). *Britain's Shield. Radar and the Defeat of the Luftwaffe*. Stroud: Sutton.

INDEX

Abel, Frederick Augustus (1826–1902) 106
Aberdeen, University of 283, 294
Aberystwyth, University of 281
Academic Assistance Council 20n, 246, 251
Acland, Henry Wentworth (1815–1900) 31–2, 35, 42, 74, 77, 120, 207
acoustics 33, 58, 189, 201, 209, 215, 284; *see also* Bosanquet, Robert Holford Macdowall; music, science of
  teaching 67–8, 33, 58, 148–9
Adam, Madge Gertrude (1912–2001) 307
Adams, John Couch (1819–92) 43
Admiralty
  Admiralty Research Laboratory 280, 284, 297
  Board of Invention and Research 228, 280
  Department of Scientific Research and Experiment 280, 284
Advisory Council for Scientific and Industrial Research 290–1
Aeronautical Research Council 280
agricultural depression 76, 85
Agriculture and Fisheries, Ministry of 302
Air Defence Experimental Establishment 284
Air Inventions Committee 280
Air Ministry's Committee for the Scientific Survey of Air Defence 297
Aldrichian chair of chemistry 26, 43n, 121
Alembic Club 168n
Alexander, Samuel (1859–1938) 194
Allen, George Dexter (f.1897–1910) 195
Allen, John (1908–2001) 297
Allen, William Douglas (b.1914) 259n, 266
Alvergniat frères instrument–makers 163
anatomy *see* physiology and anatomy
Angel, Andrea (1877–1917) 205
apparatus *see* instrumentation
Applebey, Malcolm Percival (1884–1957) 152, 166
Appleton, Edward Victor (1892–1965) 283, 289
Arms, Henry Shull (1914–72) 260, 266
Armstrong College, Newcastle *see* Newcastle, Armstrong College, and University of
Army
  Air Defence Experimental Establishment 284
  Ordnance Research Department and Signals Experimental Establishment 280, 284
  Searchlight Experimental Establishment 284, 297

Ashmolean Museum (Old Ashmolean Building) 27, 32, 37, 128
Ashmolean Natural History Society 150
Ashmolean Society 175
Asquith Commission 19
assaying 203
Astbury, William Thomas (1898–1961) 286
Aston, Francis William (1877–1945) 280, 287, 297
astronomy and astrophysics 26n, 37, 40, 43n, 62n, 74–5, 146, 156, 159, 170, 307–8; *see also* observatories; University Observatory, Oxford
Atkinson, Edmund (1831–1901) 73
atmospheric physics 19–20, 162–3, 239–40, 287–8, 306–7
atomic and nuclear physics 9, 22, 159–62, 229, 253, 257–60, 268–300, 305; *see also* Cockcroft-Walton accelerator; ion physics; radioactivity; X-rays
Austin, Herbert (1866–1941) 263, 294
Ayrton, William Edward (1847–1908) 97–102, 103

Babbitt, John David (1908–82) 254n
Bachelor of Arts (B.A.) degree, classification of 311
  syllabus 312–14
Bachelor of Science (B.Sc.) degree 152, 159, 165, 219
Baker, Herbert Brereton (1862–1935) 133, 137, 144–5, 153–6, 203, 219–20
Balliol College, Oxford 9, 17, 42, 123, 145, 149, 155, 159, 174, 182, 184, 196, 204, 240
  laboratory 7–8, 120, 127–31, 137, 165–8; *see also* Balliol-Trinity Laboratories
  relations with other colleges 130–40, 151–2, 182, 198–9
Balliol-Trinity Laboratories 9, 127–40, 145, 151–2, 154–5, 184, 189, 198–9
  closure 302
  instruments 164, 171, 199, 221
  interior, illustrations of 136, 139, 221, 241
  origins 130–2
  tradition in physical chemistry 126, 135–40, 151–3, 164, 171–2
Balliol-Trinity-St. John's Colleges, relations between 130–3, 182, 189
Barkla, Charles Glover (1877–1944) 210, 272, 276, 282–3

Barnes 218, 220
Barr & Stroud, instrument-makers 133–4
Barry, Charles (1795–1860) 75
Barton, Edwin Henry (1858–1925) 271
Bates, Leslie Fleetwood (1897–1978) 295
Bawdsey 297–8
Baynes, Robert Edward (1849–1921) 22n, 65, 73, 85, 117n, 141–5, 152, 154, 157, 159
Bazett, Henry Cuthbert (b.1885) 194–5
Bedford, Duke of see Russell, Francis Charles Hastings, ninth duke of Bedford
Belfast, Queen's College 100, 240n
Bellamy, James (1819–1909) 147
Bellevue electromagnet 252, 263
Bennett, G. A. 213
Berkeley, University of California at 254, 255n, 294
Berkeley, Randal Thomas Mowbray Rawdon, Earl of Berkeley (1865–1942) 136–7, 163–8, 170
Berkley, Evelyn Hannah 194
Berlin, University of 160, 247, 250, 258, 270, 286
    Nernst school 19, 235–8, 244–5, 286
    Physics Institute, expenditure on 78
Bernal, John Desmond (1901–1971) 292
Bethe, Hans Albrecht (1906–2005) 250, 292
Birkenhead, Earl of see Smith, Frederick Edwin
Birmingham, Mason College, and University of 252, 260, 263, 270, 272–3, 277, 279, 281, 293–6, 298
Blackett, Patrick Maynard Stuart (1897–1974) 292, 294
Blackwell, publisher 158
Bleaney, Brebis (b.1915) 21, 255, 309
Blenheim Palace 164
Bloch, Eugène (1878–1944) 214, 227
Bodleian Library 12, 54
Bohr, Niels (1885–1962) 158, 229, 278
Bolton King, Edward (1900–74) 288
Booth, Eugene Theodore (1912–2004) 259–60n
Born, Max (1882–1970) 159, 292
Bosanquet, Claude Henry (1896–1965) 159, 168, 173
Bosanquet, Robert Holford Macdowall (1841–1912) 141, 145–9, 174, 189–91, 208
Boyd, Henry (1831–1922) 223
Boys, Charles Vernon (1855–1944) 58, 81, 86, 110–14, 175–6, 199, 201, 204
Bracey, R. J. 243n
Bradley, James (1692–1762) 1
Bragg, William Henry (1862–1942) 222, 262, 276, 277, 280, 283
Bragg, William Lawrence (1890–1971) 276, 282, 284, 297
Brasenose College, Oxford 65–6, 141–2, 162, 306, 309–10

funding of chair of engineering science 204
Breslau, Technische Hochschule of 20–1, 233, 245–9, 253, 256, 293
Brewer, Alan 307
Bridges, John Henry (1832–1906) 36
Brillouin, Louis Marcel (1854–1948) 228
Bristol, Clifton College 44, 49, 183–4, 207
Bristol, University College, and University 20, 179, 183, 250n, 251, 270, 276–7, 281, 290–4, 296, 298
    Henry Herbert Wills Laboratory 20, 290–1
British Association for the Advancement of Science (BAAS)
    committee on electrical standards 6–7
    centenary meeting (1931) 267
    meeting in Australia (Adelaide, Melbourne, Sydney, Brisbane) (1914) 277
    meeting in Edinburgh (1892) 199
    meeting in Glasgow (1876) 98
    meeting in Oxford (1847) 31–2
    meeting in Oxford (1860) 35–7, 41
    meeting in Oxford (1894) 87, 132, 206–7
    meeting in Oxford (1926) 206
British Empire Cancer Campaign 295
British Medical Association 174
British Museum (natural History) 47, 64; see also Fletcher, Lazarus; Story-Maskelyne, Mervyn Herbert Nevil
British Oxygen Company 263
Brodie, Benjamin Collins, junior (1817–1880) 37, 75, 128–9, 140–1
Bröse [Brose], Henry Herman Leopold Adolf (1890–1965) 158–9, 162, 238n
Brown, O. F., recollections of 80, 81, 86, 88n, 116
Buckland, William (1784–1856) 26, 32,
Bunsen, Robert Wilhelm (1811–99) 43–4
Burch, George James (1852–1914) 8, 119, 164, 171
Burdett-Coutts scholarship 71n
Burdon Sanderson, John Scott (1828–1905) 171
Burton, Charles Vandeleur (1867–1917) 166
Busbridge, Ida Winifred (1908–88) 307
Bush, S. W. 213
Butler, Catherine Elizabeth (Mrs Clifton) (d.1917) 91

Calcutta 214
Callendar, Hugh Longbourne (1863–1930) 218, 270, 284
Cambridge, University of see also Cavendish Laboratory
    chair of experimental physics see Maxwell, James Clerk; Strutt, John William, Baron Rayleigh; Rutherford, Ernest; Thomson, Joseph John

Index

chair of mechanism 172, 202, 205
chair of theoretical chemistry 291
coaches, role of 9, 10
comparison with University of Bristol 291
comparison with University of Manchester 275, 282
comparison with University of Oxford 1–23, 72, 78–9, 83–7, 117, 161, 209–14, 219, 245–6, 268, 270, 287, 289, 295–6, 303, 305; *see also* Cavendish Laboratory, comparison with Clarendon Laboratory; Cavendish Laboratory, comparison with Electrical Laboratory, Oxford
Engineering, Department of 223
expenditure on buildings for science 60
Girton College 255–6
graduates of in Oxford *see* Clifton, Robert Bellamy; Dobson, Gordon Miller Bourne; Jackson, Derek Ainslie; Keeley, Thomas Clews; Megaw, Helen Dick; Southwell, Richard Vynne; Townsend, John Sealy Edward
hostility to experimental physics 10
ion physics 4–5, 156, 209–32, 272, 276
Mathematical Tripos 9–15, 44, 60, 74, 78
mathematics, teaching of 211
Mechanical Sciences Tripos 206
Mond Laboratory 245, 291, 293, 297; *see also* Kapitza, Piotr Leonidovich
Natural Sciences Tripos 11, 15–16, 60n, 78, 87, 288
St John's College 44, 51
Trinity College 211, 272, 281, 288
Cambridge Scientific Instrument Company 166, 213
Campbell, John Edward (1862–1924) 156
Cardiff University College 91, 273, 279
Carnegie Foundation 223
Carpenter, Henry Cort Harold (1875–1940) 194
Carroll, John (1899–1974) 294
Carroll, Lewis *see* Dodgson, Charles Lutwidge
Carse, George Alexander (d.1950) 115–16
Cavendish, Henry (1731–1810) 6–7, 84–5
Cavendish, William, 7th Duke of Devonshire (1808–91) 84–5
Cavendish Laboratory 175, 217–18, 241, 255, 267–77, 280–3, 285, 292, 294–300, 308; *see also* Maxwell, James Clerk; Rutherford, Ernest, Baron Rutherford; Strutt, John William, Baron Rayleigh; Thomson, Joseph John
annus mirabilis (1932) 294
Cavendish 'style' 5, 22, 209–14, 225–6, 231–2, 271–2, 286, 305
comparison with Clarendon Laboratory 3–5, 19, 22, 77–9, 81–6, 92, 117–18, 236, 245–6, 268, 279, 287–9, 293, 305; *see also*

Cambridge, University of, comparison with University of Oxford
comparison with Electrical Laboratory, Oxford 209–32; *see also* Cambridge, University of, comparison with University of Oxford
construction and inauguration 6, 10–11, 60–2, 84, 87
Crowther as 'press agent' 293–4
funding 60–2, 83n, 263, 300
instruments 210, 213, 220, 276
Physical Society 274
reinterpretations of 268–74, 298–300
research students 18, 83–5, 87, 209–12, 226, 269, 282, 287–8, 299
teaching 4–5, 11, 18, 108, 210, 274
Cecil, Robert Arthur Talbot Gascoyne, Marquess of Salisbury (1830–1903) 16n
Central Flying School, Upavon 236
CERN, Geneva 308
Chabaud, Victor 214
Chadwick, James (1891–1974) 292, 293–4, 296–7
Chapman, David Leonard (1869–1958) 9, 151–3
Chapman, Edward (1839–1906) 63, 121–2, 124, 126, 193–4
Chattaway, Frederick Daniel (1860–1944) 150
chemical kinetics 133, 144, 152, 196, 238
chemical physics 126, 135, 142, 144, 158, 164, 167, 291; *see also* physical chemistry as an Oxford speciality
chemistry *passim*, but *see entries for individual chemists and laboratories*
laboratory provision for 37–8, 75, 128
numbers of students 309
physical chemistry as an Oxford speciality 2, 4, 8–9, 126, 135–40, 151–5, 164–5, 171–2, 219, 263, 271, 291; *see also* chemical physics
Cherwell, Viscount *see* Lindemann, Frederick Alexander
Christ Church 22, 31–2, 37, 125, 149, 152, 205, 237, 250, 257
attendance at lectures on experimental philosophy 26
Berkeley, work of 163–7
collections 37
graduates in public life 16n
laboratory 7–9, 64, 120, 128, 132–4, 137, 140–5, 153–5, 159, 162, 165, 259, 301
Lindemann's presence 20n, 158–62
mathematics 12, 16n, 144
'new physics' 157–62
readership in chemistry 22n, 7–8, 158, 219; *see also names of holders*
readership in physics 64, 144, 160; *see also names of holders*
chronograph 114, 123, 171, 191, 195–9, 207

Church, Arthur Harry (1865–1937) 151
Churchill, Winston Leonard Spencer
    (1874–1965) 244, 261–2, 288, 297, 304–5
Citroën, André, & Co 167
Clarendon, Earl of *see* Hyde, Edward, first earl
    of Clarendon
Clarendon Building 26–8, 32, 37
Clarendon Laboratory *passim*; *see also*
    Clarendon Laboratory, new
  appearance 59
  Boys determination of G 58, 110–14
  Cockcroft-Walton accelerator 259, 293–5, 308
  comparison with Cavendish Laboratory 3–5,
    19, 22, 77–9, 81–6, 92, 117–18, 236, 245–6,
    268, 279, 287–9, 293, 305; *see also* Oxford,
    University of, Comparison with
    Cambridge, University of
  competition with Electrical Laboratory 19, 22,
    214–18, 242–4, 262–5, 303
  demonstrators 17n, 18, 22, 64–5, 80,. 81,
    85–6, 92, 97, 116, 137, 139, 143–4, 155–7,
    162, 215, 228, 236, 239–40, 243, 265, 273, 309
  disposition of space 37–40, 58–9, 61, 315–16
  foreign models 263
  electrical generation and supply 82, 94, 117,
    184–5
  funding 47n, 53–62, 75–6, 117, 215–16, 236–8,
    242–4, 251, 259, 266, 287
  hostility towards 55–8
  inauguration 35–7, 56, 87, 181
  instruments 11n, 58–9, 69, 88–90, 92–4,
    100–5, 113, 116, 159–62, 217, 236, 238, 241,
    245, 252–3, 257, 259, 288
  military research 228–9, 260–2, 297–300,
    303–8; *see also* World War, First; World
    War, Second
  'old-fashioned' physics 22
  plans for extension of University Museum
    47–53, 264
  research at 19–21, 58, 81–6, 96, 110, 116,
    158–62, 236, 241, 247–66, 272, 287
  student fees 56
  students, advanced 17, 82, 85, 107–9, 115–17,
    287, 309
  teaching at 65, 86, 92, 142, 144
  union with Electrical Laboratory 306
  workshops and other equipment and facilities
    89, 93, 94–5
Clarendon Laboratory, new 140–1, 175, 233,
    257, 295
  funding and inauguration 21–2, 262–6, 301–3
Clarendon Press *see* Oxford University Press
Clarendon Trust 53–6, 303
Clark, John Willis (1833–1910) 60n
Clifton, Catherine Elizabeth *see* Butler,
    Catherine Elizabeth

Clifton, Robert Bellamy (1836–1921) 43–79,
    80–118, 208
  appointment 44–5
  attitude to research 6–7, 47, 67, 73–4, 80–7,
    91–7, 102–4, 108, 116–17
  collection of mathematical books 44n, 90
  family estate 44, 76, 81, 85, 103, 115
  honours 76–7, 113
  importance of mathematics 8–11, 108
  interest in electricity 92–3, 97–102, 105–10
  laboratory classes 45–6, 56, 48–51, 88–92,
    95–7, 108–9
  lectures 45–6, 48–51, 56, 64, 67–8, 87–8, 93,
    107–8, 215
  letter to William Thomson 47n, 315–16
  miner's lamp 86, 104–6
  paper on voltaic action 97–102
  personality 8, 77, 79, 91, 117
  public life 81, 86, 102–9, 113
  pupils 48–51, 63–6, 85, 91–2, 95–6, 107–8
  relations with Price 62–3
  relations with Townsend 18–19, 22, 81–2,
    114–18, 154, 212, 214–18, 232
  reputation 2, 18, 23, 76–9, 80–7, 114–18,
    168, 232
  requests for funds 18, 46–53, 56–8, 62–4,
    75–6, 82–5, 92, 106–9, 184, 188, 214–15
  residences 44–5
  support for engineering 62, 76, 183
  teaching and examining duties 25–6, 45–6,
    66–74, 76–7, 86–93, 95–6, 107–8, 214–15
  testimony to Devonshire Commission 46n, 62
  testimony to 1877 Commission 67–8, 73–6,
    93, 95–7, 107
  textbooks 72–3
  working habits 66–7
Clifton, Walter Bellamy 67
Clifton College, Bristol 44, 49, 183–4, 207
coaches, role of 9–10, 149–50, 154, 179
coal-mining *see* miner's lamp; Royal
    Commission on Accidents in Mines
Cockcroft, John Douglas (1897–1967) 292,
    293–5, 298
Cockcroft-Walton accelerator 259, 293–5, 308
Coleman, F. F. 243n
Collège de France *see* Langevin, Paul
colleges *passim*; *also see entries for individual
    Oxford colleges*
  Cambridge *see entries for individual colleges*
  college laboratories 3, 7–10, 63–5, 119–208
  difficulty of securing fellowships 22, 266, 309
  end of college-based teaching in physics
    154–5
  London *see* London colleges
  mathematical lectures 10
  other, *see individual entries under city*

Collie, Carl Howard (1903–91) 9, 21, 155, 158n, 159–62, 257–9, 266, 304–5, 308
Commission, Royal
  on Accidents in Mines 82, 86, 102–9
  on Scientific Instruction and the Advancement of Science (Devonshire) 46n, 62, 181–3
  on the University of Oxford, 1877 67–8, 73–6, 81, 93, 95–7, 107, 214
Committee of Imperial Defence 261
Common University Fund 188, 204, 213
Condliffe, George E. 295–6
Conroy, John (1845–1900) 8, 48, 134, 138, 141, 142, 149, 155, 166, 170
Cooke, Arthur Hafford (1912–87) 255, 266, 303, 309
Cooke-Yarborough, Edmund Harry (b.1918) 260
Corpus Christi College, Oxford 129
Cotton, Aimé Auge (1869–1951) 252
Craig, Edwin Stewart (1865–1939) 149–50, 154
Craig-Henderson 210
Croft, William B. 74n
Cronshaw, George Bernard (1872–1928) 150
Crossley, Kenneth Irwin (1877–1957) 195
Crowther, James Gerald (1899–1983) 293–4
cryogenics *see* low-temperature physics
crystallography 17n, 138, 170–1, 255, 282–3, 284, 292, 297
Cunningham 214
Cunnold, Frank Allan (b.1912) 243n
Curie, Marie (1867–1934) 228

Dale, John Andrews (1816–88) 129
Darwin, Charles Robert (1809–82) 37
Darwin, Horace (1851–1928) 166
Daubeny, Charles Giles Bridle (1795–1867) 42, 120, 125–6, 128
  bequest to Magdalen 121, 123
  campaign for Museum and reform of syllabus 30–5
  lectures 26, 31n, 121
  primary sciences 32–3, 69
Daubeny Laboratory 120–7, 133, 150, 154, 173; *see also* Magdalen College
Daunt, John Gilbert (1913–87) 253, 254n
de Broglie, Louis-César-Victor-Maurice (1875–1960) 220, 227
Debye, Petrus Josephus Wilhelmus (1884–1966) 153
De la Rue, Warren (1815–89) 75
de Morgan, Augustus (1806–71) 44
Deane, Thomas (1792–1871) 58n
Department of Scientific and Industrial Research (DSIR) 239, 241, 266, 280–1, 283–5, 290–2, 298
  research stations 284

depression
  agricultural 76, 172
  economic (1930s) 295
Derby, Earl of *see* Stanley, Edward George Geoffrey Smith
Dermantine Company 165
Desaguliers, John Theophilus (1683–1744) 1
Devonshire, Duke of *see* Cavendish, William
Devonshire Commission *see* Royal Commission on Scientific Instruction and the Advancement of Science
Dewar, James (1842–1923) 270
Dickson, H. N. 203
Dixon, Harold Baily (1852–1930) 63, 132–5, 152, 153, 171, 182, 192, 196
Dobson, Gordon Miller Bourne (1889–1976) 19–20, 162–3, 230, 233, 236, 239–40, 266, 287–8, 306–7
Dodgson, Charles Lutwidge (1832–98) 11, 13, 16, 55–6, 143–4
Donkin, William Fishburn (1814–69) 15, 26, 37, 53, 73–4, 146,
Donkin, William Frederick (1846–88) 48, 149
Drapers' Company 5, 115, 157n, 215, 223–5, 232, 276; *see also* Electrical Laboratory
Drinkwater, John Wilson 243n
dry pile 30
Dublin, Trinity College 209
Duckworth, John Clifford (b.1916) 260n
Duncan, Philip Bury (1772–1863) 32
Dundee, University College of St Andrew's 275
dynamometers and dynamometry 179–80, 187, 191, 201, 204, 207; *see also* ergometer; precision physics
Dyson Perrins Laboratory 126–7

Earwaker, John Parsons (1847–95) 60, 326
Eddington, Arthur Stanley (1882–1944) 158, 294
Edinburgh, University of 60, 115–16, 238n, 275–7, 282–3
Edmondson, William (b.1902) 243n
educational extension 125–6, 134, 168, 171, 183–4, 194; *see also* women
Egerton, Alfred Charles Glyn (1886–1959) 233, 241–4, 287–8
Einstein, Albert (1879–1955) 21n, 158–9, 160–1, 235, 250
Electrical Laboratory 5, 119, 137, 154, 208, 209–32, 271, 276–7, 289, 299; *see also* Drapers' Company; Townsend, John Sealy Edward
  comparison with Cavendish Laboratory 209–32; *see also* Oxford, University of, comparison with University of Cambridge
  competition with Clarendon Laboratory 19, 22, 109, 117, 214–18, 236, 242–4, 262–5, 276, 303, 306

Electrical Laboratory (*cont.*)
  contribution to war effort 228–9
  demonstrators 116, 126, 127, 138, 143, 155, 157, 205, 213, 215, 243, 265, 309
  inauguration 18, 83, 109, 224
  income 244
  Moseley's research 277–9
  research at 222
  union with Clarendon Laboratory 306
electricity
  electrical standards 6–7, 102
  supply in Balliol-Trinity Laboratories 184, 189
  supply and machinery in Clarendon Laboratory 82, 94, 117, 184
  supply in Oxford 203
  teaching and research in at Oxford 8, 28–30, 58, 92–3, 97–102, 105–10, 148, 184, 187–8, 209, 213, 222–5; *see also* ion physics
Elliott, Edwin Bailey (1851–1937) 10
Elliott Brothers, instrument-makers 167, 196, 198
Ellis, Charles Drummond (1895–1980) 292, 294
EMI Company 293, 295–6
engineering science at Oxford 8, 76, 169–208, 224–5
  chair 203, 204–5
  degree 204, 206
  department 204–6
  Diploma in Scientific Engineering and Mining Subjects 203
  integration in University 207
  new laboratory 205
  numbers of students 194–5, 206, 309
  roots in Millard Laboratory 170, 185–95
English Mechanic, journal 172–3, 179, 201
entomology 40
Epstein, Hans Georg (1909–2002) 266n
ergometer 179–80
Esson, William (1839–1916) 10, 72n, 73, 122, 144, 156
Evans, Evan Jenkin (1882–1944) 281
Ewing, James Alfred (1855–1935) 202, 204–5
examination papers 70–2
Examination Schools 54
examinations *see also* Moderations; Preliminary Honour examination and students; Responsions
  candidates, numbers of 14–15
  examiners, appointment of 13–14, 188–9
  ordeal of 12–13
  questions 71–2
  statutes 16, 30–3, 40–1, 69–70, 78
  timing xviii
Exeter College, Oxford 54n, 120, 128, 131, 139, 156, 194
  laboratory 131–2
  relations with Trinity and Balliol Colleges 182

exhibitions
  Electrical Exhibition, Paris 180
  Inventions Exhibition, South Kensington 180
  South Kensington Loan Exhibition 146
  Universal Exhibitions, Paris 58
experimental philosophy, creation of readership in 1, 24–6
explosions, kinetics of 133, 152, 196

Faraday Society 291
fees
  disparity between Clifton's and Townsend's 216–17
  for laboratory work 188, 262
  for lectures 29, 33–7, 132, 188, 262
ferromagnetic resonance 307
Finsbury Technical College 103
Firth College, Sheffield *see* Sheffield, Firth College, and University of
Fisher, Walter William (1842–1920) 129
Fleming, John Ambrose (1849–1945) 87
Fletcher, Lazarus (1854–1921) 17, 63–4, 85, 91–2, 103, 133, 170–1, 182
Foster, George Carey (1835–1919) 17
Fowler, Thomas (1832–1904) 215
Foxcombe, Berkeley's laboratory at 165–7, 168
Franck, James (1882–1964) 229, 292
Freundlich, Erwin Finlay (1885–1964) 159
Fröhlich, Herbert (1905–91) 292
Froude, James Anthony (1818–94) 175
Froude, Richard Hurrell (1803–36) 13, 175
Froude, William (1810–79) 174–5, 179
Fuchs, Emile Julius Klaus (1911–88) 292

Gaisford, Thomas (1779–1855) 26
Galton, Francis (1822–1911) 42
Galway 210
Gamble, John George (1842–89) 173–4
Gamlen, William Blagdon (1844–1919) 220–1
Ganot, Adolphe (1804–87) 72
Gassiot, John Peter (1797–1877) 170
Gates, Stanley Frederick (1897–1972) 243n
Gedney, Clifton's property at 44, 76, 81, 85, 103, 115
Geiger, Hans (1882–1945) 160, 225–6, 274
General Electric Company 280, 283–4, 293, 296
geography 33
geology 26–8, 33, 71n, 121, 203
George IV (1762–1830) 1
Gerrans, Henry Tresawna (1958–1921) 194
Gill, Ernest Walter Brudenell (1883–1959) 143, 205, 221, 228, 289
Girton College, Cambridge 255–6
Gladstone, William Ewart (1809–98) 12, 16n, 52–5
Glasgow, University of 2, 98, 158, 187, 270, 272–3, 275
  laboratory 60, 77; *see also* Thomson, William

Glastonbury Abbey, abbot's kitchen, model for Oxford chemical laboratory 38, 128
Glazebrook, Richard Tetley (1854–1935) 90, 273
Goodwin, Alfred 17–18
Gordon, James Edward Henry (1852–93) 102n
Gotch, Francis (1853–1913) 171
Göttingen, University of 20n, 21, 247, 249, 270, 292
government ministries and departments *see entries for individual ministries and departments*
Graaff, Robert Jemison van de (1901–67) 289–90
van de Graaff machine 308
Gratias, Orvald Arthur (b.1909) 162, 259n
gravitational constant, measurement of 58, 81, 86, 110–14
Grazebrook, George Ward William 195
Gray, Andrew (1847–1925) 270
Gray lamp 105
'Greats'; *see* Literae Humaniores, School of
Griffith, George (1834–1902) 39, 41–2, 47n, 73, 123
Griffith, Idwal Owain (1880–1941) 66n, 139, 155, 228, 236, 238–9, 262–3, 287
Griffith, J. G. 66n
Griffiths, James Howard Eagle (1908–81) 21, 161, 257, 259, 307
Grubb, Howard (1844–1931) 75
Gunther, Robert William Theodore (1869–1940) 106, 123, 164, 172–3, 176, 193–4
Guthrie, Frederick (1833–86) 102–3

Halban, Hans von 308
Haldane, John Scott (1860–1936) 194, 203
Harcourt, Augustus George Vernon (1834–1919) 8, 28, 128, 135, 140–1, 150, 153, 165, 166, 198
Harrison, Douglas Neill (1901–87) 239n
Harrison, Edward 284–5
Harrow School 44
Hartley, Ernald George Justinian (1875–1947) 165–7
Hartley, Harold Brewer (1878–1972) 137, 138–40, 159, 167–8
Hartree, Douglas Rayner (1897–1958) 296
Harwell 161, 308
Haselfoot, Charles Edward (1864–1936) 155–7, 219
Hatton, John Leigh Smeathman (1865–1933) 156
Hawkins, Charles Caesar (1864–1938) 195
Hawkshaw, John (1811–91) 174
heat and thermodynamics 65, 66, 73, 96, 132, 141–2, 206, 215, 218–19, 287–8; *see also* low-temperature physics
Heathcote, William (1801–81) 54–5

Heidelberg, University of 42–4
helium, liquefaction of 245–6, 293
Helmholtz, Hermann (1821–94) 42–3, 72, 84, 147
Henry Herbert Wills Physics Laboratory, Bristol 20, 290; *see also* Bristol, University of
Herschel Astronomical Prize 156
Hertford College 155–7, 175, 195, 223
Hertz, Gustav Ludwig (1887–1975) 229
Hertz waves 136, 200–1, 206
Hilger, instrument-makers 167
Hinshelwood, Cyril Norman (1897–1967) 9, 139–40, 238
Hoenow, Alfred 238
Hofmann, August Wilhelm (1818–92) 128
Holbourn, Athelstan Hylas Staughton (1907–62) 266
Holmes, Muriel Catherine Canning (1894–1988) 152
Hope, Edward (1886–1953) 126–7
Hopkins, Thomas Henry Toovey (1833–85) 121, 173
Hopkinson, John (1849–98) 94
Hornsby, Thomas (1733–1810) 1
Hornsey, John 88, 93
Horton, Frank (1878–1957) 281
Hughes, L. G. 195
Hull, Richard Albert (1911–49) 21, 254, 255, 266, 309
Hume-Rothery, William (1899–1968) 292, 299
Hurst, Claude (1907–87) 21, 257, 259
Hurst, Harold Edwin (1880–1978) 154, 157, 217–19, 221, 257–8
Huxley, Leonard George Holden (1902–88) 35–7, 289
Hyde, Edward, first earl of Clarendon (1609–74) 53
Hyde, Henry, Lord Hyde, 4th Earl of Clarendon (1672–1753) 53
Hyde Institute 59; *see also* Clarendon Laboratory
hydrostatics 28–30, 32, 132–3, 188

Imperial Chemical Industries (ICI) 20–1, 233, 243n, 245–52, 258, 263, 266, 293
Imperial College 127, 243, 258, 279–81, 283, 285, 294
independent physicists 19–20, 162–7, 230–1, 233, 239–44
inductoscript 199–200
Institute of Physics 269, 292, 295
Institution of Civil Engineers 175
instrumentation *see also* Cambridge Scientific Instrument Company; Barr & Stroud; Elliott Brothers; Koenig, Rudolf; precision physics

instrumentation (*cont.*)
  astronomical 74–5, 121
  Balliol-Trinity Laboratories 171, 191, 195–9, 207
  chronograph 123, 171, 196, 198
  Cavendish Laboratory 210, 213, 220, 276
  Clarendon Building 11n, 26–30
  Clarendon Laboratory 58–9, 67–9, 75–6, 88–90, 92–106, 113, 116, 163–4, 217, 236, 241, 252, 288
  cloud chamber 220–1, 282
  Daubeny's bequest 121, 123
  Duke of Marlborough's bequest 163–4
  dynamometers and dynamometry 179–80, 187, 191, 201, 207
  electrical 103, 163–4, 200, 213, 257
  Electrical Laboratory 227, 231
  ergometer 179–80
  hydrogen liquefier 245–6, 283
  instrument making and design 88, 90, 101, 124, 133, 178, 196, 219, 282
  Lord Leigh's benefaction and trust fund 1, 46, 237, 262
  Magdalen College 121, 123, 124–6
  Millard Laboratory 185–91, 200–1
  optical 103–4, 110, 123, 201
  particle detector 226
  purchase of instruments 1, 42n, 46, 83, 141, 150, 167, 243, 253, 259, 263, 265, 301
  research, use for 82, 127, 148
  St John's College 147–9
  teaching, use for 7, 70, 81, 88, 102–4, 256, 272, 279
  University Museum 37, 46–53, 212–15
ion physics 4–5, 156, 209–32, 272, 276
Institut Henri Poincaré, Paris 251
Institute of Physics 295
Inventions and Research, Board of 228
Isaac, Florence 138

'J phenomenon' 210, 272, 276, 282–3
Jackson, organ builder 167
Jackson, Derek Ainslie (1906–82) 19–20, 230–1, 233, 239–44, 252–3, 265, 288, 304, 307
Jackson, Thomas Graham (1835–1924) 223–4
James, Reginald William (1891–1964) 284
Jamin, Jules Célestin (1818–86) 71, 72
Japan, Imperial College of Engineering 98, 100–1
Jenkin, Charles Frewen (1865–1940) 204–5, 225
Jenkin, Fleeming (1833–85) 204–5, 225
Jervis, John, Earl of St Vincent (1735–1823) 178
Jervis-Smith, Eustace John (1876–1920) 203–4
Jervis-Smith, Frederick John (1848–1911) 8, 107–9, 148–9, 156, 169–72, 175, 178–208
  chronograph 171, 191, 195–9, 207
  dynamometers and dynamometry 179–80, 187, 191, 201, 204, 207
  ergometer 179–80
  inductoscript 199–200
  lecture syllabuses 190, 191, 193
  retirement, research in 203–4
  unsuccessful applications 202–3
Jesus College, Oxford 258n
  fellows and lecturers 41, 122, 137
  laboratory 8, 126, 151–3, 166, 301
Jeune, Francis (1806–68) 53n
Johnson, Raynor Carey (1901–87) 240n
Johnson Prize 174
Jones, Edward Taylor 270
Jones, Francis Edgar (1914–88) 295
Jones, Ivor Rhys (b.1916) 260n
Jones, John Viriamu (1856–1901) 17–18, 91, 103, 133
Jones, Reginald Victor (1911–97) 261–2, 297
Joule, James Prescott (1818–89) 43
Jowett, Benjamin (1817–93) 183
Junior Scientific Club 119, 136, 164, 165, 196, 200

Kapitza, Piotr Leonidovich (1894–1984) 245–6, 291, 293
Kaye, George William Clarkson (1880–1941) 276
Keble College, Oxford 48, 59, 135, 194
  laboratory 145, 149–50, 154
Keeley, Thomas Clews (1894–1988) 160, 236, 239, 244–5, 258, 266, 287, 304–5
Keill, John (1671–1721) 1
Kelvin, Lord *see* Thomson, William
Kennedy, Alexander Blackie William (1847–1928) 186–7, 204, 206–7
Kerr, William Schomberg Robert, 8th Marquess of Lothian (1832–70) 54
Kerslake, Ralph Trevor (b.1915) 255n
King's College, London 87n, 156–7, 240n, 267, 276–9, 281, 283, 285, 294–5
Kirchhoff, Gustav Robert Georg (1824–87) 43–4
Kirkby, Paul Jerome (1869–1931) 156, 194, 213, 219–22
Koenig, Rudolf, instrument-maker 147
Kohlrausch, Friedrich Wilhelm Georg (1840–1910) 98
Kuhn, Heinrich Gerhard (1904–94) 20–1, 233, 249, 252–3, 307
Kundt, August Adolf (1839–1894) 78
Kurti, Nicholas (1908–98) 20–1, 84–5, 233, 246–9, 251–6, 257, 266, 293, 310

Laby, Thomas Howell (1880–1946) 273–4
Lady Margaret Hall, Oxford 194, 254
Lambert, Mary Georgiana (Mrs Townsend) (1890–1986) 230
Langevin, Paul (1872–1946) 210–12, 214, 217–18, 220, 225, 227–8, 231

Index

Lankester, Edwin Ray (1847–1929) 50
Larmor Joseph (1857–1942) 175
Lattey, Robert Tabor (1881–1967) 137–9, 154, 228
Laue, Max von (1879–1960) 221–2
Law, School of 34, 192
Law and Modern History, School of 16, 32–3, 34, 40–1, 314
Lawrence, Ernest Orlando (1901–58) 294
lectures
  audiences for 26–31, 36, 38, 87–8, 107–8
  Clifton's 45–6, 48–51, 56, 64, 67–8, 87–8, 96, 107–8, 215
  college lectures, *see entries for individual colleges*
  compulsory attendance 31
  Danbeny's 26, 31n, 121
  extramural 125–6
  fees 29, 33–7, 132, 188, 262
  laboratory lectures, *see entries for individual laboratories*
  Maxwell's 97–8
  Millard Laboratory 185–95
  popular 180–1, 199
  university lectures, *see entries for individual lectures*
Leeds, Yorkshire College, and University of 65–6, 133–4, 222–3, 227, 276–7, 281, 283, 286
Lee, William Bell (b.1886) 243n
Leicester, University College 289
Leigh, Lord, benefaction and trust fund 1, 46, 237, 262
Leipzig, University of 126, 153
Lennard-Jones, John Edward (1894–1954) 290–1
Lester, Laura 194
Liddell, Henry George (1811–98) 75n
Liège, University of 255
Lightfoot, John Prideaux (1803–87) 54n
Lincoln College, Oxford 9n, 77, 126, 194–5
Lindemann, Frederick Alexander, Viscount Cherwell (1886–1957) 2, 18–23, 117n, 233–66
  appointment 144, 233, 236, 286–7
  comparison with Rutherford 238, 287–8
  death 22n
  judgement of Clifton 81–7, 117, 302
  lectures 234
  'new physics', attitude to 158–62, 238, 250–1, 289
  'old fashioned physics', admiration for 296
  personal research 20, 234–8, 302
  political activity 262, 288, 299
  priorities in research 253, 257–60
  pupils and researchers 158
  refugee physicists, support for 20–1, 246–53, 291–3, 300, 304
  relations with Oxford colleagues 19, 236, 243–4
  relations with Tizard 261
  reputation 18–23, 290–1, 293
  residence and activity in Christ Church 22n, 158–62
  retirement 308
  rivalry with Cavendish Laboratory 246
  rivalry with Townsend 236
  Solvay conferences 235–6
Literae Humaniores, School of 11–16, 31–3, 34, 38–41, 312–14
Littlehailes, Richard (1878–1950) 137
Liverpool, University College, and University of 66n, 211, 260, 270–2, 276, 279, 294, 296
  George Holt Physics Laboratory 272, 298
Lodge, Oliver Joseph (1851–1940) 199, 200, 202–3, 204, 270
London, Fritz Wolfgang (1900–54) 21n, 251
London, Heinz (1907–70) 20–1, 233, 249, 251
London colleges
  Birkbeck College 294, 296
  City and Guilds of London Institute 195, 223
  East London College 156, 279
  Finsbury Technical College 103
  Imperial College 243, 258, 279–81, 283, 285, 294
  King's College 87n, 156–7, 240n, 267, 276–9, 281, 283, 285, 294–5
  Normal School of Science 102, 110; *see also* Royal College of Science
  Royal Academy of Music 148
  Royal College of Chemistry 128
  Royal College of Science 58, 86, 110, 270, 273, 284; *see also* Normal School of Science
  Royal Holloway College 281
  Royal Naval College 64, 141
  Royal School of Mines 267
  University College 17–18, 44, 174, 186–7, 270, 275, 281, 283
Lorentz, Hendrik Antoon (1853–1928) 228
Los Alamos laboratory 260
Lothian, Marquess of *see* Kerr, William Schomberg Robert, 8th Marquess of Lothian
Love, Arthur Edward Hough (1863–1940) 307–8
Lovell, Alfred Charles Bernard (b.1913) 296
low-temperature physics 19–22, 206, 233, 235–9, 244–56, 257, 266, 273, 282, 284, 287, 291, 293, 298, 300–1, 304, 306; *see also* Mond Laboratory, Cambridge, *and the names of individual physicists*
Lowe, Hubert Foster (1861–1938) 133
Luard, Charles Eckford (1869–1927) 192
Lyell, Charles (1797–1875) 42

McClelland, J. A. 210
Mac Donald, Archibald Simon Lang (1840–85) 63, 123, 182

MacDonald, James Ramsay (1866–1937) 261
Macdonell, Alexander (b.1852) 63, 182
McGill University, Montreal 212, 214, 218–19, 220, 274
McGowan, Harry Duncan (1874–1961) 246
MacKenzie, Nicol F. 205
McLennan, John (1876–1935) 214, 218, 227, 231
Madras 137
Magdalen College, Oxford 7, 46, 71n, 127, 140, 145, 150–1, 154, 161, 172–4, 182, 193–5, 250; *see also* Chapman, Edward; Daubeny, Charles Giles Bridle; Schrödinger, Erwin; Yule, Charles John Francis
  college lectures 63–4, 123, 131–2
  laboratory and lecture-room provision 7, 63, 120–7, 127, 171; *see also* Daubeny Laboratory
  relations with Merton and Trinity 123, 131–2
Magdalen College School 46, 48–51, 125–6
magnetism 28–30, 58, 114, 132, 148, 209, 213, 306
Magrath, John Richard (1839–1930) 55
Main, Robert (1808–78) 42
Makower, Walter (1879–1945) 274
Malherbe, Tielman François Tertius (1899) 238
Mallock, Henry Reginald Arnulph (1851–1933) 175–8
Manchester, Owens College, and University of 6, 18, 20, 43–5, 133, 152, 171, 221–2, 231, 259, 267, 287, 292–6, 298
  comparison with Oxford 6, 20
  refugee physicists 292–3
  research on 'modern' physics 270–85
Manhattan Project 304
Manley, John Job (1863–1946) 123–5, 127, 150, 170
March, Arthur (b.1891) 251
March, Hildegunde 251
Marconi Company 280
Marlborough, Duke of, *see* Spencer-Churchill, George Charles
Marsh, James Ernest (1860–1938) 222
Massachusetts Institute of Technology 289–90
mathematics *see also* Cambridge, University of, Mathematical Tripos
  Christ Church 12, 16n, 144
  college lectures 10, 127–8, 133, 143–4, 156
  mixed mathematics 4, 10, 14
  St Hugh's College 307
  School of, Oxford 8–18, 32–3, 34, 62–3, 91–2, 141, 192, 311, 312–14
Mather and Platt 94
Mavrogordato, Anthony E. 194
Maxwell, James Clerk (1831–79) 10–11, 65
  Ayrton and Perry, relations with 97–8, 101n
  Clifton, relations with 58n, 60–2, 87
  lectures 87–8
  precision physics 6–7, 210

scholarship, Clerk Maxwell 211
Treatise on Electricity and Magnetism 73, 77–8, 84
works of Henry Cavendish 6–7, 84–5
measurement, *see* precision physics
'mechanical physicists' 171–2
mechanics, amateur 121–2, 171–2, 172–3, 179–80
mechanics, teaching of 28–30, 32, 62, 71, 132–3, 169–208
Meetham, Alfred Roger (1910–94) 239n
Megaw, Helen Dick (1907–2002) 255–6
Meitner, Lise (1878–1968) 259
Mendelssohn, Kurt Alfred Georg (1906–80) 20–1, 233, 245–56, 257, 266, 293
Merton, Thomas Ralph (1888–1969) 19–20, 230, 233, 239–44, 287–8
Merton College, Oxford 10, 17n, 60, 76, 100, 126, 128, 132–3, 141, 143, 174, 194, 239
  grant to Townsend 212–13
  relations with Magdalen and Trinity 123
  science college, status as a 63–4, 122–3, 182–3
metallurgy 194, 290–1, 292
Meteorological Office 284
meteorology 28–30, 162, 173, 174, 239–44, 275, 284
metrology *see* precision physics and engineering
Metropolitan-Vickers Company 280, 296
Metford, William Ellis (1824–99) 178–9
Meyer, Oskar Emil (b. 1834) 142n
Michels, Mathias Johannes Friedrich (1889–1969) 252
Michelson, Albert Abraham (1852–1931) 286
microphones 179–80, 204
microphysics *see* atomic and nuclear physics
Miers, Henry Alexander (1858–1942) 165
Milford, Michael (1905–78) 243n
military research 228–9, 260–2, 266, 279–80, 283, 285; *see also* World War, First; World War, Second
Millard, Thomas (d.1871) 181
Millard endowments
  demonstrationship 137, 145
  lectureship in experimental mechanics and engineering 63, 180, 182, 185–8, 190–1
  lectureship in physics 63, 64, 123, 131–5, 145, 181–2
  scholarships 181
Millard Laboratory 8, 107–9, 130–1, 145, 148, 154, 156, 169–208
  condition, 1908 205
  equipment 185–91
  inauguration 185
  prospectuses 190, 191, 193
Millikan, Robert Andrews (1868–1953) 286
Milne, Edward Arthur (1896–1950) 250, 307–8

Milner, Samuel Roslington (1875–1958) 270, 281
mineralogy 26–8, 38, 39, 47, 64, 138; *see also* crystallography
miner's lamp, Clifton's 86, 104–6
Mines, Royal Commission on Accidents in 82,, 86, 102–9
Ministry of Agriculture and Fisheries 302
Minn, Henry (1870–1861) 172–3
Minnesota, University of 212, 217–18
mixed mathematics 4, 10, 14
Moberly, George (1803–85) 42n
Moderations 16–18, 35, 40–1, 69 78, 206, 313–14
Modern History, School of 34, 192; *see also* Law and Modern History, School of
Mond, Robert Ludwig (1867–1938) 266
Mond Laboratory, Cambridge 245, 291, 293, 297; *see also* Kapitza, Piotr Leonidovich
Montreal *see* McGill University, Montreal
Moon, Philip Burton (1907–94) 294
Moore, Judith Rachel (1911–43) 254
Moore, Tom Sidney (1881–1966) 126
Morrell, George Herbert (1845–1906) 50
Morris, Edgar Ford (b.1874) 136
Morris, William Richard, Viscount Nuffield (1877–1963) 263, 294
Morris car factory 206, 230
Moseley, Amabel 220
Moseley, Henry Gwyn Jeffreys (1887–1915) 14n, 158, 221–2, 227–9
Mott, Nevill Francis (1905–96) 292
Mullaly, John Mylne (1901–25) 243n
Mullard, Stanley Robert (1883–1979) 280
Müller, Friedrich Max (1823–1900) 42–3, 84
Munitions Inventions Department 280
Museum of the History of Science, Oxford 28, 163, 164, 28n
music, science of 145–9, 201; *see also* acoustics

Nagel, David Henry (1862–1920) 136–7, 148–51, 154, 159
National Physical Laboratory 167, 239n, 267, 269, 273, 276, 284, 290–1
Natural Science, School of; *see also* Preliminary Honour examination
 'double first' as favoured qualification 64–5
 inauguration 15
 numbers of candidates 14–15, 33–4, 38–41, 63, 68–9, 73, 87–8, 107–9, 114–17, 136, 152, 192–3, 233–4, 308–9
 Pass degree 12, 31, 33–4, 38–41, 70–1, 192, 311
 practical requirement 46, 123, 150, 154–5, 171, 188, 192
 relations with School of Mathematics 11
 syllabus 8, 31–3, 42n
 wartime degree 304

Navy Signal School, Portsmouth 284
Neate, Charles (1806–79) 55
Nernst, Hermann Walther (1864–1941) 19, 235–6, 238, 246–7, 254–5, 286
 pneumatics 28–30
New College, Oxford 22, 28, 35, 97, 150, 155–6, 159, 204, 212–14, 220, 309
Newcastle, Armstrong College, and University of 281, 292–3
Newcastle, Duke of *see* Pelham-Clinton, Henry Pelham Fiennes
Nicholls, Frank 195
Nobel prizewinners 139, 250, 259, 274, 275, 276, 282
Non-Collegiate students 126, 155, 194
Normal School of Science, London 102, 110; *see also* Royal College of Science, London
Nottingham, University College, and University 66n, 159, 271, 279, 289, 295
nuclear physics *see* atomic and nuclear physics
Nuffield, Viscount *see* Morris, William Richard

observatories 74–5, 106, 137, 244; *see also* University Observatory, Oxford
Odling, William (1829–1921) 75, 168
Oliphant, Marcus Laurence Elwin (1901–2000) 294–6
optics
 equipment 103–4, 110, 123, 201
 examination questions 71–2, 81
 papers by Clifton 103–4
 research 81, 86, 103–4, 134–6
 teaching 14, 28–30, 47, 58, 62n, 67–8, 88, 96,132, 209, 215
Oriel College, Oxford 127
Ostwald, Wilhelm (1853–1932) 126
Owen, Gwilym (1880–1940) 281
Owens College, Manchester *see* Manchester, Owens College, and University of
Oxford, University of; *see also* Clarendon Laboratory; Electrical Laboratory; Natural Science, School of; Oxford University Press; University Museum; *and entries for individual colleges, laboratories, and postholders*
 Ashmolean Museum 27, 32, 37, 128
 Bodleian Library 12, 54
 Chemistry, Department of 9, 135
 chemistry, laboratory provision for 37–8, 56, 128, 140–1; *see also entries for individual college laboratories*
 classics, prominence of in Oxford syllabus 15–16
 coaching 9, 10, 149–50
 Commission of 1850–1 31

Oxford, University (*cont.*)
  Commission of 1877 67–8, 73–4, 76, 81, 93, 95–7, 107, 214
  comparison with University of Cambridge 2, 3–7, 9–22, 72, 77–9, 83–7, 117, 161, 210, 212, 219, 245, 268, 270, 286–91, 295–6, 303, 305
  comparison with University of Glasgow 2
  Congregation, debates and decisions in 55, 69n, 70n
  Convocation 12–13, 32, 35, 40, 54n, 55–8, 74, 75n, 76n, 82, 92, 107, 115
  Dyson Perrins Laboratory 126–7
  engineering science *see* engineering science at Oxford
  Hebdomadal Council, debates and decisions in 42, 47, 54, 59, 62, 75, 78, 83, 97, 215, 224
  Law and Modern History, School of 16, 32–3, 34, 40–1, 314
  Mathematics, School of 8–18, 32–3, 34, 62–3, 91–2, 141, 192, 311, 312–14
  Museum of the History of Science 28, 163, 164
  Physical & Radio Society 153
  Physical Chemistry Laboratory 140, 152, 263
  reputation in physics 1–23, 77–9, 167–8, 169–70
  suspicion of science and engineering 43n, 55–8, 207, 306, 310
  University Observatory 74–5, 156–7, 170, 181, 215, 261
Oxford Movement 26
Oxford University Press 27–8, 35, 72–3, 78n, 228
  Clarendon Press Series of text books 72–3, 78n, 142

Palmer, Roundell, 1st Earl of Selborne (1812–95) 96
Parks, University 54
Parks Road, traffic on 58n
Pass degree 12, 31, 33–4, 38–41, 192, 311
  restructuring (1872) 70–1, 314
Pattison, Mark (1813–84) 77
Pawsey, Joseph Lade (1908–62) 336
Payne, Joseph Frank (1840–1910) 71n
Pedder, Arthur Lionel (1868–1934) 126–7
Peddie, William (1861–1946) 275
Peel, Robert (1788–1850) 12, 13, 16n
Pelham Fiennes Pelham-Clinton, Henry, 5th Duke of Newcastle under Lyme (1811–64) 53–4
Pembroke College, Oxford 62
Percival, John (1834–1918) 180–4, 207
Perkin, William Henry, junior (1860–1929) 231
Perrin, Jean Baptiste (1870–1942) 228
Perry, John (1850–1920) 97–102, 103, 168, 183–4, 207

Philips Company Laboratory, Eindhoven 295
Phillips, John (1800–1874) 37, 38n, 42, 46–7, 52, 74–5, 77
photography 58, 195–6, 199–200, 207, 296
Physical & Radio Society, University of Oxford 153
physical chemistry as an Oxford speciality 2, 4, 8–9, 126, 135–40, 151–5, 164–5, 171–2, 219, 263, 271, 291; *see also* chemical physics
Physical Chemistry Laboratory 140, 152, 263
Physical Society of London 102–4, 107, 110, 269, 283
  Clifton as President of 82, 86, 102, 104
physics *see* acoustics; atomic and nuclear physics; chemical physics; electricity; instrumentation; low-temperature physics; magnetism; optics; pneumatics; precision physics and engineering; spectroscopy; X-rays
  definitions of 4–5, 8, 71, 119, 170, 208, 268, 287, 295, 298–300, 310
  and industry 22, 105, 159, 166, 171, 204, 207, 210, 215, 231, 241–3, 267–9, 273, 280, 282–6, 288, 291–3, 295–7, 300–1
physiology and anatomy 31, 43n, 64, 107–8, 123, 171, 192–3, 194
Pickard, George Lawson (b.1913) 261–2, 297
Pidduck, Frederick Bernard (1885–1952) 205, 221
Pidgeon, L. M. 243n
Pike, William Herbert (b.1851) 63
Pilley, John Gustave (1899–1968) 238n
Plaskett, Harry Hemley (1893–1980) 307–8
Plummer, Henry Crozier Keating (1875–1946) 156–7
pneumatics 28–9, 32, 123
Pontius, Rex Bush (1909–87) 253
Potter, Richard (1799–1886) 44
Poulton, Edward Bagnall (1856–1943) 193–4
Powell, Baden (1796–1860) 13–14, 15, 26
Powell, Cecil Frank (1903–69) 294
Poynting, John Henry (1852–1914) 270, 273, 277
precision physics and engineering 90–2, 96, 110, 166–7, 170–8, 195–202, 207, 210, 217–18, 273
Preliminary Honour examination 157, 195
  introduction 9, 17–18, 69–70
  numbers of students 108–9
  practical requirements and teaching 123, 150, 154–5, 171, 188, 192
  questions 71
  teaching 115–16, 130, 188, 215–16
Presidency College, Madras 137
Press, Oxford University 27–8, 35, 72–3, 78n, 228
Pressed Steel Company 206
Price, Bartholomew (1818–98) 9–10, 13, 15, 17, 37, 62–3, 73, 103, 107, 147, 307

Price, Charles James Coverly (1838–1905) 128, 194
primary sciences, Daubeny's 32–3, 69
Prince Regent, later George IV (1762–1830) 1
Princeton 289–90
Pringle, John William Sutton (1912–82) 308
Pritchard, Charles (1808–93) 74–5

quantum physics 158–9, 161, 231
Queen's College, The, Oxford 55, 251
  laboratory 8, 150, 153–4

Radcliffe Science Library, Oxford 223
Radcliffe travelling fellowship 71n
radar 161, 260–1, 297–8, 303, 305
radio 180, 200–1, 204, 228–9, 275, 280, 283–4
radioactivity 4, 159–60, 162, 200, 218–19, 223, 257, 259, 269, 274–7, 281, 285, 288, 294
Raman, Chandrasekhara Venkata (1888–1970) 297
Ramsay, William (1852–1916) 275
Randall, John Turton (1905–84) 284, 296
Rankine, William John Macquorn (1820–72) 187, 204–5
Ratcliffe, John Ashworth (1902–87) 297
Rayleigh, Baron *see* Strutt, John William, Baron Rayleigh
Reading, University of 171
Redesdale, Lord 164
reductionist (and non-reductionist) physics 22, 274–81, 292–300, 305
refrigeration 206; *see also* low-temperature physics
refugee physicists 20–1, 245–53, 266–7, 291–8, 300, 303–4; *see also* Bethe, Hans; Einstein, Albert; Fröhlich, Herbert; Fuchs, Emil Julius Klaus; Kurti, Nicholas; London, Fritz Wolfgang; London, Heinz; Simon, Fritz Eugen; Szilard, Leo; Zeleny, John
Reiche, Fritz (1883–1969) 159
Reinold, Arnold William (1843–1921) 17, 50, 56, 64–6, 85, 91–2, 122, 141
relativity 158–9, 231; *see also* Einstein, Albert
reputation of Oxford physics 1–23, 77–9, 167–8, 169–70
Responsions 16, 179
retirement age 117n
Rhodes Scholars 158, 266, 289–90
Rice, James (1874–1936) 276
Richards, Theodore William (1868–1928) 167–8
Richardson, Owen Willans (1879–1959) 210, 276, 283, 297
Rigaud, Stephen Peter (1774–1839) 1, 24, 25, 26–8
Ritson, David M. 161
Roaf, Douglas (1911–96) 162, 259n, 266, 308
Rochester, George Dixon (1908–2001) 296

Rockefeller Foundation 20, 237, 245, 262, 291, 293
Rolleston, George (1829–81) 37, 73
Rollin, Bernard Vincent (1911–69) 21, 255
Roscoe, Henry Enfield (1833–1915) 43–4, 133
Royal Academy of Music 148
Royal Aircraft Factory (Establishment), Farnborough 19, 236, 239, 262, 280, 288
Royal Astronomical Society 76, 148
Royal College of Science, London 58, 86, 110, 270, 273, 284; *see also* Normal School of Science, London
Royal Commission on Accidents in Mines 82, 86, 102–9
Royal Commission on Scientific Instruction and the Advancement of Science (Devonshire) 46n, 62, 181–3
Royal Flying Corps 228–9
Royal Indian Engineering College, Cooper's Hill 65–6, 142–3
Royal Institution, London 267, 270, 273, 283, 286
Royal Naval College, Greenwich 64
Royal Observatory, Greenwich, Board of Visitors 106
Royal Society 99, 111, 196, 198, 201, 202, 204
  Clifton as vice-president and on council 102, 113
  elections to fellowship 24, 66n, 76, 149, 167, 175, 199
  gold medal, award of to William Froude 175
  grants 147, 214, 239, 296
  officers 240, 243, 283
  physics committee 283
  publications 68n, 86, 93, 97–102, 104, 110–14, 146, 167, 175
  reform 289
  research fellowships 292, 296
Rücker, Arthur (1848–1915) 17, 65–6, 85, 87, 91–2, 106, 270
Rugby School 183
Russell, Alexander Smith (1888–1972) 123, 159, 160, 162
Russell, Francis Charles Hastings, 9th Duke of Bedford (1819–91) 184
Russell, John Wellesley (1851–1922) 128, 133, 182
Rutherford, Ernest (Lord Rutherford of Nelson) (1871–1937) 20, 219, 235, 240–1, 290, 294–5
  appointment to Cavendish chair 238, 281
  attitude to wave mechanics 289
  career at Manchester 221, 274, 277–80
  careers of students 285, 292–7
  death 297

Rutherford, Ernest (Lord Rutherford of Nelson) (1871–1937) (*cont.*)
  'old-fashioned' physics, attitude to 218
  relations with Townsend 210–12, 225–8
  research and leadership at the Cavendish 281–2, 285–9
Ryle, Martin (1918–84) 162, 259

St Andrews, University of 292–3
St Andrews, University College of, Dundee 275
St Anne's College, Oxford 309
St Edward's School, Oxford 195
St Hugh's College, Oxford 307
St John's College, Cambridge 44, 51
St John's College, Oxford 139, 145, 166, 172, 185, 189, 194, 251; *see also* Balliol-Trinity-St John's Colleges, relations between
  laboratory 145, 147–9, 189
salaries
  Clarendon laboratory 266–7
  demonstrators 57, 64, 74
  Dr Lee's readers 141
  Lindemann 236
  London (Heinz) 251
  offer to Helmholtz 42
  refugee physicists 245, 249–52, 266
  Simon 251–2, 266
  Townsend 212
  Walker 1, 26, 41
Salters' Company 266
Sanderson, John Scott Burdon *see* Burdon Sanderson, John Scott
Savilian chairs 1, 13–15, 24, 25, 26, 40, 74; *see also* entries for individual professors
Sayce, Archibald Henry (1845–1933) 146
Schlick, Moritz (1882–1936) 159
scholarships
  Brackenbury, Balliol 17–18
  Burdett-Coutts 71n
  Clerk Maxwell 211
  1851 Exhibition 209, 210, 266, 270, 273, 281, 284–5
  mathematical 17n
  Radcliffe travelling fellowship 71n
  Rhodes Scholarships 158, 266, 289–90
schools, teachers and science teaching in 6, 32, 46, 88, 95–6, 107, 142, 149, 183–4
Schrödinger, Erwin (1887–1961) 21n, 158–9, 161, 250–1
Schuster, Arthur (1851–1934) 274
Science and Art Department 186n
Science Area, planning of 21–2, 184, 263–5
Searle, George Frederick Charles (1864–1954) 271–2
Sedleian chair of natural philosophy 62, 307–8; *see also* Love, Arthur Edward Hough; Price, Bartholomew
Selborne, Lord *see* Palmer, Roundell, 1st Earl of Selborne
Selby, Arthur Laidlaw (1861–1942) 273
Seward, Margaret (1864–1929) 134
Sheffield, Firth College, and University of 18, 91, 223, 279, 281, 292–3
Sherwood, William Edward (1851–1927) 125
Shirley Institute, Manchester 293
Short, Walter Francis 55n
Sidgwick, Nevil Vincent (1873–1952) 9n, 126, 151, 158, 166
Simon, Franz Eugen [Francis] (1893–1956) 21, 233, 244–9, 251–6, 257, 266
Skinner, Herbert (1872–1931) 288
Smirke, Robert (1781–1867) 27
Smith, Frank 284
Smith, Frederick Edwin, Earl of Birkenhead (1872–1930) 82–4
Smith, Frederick Jeremiah (1809–84) 179
Smith, Frederick John *see* Jervis-Smith, Frederick John
Smith, Frederick Llewellyn (1909–88) 243n
Smith, Henry John Stephen (1826–83) 14, 17, 42, 46n, 47n, 81–2, 103, 107, 129–30, 140, 145, 174
Smith, Samuel Walter Johnson (1871–1948) 281
Soddy, Frederick (1877–1956) 9, 19, 124, 137, 158–9, 160, 194, 231, 240, 275, 288–9, 299
Solvay Institute and conferences on physics 226–8, 235, 275, 277, 286, 289
Somerville College, Oxford 134, 309
Sorbonne 256
South Kensington Loan Exhibition 146
Southampton, University of 292–3
Southwell, Richard Vynne (1888–1970) 206, 262–3
spectroscopy 9, 163–4, 170–1, 227, 233, 239–40, 249, 252–3, 270, 273–5, 277, 278, 283, 285–8, 295, 299, 305, 307
  Jackson and Kuhn, contributions of 21
Spencer-Churchill, Edward George (1876–1964) 145
Spencer-Churchill, George Charles, Duke of Marlborough (1844–92) 163–4
Stainer, John (1840–1901) 179
Stallybrass, William Teulon Swan (1883–1948) 306, 310
Stanley, Edward George Geoffrey Smith, 14th Earl of Derby (1799–1869) 35
Stapleton, Henry Ernest (1878–1962) 172–3
statutes, examination 16, 30–3, 40–1, 69–70, 78, 311–14
Stewart, Balfour (1828–87) 72n, 142
Stocker, William Nelson (1851–1949) 65–6, 92, 142–4, 310

Stokes, George Gabriel (1819–1903) 17, 43, 45, 88, 92
Stoletow 211, 226
Stoner, Edmund Clifton (1899–1968) 283
Story-Maskelyne, Mervyn Herbert Nevil (1823–1911) 37, 39, 42, 47, 73, 128, 170–1
Strasbourg, University of 78
Stroud, William (1860–1938) 133–4, 276
Strutt, John William, 3rd Baron Rayleigh (1842–1919) 11, 84–5, 87, 102, 106, 111, 147, 175, 210
submarine detection 280
superconductivity 253
surveying 203, 205
Swansea, University College 281
syllabus, debates on 5, 8, 12–18, 31–3, 213, 232
Sylvester, James Joseph (1814–97) 10n,
Symm, Joshua Robinson (c.1809–87) 56
Szilard, Leo (1898–1964) 21, 161, 258–9, 294–5

Tait, Peter Guthrie (1831–1901) 60, 73, 275
Talbot, Edward Stuart (1844–1934) 149
Tassel, Emile 228
Taunton, Jervis-Smith's workshop in 179–80, 185
Taylor, Geoffrey Ingram (1886–1975) 297
Taylor Institution, Oxford 55
teaching *see entries for individuals' laboratories and colleges*; *see also* schools, teachers and science teaching in
telephony and telegraphy 179–80, 195–6, 201–2, 204, 207, 224, 280
terms and residence, University xviii
textbooks 13–14, 24, 32, 65, 72–3, 78, 142, 154, 159, 195, 275
textile physics 286
Theology, School of 34
theoretical physics 158, 161, 250–1, 290
 reallocation of Wykeham chair 262
Thompson, Silvanus Phillips (1851–1916) 199
Thomson, Arthur (1858–1935) 107–8
Thomson, George Paget (1892–1975) 280, 283
Thomson, Joseph John (1856–1940) 84, 87, 209–11, 217–18, 226, 228, 235, 267–83, 285, 286
Thomson, William (Lord Kelvin) (1824–1907) 43, 47n, 55, 60, 77–8, 97–8, 199, 211, 214, 315–16
Tizard, Henry Thomas (1885–1959) 82, 84n, 126, 127, 261, 285
Todd, George William (1886–1950) 281
Todhunter, Isaac (1820–84) 43
Tootell, Henry 179
Toronto, University of 63n, 214, 217, 227
Torricelli tercentenary celebrations 205
Touch, Arthur Gerald (1911–94) 259n, 297–8

Townsend, John Sealy Edward (1868–1957) 4–5, 9n, 23
 appointment to Wykeham chair 81, 114, 203, 209, 212, 271
 marginalization 226–32, 289, 303
 personality 225, 230–1
 precision physics, attitude towards 217–18
 relations with chemists 218–22
 relations with Clifton 18–19, 22, 81–2, 114–18, 154, 212, 214–18, 232
 relations with Lindemann 19, 236, 262, 289
 relations with Rutherford 210–12, 226, 228
 relations with J. J. Thomson 210–11, 226, 228
 reputation of 23, 225–6, 262
 resignation 304
 strategies for physics in Oxford 149–50, 153–5
Townsend, Mary Georgiana *see* Lambert, Mary Georgiana
Translation of textbooks and monographs 65, 72, 72–3n, 142n, 159
Trinity College, Cambridge 211, 272, 281, 288
Trinity College, Dublin 209
Trinity College, Oxford 137–8, 140, 147–9, 168, 169–208, 221, 309
 funding of science and engineering 9, 123, 145, 181–5
 laboratories, 169–208; *see also* Millard Laboratory; Balliol-Trinity Laboratories
 relations with Balliol and Exeter Colleges 182
 relations with Balliol and St John's Colleges 130–3, 182
 relations with Magdalen and Merton Colleges 63, 123, 131–2
Tube Alloys project 152, 259–60, 298, 303–5
Tuck, James Leslie (1910–80) 259
Tucker, William Sansome (c.1876–1955) 284, 297
Tuckwell, William (1829–1919) 28
Turner, Herbert Hall (1861–1930) 106, 159
Tyndall, Arthur Mannering (1881–1961) 102, 276, 290–2, 294
Tyndall, John (1829–93) 72, 147

Ubbelohde, Alfred Rene Jean Paul (1907–88) 243n
universities *see individual entries under city*
University College, London 17–18, 44, 174, 186–7, 270, 275, 281, 283
University College, Oxford 15, 149–50
University Grants Committee 236, 266
University Museum, Oxford 64, 130, 145–6
 building and cost 7–8, 35, 60
 chemical laboratory 37–8, 128, 140–1
 impracticality 37
 inauguration 35–7
 interior arrangement 37–40, 47–53

University Museum, Oxford (*cont.*)
  opposition to 55
  'philosophy', guiding 38, 120
  plans for extension 47–53, 264
  premises for engineering 205
  premises for physics 38–40
  Townsend in 115, 154, 212–15
University Observatory, Oxford 74–5, 156–7, 170, 181, 215, 261, 264
University Parks 54
Urbain, Georges (1872–1938) 227

Veley, Victor Herbert (1856–1933) 124, 135, 137, 142, 150, 154, 164, 194
Vernon Harcourt, Augustus George *see* Harcourt, Augustus George Vernon
Vernon-Harcourt, Leveson Francis (1839–1907) 174
Villard, Paul (1860–1934) 335
Vincent 211
voltaic action, Clifton on 97–102
Von Engel, Alfred (1898–1990) 230

Wadham College, Oxford 15, 22, 24, 41–3, 76, 122, 139, 141, 155–6, 195, 243
Walden, Allan Frederick (1871–1956) 125, 150, 159, 194
Walker, James (1857–1929) 143–4
Walker, Robert (1801–65) 15, 24–41, 72, 117n
  appointment to readership 1, 24
  illness 41, 46
  lectures 28–30, 33–8, 41
  mathematical interests 15
  reform of syllabus 30–5
  textbooks 32, 72
Walter, Henry 56–7, 93
War Office X-ray Committee 280
wartime research 228–9, 260–2, 266, 279–80, 283, 285; *see also* World War, First; World War, Second
Watkins & Hill, instrument-makers 28
Watson-Watt, Robert Alexander (1892–1973) 259n, 275, 284, 297–8
Watts, John (1843–1933) 126, 194
Waynflete chair of mathematics 10
Wells 179–80
Westminster, Duke of 237, 262, 266
Westwood, John Obadiah (1805–93) 37, 40
Whewell, William (1794–1866) 43
Whiddington, Richard (1885–1970) 280, 281
White, Stuart Arthur Frank (1870–1951) 156–7
Whiteside, John (1679–1729) 1
Whitley, Edward (1879–1945) 194
Wilberforce, Lionel Robert (1861–1944) 271–3
Wilberforce, Samuel (1805–73) 35–7

Wilde, Henry (1833–1919) 95
Wilkinson, Denys Haigh (b.1922) 308
Williams, David (1786–1860) 35
Williamson, Alexander William (1824–1904) 73
Willis, Robert (1800–75) 60n
Wilsdon, Bernard Howell (b.1888) 166
Wilson, Charles Thomson Rees (1869–1959) 211, 220
Wilson, Duncan Randolph (1875–1945) 126
Wilson, James Maurice (1836–1931) 44
Wimperis, Harry Egerton (1876–1960) 261
Winchester College 42, 74n
women, *see also* Adam, Madge Gertrude; Berkley, Evelyn Hannah; Busbridge, Ida Winifred; Butler, Catherine Elizabeth; Curie, Marie; Holmes, Muriel Catherine Canning; Isaac, Florence; Lady Margaret Hall; Lester, Laura; Megaw, Helen Dick; Moore, Judith Rachel; St Anne's College; St Hugh's College; Seward, Margaret; Somerville College
  education, campaign for women's 183
  fellows in physics and mathematics 307, 309
  students 14–15, 116, 134, 152, 192, 194, 234, 254, 307
women's colleges, opening of 134, 194
Woods, Henry George (1842–1915) 184–5, 195
Woodward, Benjamin (1815–61) 55n
Woodward, Leonard Ary (1903–76) 152–3
Woolwich 198, 283, 284, 297
Worcester College, Oxford 194–5
World War, First *see also* military research
  Admiralty's Board of Invention and Research 228, 280
  Admiralty's Department of Scientific Research and Experiment 280, 284
  Admiralty's Research Laboratory 280, 284, 297
  Aeronautical Research Committee 175
  Air Inventions Committee 280
  Clarendon Laboratory 228–9
  Electrical Laboratory 228–9
  Munitions Inventions Department 280
  Ordnance Board and Committee 175
  Royal Aircraft Factory, Farnborough 19, 236, 239, 280, 287–8
  Royal Flying Corps 228–9
  Royal Naval Air Service 229
  submarine detection 280
  X-rays, research on 279–80
World War, Second
  radar 161, 260–1, 297–8, 303, 305
  Royal Aircraft Establishment, Farnborough 262, 280
  Tube Alloys Project 152, 259–60, 298, 303–5

Wykeham chair of physics 5, 81, 97, 114–15, 212, 229–30, 271, 304
   creation 202–3, 209, 214
   reallocation to theoretical physics 262
Wyndham, Thomas Heathcote Gerald (1842–76) 63
Wynn-Williams, Charles Eryl (1903–79) 294

X-rays 200–1, 221–2, 272, 276–80, 286, 295, 299; *see also* Moseley, Henry Gwyn Jeffries
   crystallography 282–3, 284, 292
   wartime research 279–80

Yule, Charles John Francis (1848–1905) 51, 52, 63, 123, 171, 196
Yorkshire College *see* Leeds, Yorkshire College, and University of

Zeleny, John (1872–1951) 210–12, 217–18